HEAVEN'S MIRROR

Heaven's Mirror

QUEST FOR THE LOST CIVILIZATION

Graham Hancock *and* Santha Faiia

Three Rivers Press

New York

Acknowledgements

Donald and Muriel Hancock, James Macaulay and Solomon Lorthu for being with us all the way as ever. Lew and Linda Jenkins, who put us on to *Hamlet's Mill* years ago – and for looking after us in California. Tom Weldon our editor and old friend. Julie Martin, who designed this book. Zahi Hawass for surprising us at the Sphinx. Oswaldo Rivera for opening the way in Tiahuanaco. John Grigsby for his brilliant research work and the discovery of the Draco correlation. Yasuo Watanabe for his generosity, breadth of spirit and openness of mind. Shun Daichi, our friend, fellow researcher and bridge to Japanese culture. Tim Copestake, Stefan Wickham, Colin Clarke, Catherine Brandish, Danny Hambrook and all the other members of the great team who filmed the television series *Quest for the Lost Civilization* with us. Eileen Warren, Barry Jeffries and Jan Mathews for looking after things at home.

Page i: Southern gateway, Angkor Thom, Cambodia.
Page ii: The Great Sphinx, Giza, Egypt.
Page iii: Colossal heads, Rano Raraku, Easter Island.
Page vi: Falcon-headed Soul of Pe, tomb of Seti I, Valley of the Kings, Egypt. The Souls of Pe were believed to assist in the rebirth of the deceased.
Page vii: Ground and sky: Mayan pyramids at Tikal.

Published by Three Rivers Press, 201 East 50th Street, New York, New York 10022. Member of the Crown Publishing Group.

Originally published in Great Britain by Michael Joseph, Ltd., and in the U.S. by Crown Publishers in 1998.

Random House, Inc. New York, Toronto, London, Sydney, Auckland
www.randomhouse.com

THREE RIVERS PRESS is a registered trademark of Random House, Inc.

Printed in Great Britain by Butler & Tanner Ltd, Frome and London

Library of Congress Cataloging-in-Publication Data

Heaven's mirror: quest for the lost civilization / Graham Hancock and Santha Faiia.
      p.   cm.
   Includes bibliographical references and index.
    1. Lost continents.   2. Civilization, Ancient.   3. Megalithic monuments.
I. Faiia, Santha. II. Title.
GN751.H295   1998
930—dc21   98-16028
                                                   CIP

ISBN 0-609-80477-4

10 9 8 7 6 5 4 3 2 1

First U.S. Paperback Edition

Contents

To our children, Sean, Shanti, Ravi, Leila, Luke and Gabrielle. We hope one day you'll tread all of the amazing monuments this book has taken us to. Thank you for your patience and understanding. Love you all.

WHAT IS IN THE GREAT BEYOND?

'Heaven above, Heaven below;
Stars above, Stars below;
All that is over, under shall show.
Happy thou who the riddle readest.'

Tabula Smaragdina[1]

DEATH IS THE FUNDAMENTAL mystery of life. It is certain for all and yet we do not know what it means.

The mystery can be reduced to simple dilemmas. When we die does everything end for us, or is there some way that we go on? Is there nothing more to us than the sum of our material parts, or does the soul exist? Is the notion of the soul a figment of human psychology, or perhaps an invention of religion? Or could it be something wonderfully real?

Science, which can weigh, measure and assess the corpse of a dead person, is powerless to tell us whether anything spiritual occurs after death. A widely held assumption of science, though it is by no means unanimous, tends to be that there is no soul, and that 'dead means dead'. Some scientists promote such views as though they are facts that have been empirically tested. Yet there are no facts here, only assumptions that cannot be proved. Indeed, the scientific case in this area is religious in nature since it expresses a passionate *belief* in the non-existence of the soul, but has no evidence to support that claim.

Religion presents the opposite case, equally passionately, on equally flimsy grounds. There is no scientific proof of the existence of the various religious heavens and hells and afterlife realms. Nevertheless, the religious or spiritual point of view strongly asserts that the soul does exist, and will undergo a judgement after death, and can transmigrate through many forms, and can be reborn.

A QUESTIONER LIKE THEE ...

Confronted by these two polar opposite opinions, it is natural for thinking people to consider the universal law of physical death and to wonder what their own fate might be. One such person, according to an Indian sacred text called the

OPPOSITE: *Granite stela of Axum, Ethiopia. This megalith is over 20 metres high and weighs 300 tonnes. It is at least 2,000 years old, but could be much older. There is evidence that its apex was once decorated with a metal plaque showing the stars, the sun and the moon. Although nothing is known about the religion of its builders it has much in common with other gigantic structures – frequently megalithic, frequently connected to astronomy – that are scattered all around the world.*

Katha Upanishad, was Nachiketas, a brave and inquisitive young man who had found his way to the 'House of the Dead' and earned the right to demand a wish of Yama, the Hindu god of death.

> *Nachiketas*: This doubt there is of a man that has died: 'He exists' say some; and 'He exists not', others say. A knowledge of this, taught by thee, this is my wish . . .
>
> *Yama*: Not easily knowable and subtle is this law . . . Choose Nachiketas, another wish, hold me not to it . . . Choose sons and grandsons of a hundred years, and much cattle and elephants and gold and horses . . . Choose wealth and length of days . . .
>
> *Nachiketas*: Tomorrow these fleeting things wear out the vigour of a mortal's powers. Even the whole of life is short . . . Not by wealth can a man be satisfied. Shall we choose wealth if we have seen thee? Shall we desire life while thou art master . . .? This that they doubt about . . . what is in the great Beyond, tell me of that. This wish that draws near to the mystery, Nachiketas chooses no other wish than that.
>
> *Yama*: Thou indeed, pondering on dear and dearly loved desires, O Nachiketas, hast passed them by. Not this way of wealth hast thou chosen, in which many men sink . . . The great Beyond gleams not for the child led away by the delusion of possessions. 'This is the world, there is no other,' he thinks, and so falls again and again under my dominion . . . Thou art steadfast in the truth; may a questioner like thee, Nachiketas, come to us.[2]

Yama, the god of material death, then goes on to utter one of the luminous statements of Indian scripture, a statement concerning the nature of the soul 'which is never born nor dies, nor is it from anywhere, nor did it become anything':[3]

> Unborn, eternal, immemorial, this ancient is not slain when the body is slain . . . Smaller than small, greater than great, this Self is hidden in the heart of man . . . Understanding this . . . the wise man cannot grieve . . . Bodiless in bodies, stable among unstable . . . he is released from the mouth of death.[4]

We will present evidence in the chapters that follow that the revelations attributed to Yama in the Upanishads did not originate in Hindu religious philosophy. They are part of an ancient spiritual teaching that was promulgated not only in India but also as far afield as Mexico, Egypt, Indochina, the Pacific and South America. Hinting at the former existence of an important civilization not spoken of in any history books – a lost 'common source' that influenced all these regions – this mysterious system of ideas used an esoteric form of astronomy as its principal methodology, and built great works of architecture on the ground to reflect the patterns and movements of the heavens. The system was a kind of 'science of immortality' designed to release mankind 'from the mouth of death'. Its origins are forgotten in prehistory. And 'prehistory' itself is just the name that we give to the almost total amnesia that our species has suffered concerning more than 40,000 years of our own past. This amnesia covers the entire period from the emergence of anatomically modern humans until the first 'historical records' began to be written down in Sumer and in Egypt in the third millennium BC.

Gavrinis, megalithic passage grave near Carnac with an alignment to the winter solstice sunrise. The meanings of the patterns carved into the stones are unknown. Could they relate to the twin concerns of astronomy and the immortality of the soul that are expressed in the orientation of the passage?

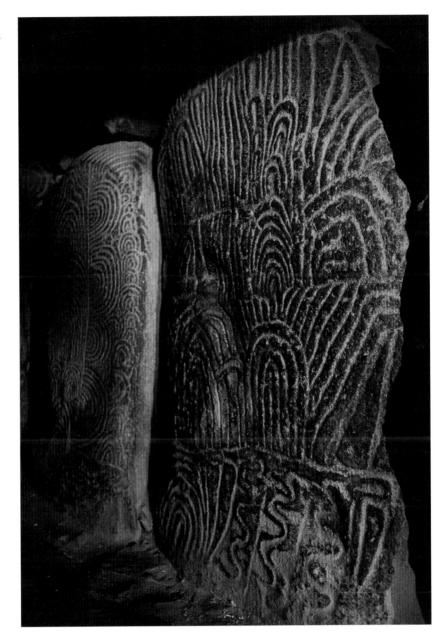

MEMORIALS MIGHTY

Out of that long period of amnesia, and from its borders with history, a number of mighty monuments have come down to us. These include rock-hewn temples, circles of megaliths, and sacred sites arranged in dead straight lines over vast distances, such as the avenues of standing stones at Carnac in northern France. One earthen mound there, which contains a megalithic passageway oriented to the winter solstice sunrise, has been carbon-dated to 4700 BC.[5]

In the British Isles the oldest stone circles, such as Callanish in the Outer Hebrides, are believed to date from about 3000 BC, but could be older: nobody is really sure.[6] There are megalithic circles in Japan that have never been excavated.[7] Megalithic temples in Malta may be as old as 4000 BC.[8] In Ethiopia the

rock-hewn 'churches' of Lalibela and the 300-tonne granite stele of Axum are of unknown origin and are undatable by any objective technique.[9] The Pacific islands are littered with dozens of mysterious megalithic constructions.[10] We must consider, too, the monuments of Egypt, Mexico and South America, where individual stone blocks sometimes weigh as much as 200 tonnes.[11]

What many of these structures have in common is a degree of uncertainty about when they were built, why they were built, how they were built, and who they were built by. They also share advanced engineering techniques and demonstrate finely calibrated astronomical alignments.

Thus some of the avenues of megaliths at Carnac are thought to have been used for lunar observations. Similarly the stone circle at Callanish seems to have been purposefully oriented to draw attention to an obscure lunar phenomenon

The megaliths sometimes seem like hieroglyphs expressing a mysterious language of symbol and form.
ABOVE: *Callanish, Outer Hebrides.*
RIGHT: *Carnac, Brittany.*

The stone circle at Callanish.

known to astronomers as 'the southern extreme of the moon's major standstill' – an event that occurs only once every nineteen years.[12] In addition, one of the principal axes running through Callanish is aligned to sunrise and sunset on the spring and autumn equinoxes.[13] By contrast the principal axis of the world-famous megalithic circle at Stonehenge in the English county of Wiltshire is firmly oriented, through the sighting device of the so-called 'heelstone', to the summer solstice sunrise (the furthest north of east of the sun's annual range along the horizon) and to the winter solstice sunset (furthest south of west).[14]

THE DRUID CONNECTION

Stonehenge was for a long while thought to have been built slowly – over about 1000 years between 2100 and 1100 BC.[15] This chronology was called into question in 1996 by new archaeological evidence. Following a two-year study commissioned by the English Heritage Foundation, researchers concluded that the great circles of bluestones and sarsens had in fact been put up between 2600 BC and 2030 BC – with the bulk of the work being completed in just three centuries between 2600 BC and 2300 BC.[16] Less than a year after these results were published another study showed that the stone circles had been preceded by wooden circles of 6-metre pine 'totem poles' dated to 8000 BC.[17]

Nothing is known about the religious ritual that Stonehenge was built to serve. Indeed, it is only an assumption (though probably a very good one) that religious rituals were conducted here at all. The earliest written accounts of the site date to

Stonehenge at the summer solstice an hour after dawn. The functions of the monument as an observatory and calendar are not in doubt, but the purpose to which these functions were put has never been explored. Could it have been a part of an ancient quest for immortality?

the time of the Romans and by then almost all knowledge of its original function had been lost. Only the Druids, who did not build Stonehenge, were rumoured to know its secrets – but they were not talking.

The Druids first appear in the history books in Julius Caesar's *De Bello Gallico* written at around 50 BC. The reference is relatively concise – less than 1000 words – but in it the Roman Emperor gives us a number of very important clues about the Druid religion:

> In particular they wish to inculcate this idea, that souls do not die, but pass after death from one body to another ... They also dispute largely concerning the stars and their motion, the magnitude of the world and the earth, the nature of things ...[18]

The precise alignments and geometry of Stonehenge tell us that astronomy in general, and the summer solstice sunrise in particular, must have been very important to its builders, but they do not tell us *why* these things were important. The Druid beliefs, which were connected to solstitial ceremonies,[19] could help fill in the gap. They raise the possibility that Stonehenge was not merely the venue of some 'primitive' British solar cult but was once connected to a spiritual quest for reincarnation and the immortality of the soul, in which empirical inquiries about the measurements of the earth and the 'movements' of the stars formed part of the initiatory process.

Scholars do not dispute that the Druids had a special interest in numbers. For some reason they particularly venerated the number 72,[20] which, as we shall discover in later chapters, is derived from astronomical observations. This number appears everywhere in Druid traditions, and there are even 72 letter-strokes required to write out the 22 characters of the 'Ogham' alphabet that the priesthood used for secret communications.[21] The Ogham script also contains a code of its own. As the poet Robert Graves shows in *The White Goddess*, his erudite study of Celtic myths, 'the proportion of all the letters in the alphabet to the vowels is 22 to 7 . . . which is the mathematical formula, once secret, for the relation of the circumference of a circle to its diameter'.[22]

Today, with a pocket-calculator, we can easily work out the circumference of any circle by multiplying its diameter by the transcendental number *pi* – which has a value of 3.141592 . . . To two decimal places, this figure is expressed in the ratio 22:7 (since 22 divided by 7 equals 3.142857).

The sense that the Druids must have been geometers and mathematicians is therefore very strong. Yet it is only an intuition. The truth is that nothing is known about their origins – not even how long they had existed before Caesar first described them.[23] Moreover, although they are associated with the Celts, who entered Britain at around 600 BC, there are suggestions that this charismatic priesthood may have established itself in the British Isles centuries, or even thousands of years, before the Celtic migration.[24] At the very least the Druids must be regarded as the inheritors of genuinely ancient traditions about Stonehenge which they are likely to have preserved and honoured.

NOT IN ISOLATION

Stonehenge and its mysteries do not stand in isolation. Megaliths on an even larger scale were used in ancient Peru and Bolivia to create gigantic monuments such as the zig-zag walls of Sacsayhuaman and the temples of Ollantaytambo and Tiahuanaco with their 200-tonne ashlars. Equally huge blocks were used to build the anonymous 'valley' and 'mortuary' temples of Giza in Egypt and the temple of the Sphinx. The Sphinx itself is a rock-hewn monolith almost 87 metres long.

Sphinx Temple at Giza in a veil of morning mist, with the outer wall of the Valley Temple in the background.

The Pyramids and the Sphinx of Giza. Could these monuments have had a far earlier genesis, a far more extended history and a far more mysterious purpose than historians imagine?

Gateway to the Osireion, Abydos. A mighty memorial to the wisdom of a former civilization?

The Great Pyramid incorporates a number of blocks in the 100-tonne range that were somehow manoeuvred into position at altitudes in excess of 45 metres above the ground. In Upper Egypt there is the problem of the Osireion at Abydos, a truly astonishing semi-subterranean temple made of 100-tonne blocks, seeming to date from a far earlier epoch than the other structures that surround it.[25]

There is no doubt that the Pyramids of Giza, the best known of all the Egyptian monuments, have a strong connection to the epoch 2600 to 2300 BC, the same dates as Stonehenge. There is also no doubt that they show many signs of a far earlier genesis – again like Stonehenge.[26] Indeed, the same geometrical and astronomical concerns that the megaliths express, linked to the same quest for immortality (and frequently to the number 72), are found not only in Egypt but in a great band of cultures encircling the globe and extending back in time to the remotest antiquity.

The common project of all these cultures was to use intelligence and insight to fathom out the mystery of the soul – exactly as Caesar's account suggests that the Druids may have done. Whether in Mexico, as we shall see in Part I, or in Egypt, Cambodia, Easter Island, South America or Celtic Britain, this was a worldwide spiritual inquiry. Very frequently it was associated with the figure of a 'god' or a 'civilizing hero' who was said to have been a great teacher and the founder of religion. And at all times, in all places, it was pursued in the ambience of sacred monuments designed to maintain a link between earth and sky.

The monuments were believed to be gateways to the afterlife realms – both heaven and hell – and it was understood that the ultimate destination of those who passed through would be determined by their own choices.

In Mexico they chose hell.

PART I

MEXICO

CHAPTER I

THE FEATHERED SERPENT AND THE FLAYED MAN

TRAVELLERS IN CENTRAL AMERICA who have attempted to explore its monuments and its past have come away haunted by the intuition of a great and terrible mystery. A dark sorrow overhangs the whole land like a pall, and what is known of its history is filled with inexplicable contradictions.

On the one hand there is tantalizing evidence of lofty spiritual ideas, of a deep philosophical tradition, and of astonishing artistic, scientific and cultural achievements. On the other hand we know that repulsive acts of psychopathic evil had become institutionalized in the Valley of Mexico by the beginning of the sixteenth century and that every year, amidst scenes of nightmarish cruelty, the Aztec empire offered up more than 100,000 people as human sacrifices.[1] Two wrongs do not make a right, and the Spanish Conquistadores who arrived in February 1519 were pirates and cold-blooded killers. Nevertheless, their intervention, motivated exclusively by material greed, did have the happy side-effect of bringing the demonic sacrificial rituals of the Aztecs to an end.

Before the Spanish had completely established their rule a number of the Conquistadores, and of the Roman Catholic priests who came after them, witnessed these rituals. Amongst the witnesses was the conqueror himself, Hernán Cortés, the veteran soldier and swordsman Bernal Diaz de Castillo, and Father Bernadino de Sahagun (1499–1590), an 'extremely wise Franciscan'[2] whose *History of the Things of New Spain* is an unrivalled source of information on pre-Conquest Mexico.[3] Their accounts reveal the dark side of a schizophrenic culture, addicted to murder, which also, with apparently quite staggering hypocrisy, claimed to venerate ancient teachings concerning the immortality of the human soul – teachings that urged initiates to seek wisdom and to be 'virtuous, humble, peace-loving . . . and compassionate' towards others.[4]

The Aztecs reported that the source of this doctrine of non-violence and cosmic gnosis was a god-king known as Quetzalcoatl – 'the plumed serpent' (*quetzal* means, literally, 'feathered' or 'plumed', *coatl* means 'serpent'). He had ruled, they said, in a remote golden age, having come to Mexico from a far-off land with a group of companions. He had taught, quite specifically, that living things were not to be harmed and that human beings were never to be sacrificed, but only 'fruits and flowers of the season'.[5] His cult was absorbed with the mysteries

PREVIOUS PAGE: *'Olmec' head from La Venta, Gulf of Mexico, approximately 1500 BC. The African features of this 4-tonne megalith cannot be explained by the prevailing theory of the peopling of the Americas.*

OPPOSITE: *Sacrificial victims with necklaces of jaw-bones, Valley of Mexico.*

Teotihuacan: emblem of Quetzalcoatl. The god's name means literally 'feathered serpent' and he is frequently depicted as such in pre-Columbian Mexican art. But in legend he also manifested as a 'fair and ruddy-complexioned man with a long beard', a great civilizer, an astronomer and a builder, who came to Mexico to teach compassion and a religion of intelligence.

of life beyond death and he was said to have made a journey to the underworld and to have returned to tell the tale.[6]

GOD OF THE PYRAMIDS

Could there have been a real historical figure behind the Quetzalcoatl story?

During the two or three hundred years before the Aztecs' rise to prominence in the thirteenth and fourteenth centuries AD a number of kings, particularly amongst the Toltecs of the Valley of Mexico, are known to have called themselves 'Quetzalcoatl'. However, they did not claim to be *the* Quetzalcoatl, but rather to be his successors, and used his name in the manner of a title or honorific.[7]

Far earlier than the Toltecs, as we shall see in the next chapter, the distinctive symbolism of Quetzalcoatl was known to the Olmecs of the Gulf of Mexico. Their culture flourished 3500 years ago. Later, although still more than 2000 years ago, work began at Cholula in central Mexico on a gigantic monument, named in honour of the same god, which continued to be extended and increased in size by all subsequent cultures occupying the site until finally being halted for ever by the firestorm of the Conquest. The result of this amazingly sustained project of engineering and sacred architecture is the pyramid-mountain of Quetzalcoatl.

The Great Pyramid of Cholula at sunset – a gigantic structure built over thousands of years in honour of the cult of Quetzalcoatl. After the Conquest the sacred mound was spiritually 'capped off' by a Roman Catholic chapel

Today spiritually 'capped-off' by a Roman Catholic chapel, its base area of 18 hectares and height of almost 70 metres make it three times more massive than the Great Pyramid of Egypt.[8]

Also around 2000 years ago the mysterious culture of Teotihuacan, 35 kilometres to the north-east of modern Mexico City, was strong in its veneration of Quetzalcoatl, constructing pyramids and other monuments in his honour. And from at least 1500 years ago until the time of the Conquest, the Maya of the Yucatan, Chiapas and Guatemala worshipped him under the names Kukulkan and Gucumatz (both of which, in different dialects, mean 'feathered serpent').[9] Between AD 900 and AD 1200, a nine-tiered pyramid of Quetzalcoatl/Kukulkan was completed at Chichen Itza. Like many of the monuments dedicated to this deity it was constructed on top of an earlier sacred mound that had occupied the same site.[10]

So the Aztecs, who only declared their empire in the 1320s AD, were merely passing on an ancient tradition picked up from their predecessors when they spoke of the 'feathered serpent' and described him, most unambiguously, as:

> a fair and ruddy-complexioned man with a long beard ... A mysterious person ... a white man with strong formation of body, broad forehead, large eyes, and a flowing beard, who came from across the sea in a boat that moved by itself without paddles. He condemned sacrifices, except of fruits and flowers, and was known as the god of peace.[11]

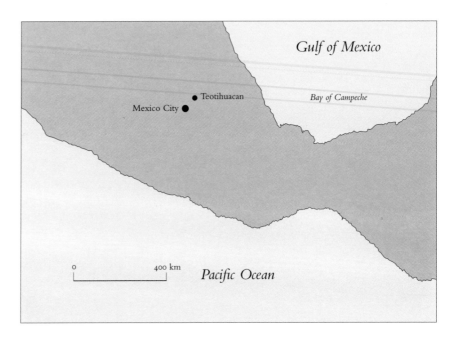

The step-pyramid of Quetzalcoatl/
Kukulkan at Chichen Itza, a
monument to the Feathered Serpent.

It is one of the great unexplained puzzles of Central American history that the murderous Aztecs worshipped and honoured this benign figure in all manner of rituals, and always spoke with awe of his peaceful and life-giving ways. They believed that he and his followers had been driven out of Mexico long ages previously, but that they would one day return, coming from the west, by boat. They also believed that Quetzalcoatl would punish them for having reverted to human sacrifice, that he would put an end to evil and fear, and that he would restore the golden age of peace and plenty over which he had presided in the mythical past.[12]

As is well known this tradition of the white-skinned, bearded god-king worked greatly to the advantage of the Spaniards when they arrived in Mexico in 1519, in boats that moved by themselves without paddles, suffering from that 'disease of the heart' for which looted gold was the only specific remedy.[13]

BROKEN FLUTES

The Conquistador Bernal Diaz de Castillo was a hard man and no stranger to violence. Nevertheless, he was badly shaken by his first experience of an Aztec temple:

> In that small space there were many diabolical things to be seen, bugles and trumpets and knives, and many hearts of Indians that they had burned in fumigating their idols, and everything was so clotted with blood, and there was so much of it, that I curse the whole of it, and as it stank like a slaughterhouse we hastened to clear out of such a bad stench and worse sight.[14]

The physical destruction of sacrificial victims, most of whom had been seized in battle, usually involved ripping out their hearts, and was frequently followed by ceremonies in which 'the celebrants flayed and dismembered the captives'.[15]

Sometimes victims would be sacrificed immediately after they had been captured and sometimes much later after long-drawn-out and excruciating anticipation. For example, Sahagun tells us of a 'feast' in the Aztec capital of Tenochtitlan at which:

> they killed a youth of very docile temperament, whom they had kept for the space of a year in pleasurable activities ... When this youth who had been cherished for one year was killed, they immediately put another in his place to be regaled throughout the next year ... From among all the captives they chose the noblest men ... they took pains that they should be the ablest and best-mannered they could find, and with no bodily blemish.[16]

Sahagun further reports that 'the youth who was reared in order to be killed at this feast [was] taught very diligently how to sound the flute well'.[17] When his year was up:

> they took him to a small and ill-furnished temple that was beside the road and away from any settlement ... Having reached the temple steps, he climbed them himself, and on the first step he broke one of the flutes he had played in the time of his prosperity, on the second he broke another, on the third another, and so he broke them all as he climbed the steps. When he had come to the top, to the highest part of the temple, there were the priests who were to kill him, standing in pairs, and they took him and bound his hands and head, lying him on his back upon the block; he that had the stone knife plunged it into his breast with a great thrust, and drawing it forth, put his hand into the incision the knife had made, and pulled out the heart and offered it at once to the sun.[18]

The cult of death.

Coatlicue, the Aztec mother of gods and men, who was believed to feed on human corpses. Her head is shown severed from her body. In its place, two serpents form the illusion of a monstrous face. She wears a necklace from which is suspended a skull, strung together with human hearts and hands. The figure bears a curious similarity to the Hindu death goddess Kali with her necklace of skulls.

When the priests were questioned by the bemused Spaniards about the reasons for this horrible annual ritual, they are reported to have replied, as though they thought it should have been obvious to all, that the tragic story of the prisoner was 'the type of human destiny'.[19]

STIFLING THE VOICE OF NATURE

The Aztecs were relentless murderers of children – as Sahagun's bare catalogue of certain 'feasts' reveals:

On the calends of the first month . . . they killed many children, sacrificing them in many places and on the hill tops, removing their hearts in honour of the gods of water . . . On the first day of the fourth month they held a feast in honour of the corn god . . . and killed many children . . .[20]

Tlaloc, the god of rain was especially insatiable. To him:

children, for the most part infants, were offered up. As they were borne along in open litters, dressed in their festal robes, and decked with the fresh blossoms of spring, they moved the hardest heart to pity, though their cries were drowned in the wild chant of the priests, who read in their tears a favourable augury for their petition. These innocent victims were generally bought by the priests of parents who were poor, but who stifled the voice of nature probably less at the suggestions of poverty, than of a wretched superstition.[21]

Or was it perhaps *fear* that was the main instrument in this 'stifling of the voice of nature'? In her classic study *Burning Water* the archaeologist Laurette Sejourne argues compellingly that the Aztecs had created:

a totalitarian state, the philosophy of which included an utter contempt for the individual . . . Death lurked ceaselessly everywhere, and constituted the cement of the building in which the individual Aztec was prisoner. There were those who, by their social status, were by law destined to extermination: the slaves – and anyone might become one through losing his fortune or civil rights; captive warriors; children born under a sign favourable for sacrifice and offered to the gods . . . Capital punishment was another constant threat: to anyone who dared without authority to wear a garment that reached below the knee; to the official who ventured into a forbidden room of the palace; the merchant whose riches had made him too proud; the dancer taking a false step . . .[22]

So important was human sacrifice within this monstrous 'mechanism for breaking men down' that any member of the public who was reluctant to spectate was 'held to be despicable, declared unworthy of all public office, and converted into a wretch without the law'.[23] Chiefs and lords, meanwhile, were frequently obliged to witness the sacrifices 'on pain of being sacrificed themselves if they did not attend'.[24]

136,000 SKULLS

What they witnessed was awful beyond imagination. According to the chronicler Munoz Camargo:

one who had been a priest of the devil told me . . . that when they tore the heart from the entrails and side of the wretched victim, the strength with which it pulsated and quivered was so great that [it] used to lift . . . three or four times from the ground [before] the heart had grown cold . . .[25]

Relief showing a skull-rack, Chichen Itza.

Sahagun recalls a feast at which:

> The owners of the prisoners dragged them by the hair, each one his own, up the steps ... They dragged them to the block of stone where they were to kill them, and taking from each one his heart ... they hurled their bodies down the steps, where other priests flayed them.[26]

On the occasion of a ritual in honour of the god of fire, the owners:

> took and bound [their captives] hands behind, and also their feet; then they threw them over their shoulders, and carried them to the top of the temple, where was a great fire and a great heap of coals; and when they had arrived at the top, they gave them to the fire ... and there in the fire the unhappy prisoner began to twist and to retch ... and he being in such agony, they brought him out with grappling irons ... and placed him on the stone ... and opened his breast ... and took out the heart and threw it at the feet of the statue of Xiuhtecutli, god of fire.[27]

It was customary for temples to preserve the skulls of sacrificial victims spiked on the horizontal cross-beams of specially designed wooden racks called *tzompantli* 'in buildings appropriated to the purpose'.[28] In one such building, soldiers with Cortés counted a total of 136,000 skulls, which were 'so arranged as to produce the most hideous effect'.[29]

ALLEGORIES TAKEN LITERALLY

As we have already observed, the great mystery of Central America is that a culture capable of such unmitigated ferocity was also a vehicle for profound religious ideas. We have to thank in particular Bernadino de Sahagun for documenting these ideas and passing them down to us – since without his intervention it is unlikely that we would know anything about them today. His *History of the Things of New Spain* has been described by modern scholars as 'the most complete ethnographic investigation of any people' and has been commended for its use of 'the most rigorous and exacting of scientific anthropological methods'.[30]

In 1956 Laurette Sejourne, drawing on the rich fund of ethnographic and religious material gathered by Sahagun, put forward a remarkable theory about the Aztecs.[31] Their entire cult of human sacrifice, she argued, had come about because an ancient system of purely spiritual initiation, linked to a quest for immortal life, had been grotesquely misunderstood. All the gruesome physical aspects of Aztec sacrifices – flaying, cutting out the heart, burning, etc. – were originally *metaphors* for spiritual processes to be undergone by initiates. 'Flaying' meant the disciplines that enabled the initiate to become detached from his physical body. The 'heart' signified the soul which was to be 'cut out' of the body at death and liberated into the land of light (this latter of course symbolized by the sun). 'Burning' was the fire of renewal, in which the eternal spirit, rising phoenix-like from the ashes of its own previous existence, would cast off the worn-out physical form of one life and be reborn again into another.

With such metaphors in mind it is easier to understand why the Aztec sacrificers described the plight of the victim so brutally killed after being 'cherished for one year' as 'the type of human destiny'. The allegory that he re-enacted as he made his way to the temple where he was to be murdered, casting off all the material finery that he had accumulated and finally breaking the flutes on which he had once played sweet music, was intended to symbolize one of the ultimate truths learned by initiates – that on death the soul must leave everything of the material, terrestrial world behind. This truth is simply encapsulated in a fragment of an ancient teaching in the Nahuatl language spoken by the Aztecs that was preserved by Sahagun in the sixteenth century:

> My well loved and tender son ... know and understand that thy house is not here ... This house wherein thou art born [i.e. the physical body] is but a nest, an inn at which thou has arrived, thy entry into this world; here dost thou bud and flower ... thy true house is another.[32]

Similarly we read:

> Birth comes, life comes upon earth.
> For a short while it is lent us,
> the glory of that by which everything lives.
> Birth comes, life comes upon earth.[33]

Sejourne makes a strong case that the system of thought to which these profound ideas belong had been present in the Valley of Mexico for an unknowable length

Aztec calendar stone depicting Tonaituh, the Fifth Sun, the face and symbol of our current epoch of the earth. His tongue, an obsidian knife, juts out hungrily, signalling his need for the nourishment of human blood and hearts. His features are wrinkled to indicate his advanced age and he appears within the symbol ollin which signifies 'movement' (see discussion below, page 15). As well as conveying a gruesome message, the calendar incorporates advanced astronomical knowledge which the Aztecs inherited from earlier civilized peoples who had inhabited the Valley of Mexico for thousands of years before their arrival.

of time, perhaps thousands of years, before it was first encountered by the Aztecs. They were simply the most ferocious amongst a large group of migrant Nahuatl-speaking tribes known as the Chichimeca (a word that signifies 'barbarians') who poured into Central America from the north in the twelfth and thirteenth centuries AD. Having no culture of their own, they set about appropriating the fading remnants of the once-great Mexican civilization that they had overwhelmed, learning its astronomy, its agriculture, its engineering and its architecture, and acquiring aspects of its religious paraphernalia. They were particularly transfixed by its colourful and dramatic initiation rituals which they co-opted wholesale. Tragically, however, they did not understand – or did not want to understand – that the rituals were metaphorical dramas that were only intended to be played out symbolically. They took them literally, with horrific consequences.

ECHOES OF THE FIRST MEN

The Aztecs believed that their tribe had been born from womb-like caves in the heart of a mountain, that Aztlan, their first homeland, had been established on an island,[34] and that they had been ordered to leave this island by their god Huitzilpochtli, who had prophesied: 'The four corners of the world shall ye conquer, win and subject to yourselves ... it shall cost you sweat, work and pure blood.'[35]

The god also foretold that in their wanderings they would one day come across an eagle perched amongst the thorns of a cactus sprouting from a rock.³⁶ On this spot they were to build the capital of their empire.

Like the Nazis mesmerized by Hitler, the Aztecs set about fulfilling the vision of Huitzilpochtli, easily overrunning the incumbent peoples of the Valley of Mexico and using war systematically to weaken and subjugate the other Chichemec tribes. In the early fourteenth century, in the swamps of Lake Texcoco, a rock was seen from which sprouted a cactus with an eagle perched amongst its thorns. The prophecy was remembered and work was begun on the city of Tenochtitlan, which became the Aztec capital in AD 1325.³⁷

In the two centuries that remained until the Conquest, the Aztec empire centred on Tenochtitlan continued to expand its power and the city itself grew correspondingly in size to become a great metropolis of 300,000 people. Built on an artificial island, surrounded by a series of square and circular canals, it was oriented to the cardinal directions with masterfully engineered causeways running across the waters into its four principal entrance gates. The central plaza was dominated by a gigantic four-tiered step-pyramid, which the Spanish, who greatly admired its architecture, called the Templo Mayor. Cortés asserted that 'no human tongue could describe its size and characteristics',³⁸ while Bernal Diaz reported: 'there were soldiers that had been in many parts of the world, in Constantinople and in all Italy and Rome, and they said they had never seen a plaza so well proportioned and so orderly, and of such a size and so full of people'.³⁹

What is particularly notable is that the Aztecs did not take credit for the high level of civilization attested to by these and many of the other wonders of their booming empire. Instead they frankly admitted that their entire system of knowledge had come to them in one piece, as a legacy, from the god-king Quetzalcoatl and his companions, those who were 'the first inhabitants of this land, and the first that came to these parts called the land of Mexico ... those who first sowed the human seed in this country'.⁴⁰

Anyone seeking to discover the identity of these 'first men', and in which epoch their civilization might have flourished, will quickly realize that there are tremendous problems with Central American history prior to the period of Aztec expansion. Indeed, before about AD 1000 there is almost no 'history' at all – and less and less the further back in time we look. In consequence, scholars know next to nothing about the origins of the three earliest and oldest high civilizations that have been identified in the region: the Olmecs, who supposedly flourished, mainly along the coast of the Gulf of Mexico, from well before 1500 BC until about the time of Christ; the Maya, who paralleled and then succeeded them and whose living descendants are still found throughout Central America today; and the civilization that built the awe-inspiring sacred domain of Teotihuacan almost 2000 years ago. Laurette Sejourne, who conducted extensive excavations at the latter site, remarked in 1956: 'The origins of this high culture are a complete mystery.'⁴¹ In 1995, after almost 40 years of further excavations, University of California archaeologist Karl Taube was obliged to admit: 'We still don't know what language the Teotihuacanos spoke, where they came from, or what happened to them.'⁴²

The view over Teotihuacan from Cerro Gordo.

THE DEATH AND REBIRTH OF WORLD AGES

In September 1996, a few days before the autumn equinox, we climbed the mountain known as Cerro Gordo and looked down from its summit over Teotihuacan, a Nahuatl name given to the city by the Aztecs and meaning literally 'the place where men became gods'.[43] Stretched out below us was the arrow-straight axis of the so-called 'Way of the Dead', which runs 15 degrees and 30 minutes east of true north and west of true south. At its northern extremity, towering 46 metres high, we could see the five-tiered hulk of the Pyramid of the Moon. A kilometre to the south of it, on the east side of the Way of the Dead, lay the more massive Pyramid of the Sun, more than 70 metres high and measuring 222 metres along each of its four sides.[44] Beyond that, set back within a spacious rectangular enclosure, was the smaller Pyramid of Quetzalcoatl.

Professor Michael D. Coe of Yale University has pointed out that the Pyramid of the Moon and Pyramid of the Sun are both 'explicitly named in old legends'[45] and concludes: 'there is no reason to doubt that they were dedicated to those divinities'.[46] The same names were adopted by the Aztecs when they stumbled upon Teotihuacan soon after they had begun to infiltrate the Valley of Mexico in the twelfth century AD. The gigantic geometrical city, already remotely ancient, had by then fallen into ruins. Nevertheless, the newcomers preserved distinct traditions concerning it, suggesting either that its reputation may have reached them before they migrated from their former homeland or more likely that they had learned about it from peoples whom they conquered on their route to power. At any rate they respected it so highly that Montezuma, the Aztec emperor at the time of the Conquest, is reported to have made frequent pilgrimages on foot to the Pyramid of the Sun during the last years of his reign.[47]

In common with all his subjects, Montezuma believed the pyramid to be the original primeval mound marking the spot where creation had been set in motion at the beginning of the present epoch of the earth. Traditions that were widely distributed throughout Central America, notably amongst unrelated neighbouring cultures such as the Maya, asserted that there had been four previous epochs, or 'Suns', each of which had ended in a cataclysm that had wiped the face of the earth clean. The fifth epoch was said to have begun in darkness on 4 Ahau 8 Cumku, a date in the Mayan calendar corresponding to 13 August 3114 BC, and was expected to come to a catastrophic end as a result of a 'great movement of the earth' on 4 Ahau 3 Kankin – which corresponds in the modern calendar to 23 December AD 2012.[48]

The Aztecs claimed that their motive for carrying out human sacrifices was to postpone, or if possible to prevent, the predicted end of the world – by offering up an endless supply of hearts and blood to 'rejuvenate' the aged Fifth Sun. When all things are considered, however, it seems most unlikely that they were ever sincere in this belief. As Sejourne argues, their repressive political system 'was founded on a spiritual inheritance which it betrayed and transformed into a weapon of worldly power'.[49] In other words they made gross material travesties of subtle symbolic rituals and exploited the old prophecy of the end of the Fifth Sun as a transcendant justification for the policies of terror that so many of them appear to have relished and that were used 'to prop up their bloody state'.[50]

View down on the apex of the Pyramid of the Sun, believed by the Aztecs to be the original primeval mound marking the spot where creation itself was set in motion.

Sunrise over the Pyramid of the Sun, meeting place of the gods.

BIRTH OF THE FIFTH SUN

Recognition of the truth behind the Aztec propaganda should not blind us to the fact that the traditions concerning Teotihuacan incorporate a cosmic vision, both ancient and sophisticated, that sees the passage of time not as a linear process but as an endless series of immense cycles which witness the rise and fall of world systems.

Fragments of this vision are found in the oldest Nahuatl songs and poems. They proclaim that after the destruction of the Fourth Sun by a universal flood ('there was water for 52 years and then the sky collapsed'[51]), the gods gathered together on the summit of the primeval mound at Teotihuacan to decide who was to sacrifice himself so as to become the new, Fifth Sun and bring light again to the world:[52]

> Even though it was night,
> even though it was not day,
> even though there was no light they gathered,
> the gods convened
> there in Teotihuacan.[53]

Two of them competed for the honour to throw themselves into the sacred fire from which the Fifth Sun would be kindled – the handsome and worldly Tecciztecal, 'who was arrogant and greedy for glory',[54] and humble, self-effacing Nanahuatzin, 'the bubonic god, ailing and covered with pustules'.[55] At the last moment Tecciztecal baulked before the fierce heat of the flames. Nanahuatzin, however, 'made an effort and closed his eyes, and rushed forward and cast himself into the fire, and then he began to crackle and burn in the fire like one roasting'.[56] As a result of this act of cosmic altruism the Fifth Sun finally arose, ushering in our present epoch: 'it took sight from the eyes, it shone, and threw out rays splendidly, and its rays spilt everywhere'.[57]

THE QUEST FOR IMMORTALITY

Sejourne places particular importance on the fact that the god whose sacrifice gives birth to the Fifth Sun is:

> the scabby one, he whose body is disintegrating, that is the one who having completed the task of reconciling the opposites has begun to be detached from his fragmentary self ... This tale, with its ritual details and secret formulae, seems to constitute the model for the final trial of Initiation, which leads through death to eternal life.[58]

It is striking that many of the myths, traditions and liturgies that have come down to us concerning Teotihuacan are deeply concerned with the immortality of the human soul and the hard work that the initiate must do in order to attain it. 'Oh brother,' warns the inquisitor in one remnant of the ancient rituals, 'thou hast

Nanahuatzin, the bubonic god, whose sacrifice ushered in the Fifth Sun.

The Pyramid of the Moon, 46 metres high, overlooks a broad plaza at the northern end of the Way of the Dead.

Hieroglyph for a cycle of time surmounting a pillar in the so-called 'Palace of the Quetzal Butterfly' at the south-western end of the Moon plaza. Ancient legends link Teotihuacan to cyclical notions of life, death and rebirth.

come to a place of great danger, and of much work and terror . . . thou hast come to a place where snares and nets are tangled and piled one upon another, so that none can pass by without falling into them . . . These are thy sins.'[59]

This leaves us in little doubt that one aspect of the rituals conducted at Teotihuacan must have been to do with penance and a spiritual ordeal. The same sources, however, make it equally clear that *the sacred city itself was also seen as an intrinsic part of the initiatory mechanism.* Some special and mysterious quality adhered to its monuments which provided the numinous setting where physical death was believed to lose its sting and mortal men could be transformed into immortal gods:

And they called it Teotihuacan
because it was the place
where the lords were buried.
Thus they said:
'When we die,
truly we die not
because we will live, we will rise,
we will continue living, we will awaken . . .'
Thus the dead one was directed,
when he died:
'Awaken, already the sky is rosy . . .'
Thus the old ones said
that he who has died has become a god,
they said: 'He has been made a god there.'[60]

Nahuatl traditions further tell us that the process of turning men into gods was overseen by a body of priests – the 'old ones', who directed the deceased to awaken. These priests are said to have been 'wise men, knowers of occult things, possessors of traditions'.[61] Sometimes referred to as the 'Followers of Quetzalcoatl',[62] they are described in the *Popol Vuh*, the sacred book of the ancient Quiche Maya, as 'Plumed Serpents ... Great knowers, great thinkers in their very being'.[63] It was they who taught the extraordinary rituals of spiritual initiation that were ultimately to be hijacked by the Aztecs during the 200 years before the Conquest.

Quetzalcoatl

The central figure in the ancient Mexican rebirth rituals was the 'Sovereign Plumed Serpent' Quetzalcoatl – the god-king of the golden age who had died, the legends said, but who would one day return. The well-known physical descriptions of this 'once and future king' which the myths relay – tall, white-skinned, yellow-haired and bearded – make him sound like a Caucasian. Yet no Caucasians are supposed to have visited the New World before the time of Columbus, and Quetzalcoatl is far older than that. Scholars therefore for a long while played down the deity's distinctly Caucasian features, which did not accord with prevailing theories about the peopling of the Americas and were also felt by some parts of the native American lobby to be politically incorrect.

Startling new evidence has recently emerged which calls for a complete re-evaluation of such thinking:

> *Washington Post, Final Edition, Tuesday April 15 1997:*
> Skeletons unearthed in several western states and as far east as Minnesota are challenging the traditional view that the earliest Americans all resembled today's Asians. The skeletons' skulls bear features similar to those of Europeans, suggesting that Caucasoid people were among the earliest humans to migrate into the New World more than 9000 years ago. Anthropologists have known of such bones for years, but did not fully appreciate their significance until reappraising them over the last few months. The new analyses were prompted by the discovery last summer of the newest addition to the body of evidence – the unusually complete skeleton of an apparently Caucasoid man who died about 9300 years ago near what is now Kennewick, Washington ... The man's head and shoulder were mummified, preserving much of the skin in that area ... Those who examined [him at first thought the bones] were the remains of a European settler [until radiocarbon dating revealed their great age]. 'It's an exciting time, and I think we're going to see some real changes in the story about the peopling of North America', said Dennis Stanford, an authority at the Smithsonian Institution's National Museum of Natural History.[64]

How old are the myths of Quetzalcoatl? Could they go back to these prehistoric Caucasians, 'similar to Europeans', who were in the Americas in the Stone Age, at least 9000 years ago?

CHAPTER 2

ON EARTH AS IT IS IN HEAVEN

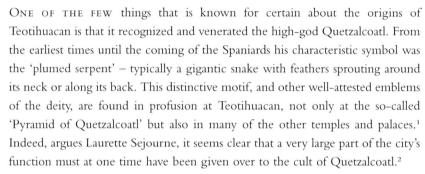

Western façade of the Pyramid of Quetzalcoatl, Teotihuacan.

OPPOSITE: *Inner chamber of the Pyramid of Kukulkan, Chichen Itza. Kukulcan, meaning 'feathered serpent', was the Mayan form of Quetzalcoatl. His pyramid at Chichen Itza has powerful astronomical characteristics (see pages 26–7) and surmounts an earlier structure on the same site of which this eerie chamber is a part. The jaguar's spots are formed out of 72 separate pieces of jade. This number is related to the astronomical phenomenon known as the precession of the equinoxes and is found in the dimensions of ancient structures all around the world.*

ONE OF THE FEW things that is known for certain about the origins of Teotihuacan is that it recognized and venerated the high-god Quetzalcoatl. From the earliest times until the coming of the Spaniards his characteristic symbol was the 'plumed serpent' – typically a gigantic snake with feathers sprouting around its neck or along its back. This distinctive motif, and other well-attested emblems of the deity, are found in profusion at Teotihuacan, not only at the so-called 'Pyramid of Quetzalcoatl' but also in many of the other temples and palaces.[1] Indeed, argues Laurette Sejourne, it seems clear that a very large part of the city's function must at one time have been given over to the cult of Quetzalcoatl.[2]

This cult sought the spiritual transfiguration of its initiates – a process of hard mental work and self-denial aimed at 'flaying away' attachments to the sense-world and thus releasing the soul, envisaged as 'a precious stone and a rich feather',[3] from the encumbrance of matter. Part of the work appears to have involved the enactment of rituals in which the initiate would undergo a symbolic death and rebirth like the god Quetzalcoatl, the once and future king, who was said to have voluntarily enclosed himself for four days within a sarcophagus – 'a box made of stone'[4] – from which he then arose, making his way to 'the celestial shore of the divine water'.[5] There:

> he stopped, cried, seized his garments, and put on his insignia of feathers ... Then when he was adorned he set fire to himself and burned ... It is said that when he burned his ashes were at once raised up and that all the rare birds appeared when Quetzalcoatl died ... for which reason in eight days there appeared the great star called Quetzalcoatl.[6]

Two potent themes, which scholars have hitherto associated only with the Old World and not with the New, are apparent in this pre-Colombian Central American tale – the theme of the phoenix rising again from the ashes of its own funeral pyre and the theme of stellar rebirth which, as we shall see in Part II, makes its earliest recorded appearance far away from Mexico in the ancient Egyptian Pyramid Texts of the third millennium BC. As in ancient Egypt also, where the sun's apparently infinite capacity for self-renewal served as an eloquent symbol for reincarnation, the ancient Mexicans spoke of the sun as 'the king of those who return'.[7]

Teotihuacan: stellar Venus symbol dispensing influence downwards towards the earth. Such symbols were part of an ancient and widely disseminated religious idea that 'all the world which lies below was set in order and filled with contents by the things which are placed above'. Compare with the second shrine of Tutankhamun, page 88.

AFTERLIFE REALMS

Such parallels exist across the entire spectrum of religious beliefs. For example, the ancient Egyptians envisaged a terrifying afterlife realm, located in a region of the sky that they referred to as the 'Duat' (usually translated as 'Netherworld') in which it was believed that the souls of the deceased must undertake a perilous journey.[8] The ancient Mexicans spoke of an afterlife realm, equally terrible, called 'the Land of the Mystery':

> The more I weep, the more I am afflicted,
> the more my heart may not desire it,
> have I not, when all is said, to go to the Land of the Mystery?
>
> Here on earth our hearts say:
> 'Oh my friends, would that we were immortal,
> oh friends, where is the land in which one does not die?'
>
> Shall it be that I go? Does my mother live there?
> Does my father live there?
>
> In the Land of the Mystery . . . My heart shudders:
> if only I had not to die, had not to perish . . . !
> I suffer and feel pain.[9]

The Mexicans taught that the deceased would have to overcome seven difficult ordeals in the Land of the Mystery, the last of which was his final judgement in the terrifying presence of the death god.[10] There was also a tradition that

Quetzalcoatl himself had opened the way to possible triumph for future travellers in the netherworld by bringing back from it the bones of the ancestors that lay hidden there and restoring them to life.[11] An almost identical function was attributed by the ancient Egyptians to their high god Osiris, lord of resurrection and rebirth.[12]

Very much like the ancient Egyptians, the peoples of ancient Central America located their netherworld in a region of the sky through which ran the Milky Way.[13] Another curious similarity is that both peoples appear to have believed that the gates of the afterlife realm swung open 'in the red glow of twilight which precedes the dawn'.[14] What is most striking, however, is the extent to which both systems of initiation focussed on astronomy, particularly upon an esoteric knowledge of the cycles of the heavens, and aspired to immortality amongst the stars. For this reason, the Aztec wise men, when asked to contemplate the meaning of death, 'said that they did not die but woke from a dream they had lived ... and became once more spirits or gods ... They said, too, that some were transformed into the sun, others into the moon.'[15]

Such an apotheosis was the ultimate goal of initiates on the path of Quetzalcoatl, 'he who knows the mystery of all enchantments',[16] of whom the myths said: 'above all he taught man science, showing him the way to measure time and study the revolutions of the stars'.[17]

CELESTIAL IMITATION

Against the background of ideas such as these, it is not surprising that the sacred city of Teotihuacan, with its pyramids dedicated to Quetzalcoatl, to the sun and to the moon, has a complex astronomical design that connects it intricately to the heavens.

We have seen that the principal axis of the city, the Way of the Dead, is deliberately offset 15 degrees and 30 minutes to the east and west of true north–south.

BELOW RIGHT: *The axis of the Way of the Dead, looking west of south, with the Pyramid of the Moon in the foreground and the Pyramid of the Sun in the background to the left. The American engineer Hugh Harleston Jr. (The Keystone: A Search for Understanding) has shown from more than 9,000 on-site measurements that the great monuments of Teotihuacan are laid out on a grid system along and around this axis and that they were designed by architects using a standard unit of measure, 1.05946 metres – the 'Standard Teotihuacan Unit' (STU). When expressed in STU, key dimensions within and between the different monuments all turn out to belong to a definite sequence based on the number 72, itself derived from the rate of the precession of the equinoxes (see page 29 ff). Thus, the centres of certain structures are 72 STU apart, or 36 STU (half of 72), or 108 STU (72 + 36), or 216 STU (108 × 2), or 54 STU (half of 108), or 540 STU (54 × 10), etc. The sequence expands outwards in multiples to 2,160 STU, 4,320 STU, 5,400 STU, etc. The length of one side of the Pyramid of the Sun at the base is 216 STU. The east–west axis of the Pyramid of the Moon is 144 STU. The centre of the Pyramid of the Sun lies 720 STU south of the centre of the Pyramid of the Moon. Is it a coincidence (see Part III) that the standard unit of measure used in the temples of Angkor in Cambodia yields the same sequence of numbers?*

The explanation of this offset is not to be found in the Way of the Dead itself but in Teotihuacan's dominant structure, the Pyramid of the Sun, which was built with its west face oriented 15 degrees and 30 minutes north of due west and its east face oriented 15 degrees and 30 minutes south of due east. The trajectory of the Way of the Dead, in other words, is determined by the orientation of the west face of the Pyramid of the Sun.

This orientation is not random. It targets the point of sunset on the western horizon on two astronomically significant days, 19 May and 25 July – the only two days of the year on which, at noon, the sun passes through the zenith vertically overhead at the latitude of Teotihuacan (19.5 degrees north of the equator).[18] A strong alignment to the setting of the Pleiades star-cluster in the constellation of Taurus has also been identified at Teotihuacan – and at around AD 150, a date that accords well with the archaeology of the site. In that epoch, on the first of the sun's two annual days of zenith passage, computer simulations reveal that the Pleiades would have performed what astronomers call their 'heliacal rising' – i.e. they would have been visible rising low in the east in the rosy skies just before dawn.[19]

For the ancient Maya of Central America, who are known to have been in regular and extensive contact with Teotihuacan at the height of its powers,[20] the Milky Way was a particularly important feature of the heavens. They conceived of it as the road that led to their netherworld, Xibalba,[21] which, in common with other Central American peoples, they located in the sky.[22] One scholar who was aware of this cosmology, and who also undertook an extensive archaeoastronomical

West face of the Pyramid of the Sun, around sunset. The pyramid's orientation, which defines the curious offset of the Way of the Dead from true north–south, targets the point of sunset on two astronomically significant days.

Teotihuacan, 'Palace of the Jaguars': celestial puma against a background of stellar symbols blowing a feathered conch shell. It was a widespread convention of the ancient world that the December solstice should be symbolized by a conch shell.

investigation at Teotihuacan, was Stansbury Hagar, Secretary of the Department of Ethnology at the Brooklyn Institute of the Arts and Sciences. In academic papers published in the 1920s he presented powerful evidence to suggest that the Way of the Dead at Teotihuacan – which was also known in certain traditions as the 'Way of the Stars' – might have been intended by its designers 'to represent the Milky Way', serving as a symbolic path on which 'spirits were believed to pass ... between earth and the land of the souls amid the stars'.[23]

At the heart of this provocative idea is an image of the earth as a 'mirror of heaven' – or rather of works of architecture being built 'below', on earth, to mimic specific sky-features, 'above', and to align themselves with important celestial events. As we shall see in later chapters, the three great Pyramids and the Great Sphinx of Egypt's Giza plateau were laid out according to such a plan and, moreover, in relation to the river Nile, which was seen as the terrestrial reflection of the Milky Way. How likely is it to be an accident that precisely the same system of ideas is found at Teotihuacan, which, according to Hagar, 'reproduced on earth a supposed celestial plan of the sky-world where dwelt the deities and spirits of the dead'.[24]

Hagar concludes his investigation by observing: 'we have not realized either the importance, or the widespread distribution throughout ancient America, of the astronomical cult of which the celestial plan was a feature, and of which Teotihuacan was at least one of the principal centers'.[25]

If he is right then this 'cult', whatever its origins, must be very old. We will find it in Egypt as early as the third millennium BC. In Mexico, archaeologists are reasonably sure that the monuments of Teotihuacan in their present form were built during the period AD 150 to 300. Underneath the façades of all three of the pyramids, however, have been found the traces of earlier layers of construction, marking earlier pyramids that stood on the same sites, going back to primordial sacred mounds – some of which contained caves – that were venerated in far antiquity.[26]

SEVENTY-TWO PIECES OF JADE

On 21 September 1996, the autumn equinox, we were at Chichen Itza in the Yucatan, a Mayan site thought to date to the sixth century AD and to have been continuously developed thereafter until at least the thirteenth century.[27] The central pyramid is dedicated to Kukulkan (Quetzalcoatl).[28] Consisting of nine superimposed platforms, it towers 30 metres high and measures 55.3 metres along each side. Like the Pyramid of the Sun at Teotihuacan its principal axis is deliberately tilted east of north and west of south. Here, however, the angle selected is not designed to signal the days of the sun's zenith passage but instead the spring and the autumn equinoxes, when the sun rises precisely due east and the hours of light and darkness are equal.

By about 5.15 in the evening it was clear what was happening. So skilfully was this magnificent pyramid aligned to the trajectory of the setting equinoctial sun

The ancient magnetism of the Pyramid of Kukulkan at Chichen Itza still draws people towards it. This picture was taken at around 3.40 p.m. on 21 September 1996 and shows the pool of shadow out of which the illusion of an undulating serpent is formed along the balustrade of the northern stairway.

The epiphany of the feathered serpent in its complete form at around 5.30 p.m. Within ten more minutes the shadowed area has filled in entirely and the illusion is gone.

that it had been possible for the ancient builders to contrive a pattern of light and shadow on the western side of the northern stairway. Very gradually, as the minutes ticked by and the sun fell lower in the sky, this pattern, which was projected by the north-western corner of the pyramid, gained in shape and substance. By around 5.30 p.m., it had manifested itself fully as a gigantic undulating serpent with seven coils of shadow defined by seven triangles of light. The tail of the serpent reached the top platform of the pyramid, with its body extending down the balustrade all the way to the ground where a huge sculpted serpent's head with gaping jaws completed the illusion at the base of the stairway.

This almost magical epiphany of an ancient deity as a signal of the equinoxes indicates that an advanced geodetic and astronomical science must at one time have been practised at Chichen Itza – for only a civilization with highly skilled surveyors, astronomers and setting-out engineers could have achieved the razor-sharp alignments necessary to materialize such an image in such a way at precisely the desired moment. It is impossible to say for sure when this science might have first begun to express itself in the architecture of Chichen Itza, because the Temple of Kukulkan – like the pyramids of Teotihuacan – surmounts an earlier structure that stood on the same site and had the same orientations.

Much of this 'ancestral' pyramid is still intact, covered up by the façade of the present pyramid, and has been extensively excavated since the 1930s. We climbed to reach its summit up a steep, dark tunnel-like stairway cut by archaeologists beneath the existing northern stairway. At the top (photo, page 21) we came to a rectangular chamber containing a reclining idol of a type known as Chacmool, which was frequently used as a prop in human sacrifices. Behind this figure, in the shadows of a second rectangular chamber, we could make out a magnificent sculpted puma carved from a single block of stone and painted red. Its mouth was open in a manner considered by the ancient Maya to represent 'an entrance to the underworld'.[29] Its spots were simulated by 72 pieces of jade embedded into its body.[30]

*The Observatory, Chichen Itza.
Its narrow windows and doors include
precise sight-lines to sunset at the
equinoxes, the setting of Venus at its
northerly and southerly extremes, sunset
on the days of the zenith passage,
summer solstice sunrise, and the south
meridian of the sky. It is possible that
the Maya astronomers were the
inheritors of a far more ancient tradition
of observing the sky than historians
have hitherto recognized.*

As we shall see in the chapters that follow, the puma has its leonine counter-part in ancient Egypt in the form of the Great Sphinx (which was also painted red) and the number 72 is 'hard-wired' into the design of the Great Pyramid.

THE MYSTERY OF THE STARS IN THE CAVES

The use of the number 72 in Egypt and Mexico – as well as in many other ancient cultures – derives from a cosmological process, known as 'precession', that can only be detected through long-term and extremely accurate astro-nomical observations and records.

For how long have human beings made such observations?

In 1972 Alexander Marshack used carbon-dated fragments of engraved bone from Stone Age caves to demonstrate that the phases of the moon, and per-haps also the winter and summer solstices, were being precisely observed and recorded 40,000 years ago in northern Europe.[31]

Some of Marshack's bone fragments, bearing particularly curious spiral pat-terns of incised dots, were from the Lascaux region of France. In December 1995 the cosmologist Frank Edge circulated a research paper making a case for a 'celes-tial interpretation' of the Hall of Bulls in the world-famous cave of Lascaux.[32] Scientists are generally agreed that the Hall of Bulls was painted some 17,000 years ago, i.e. *more than 14,000 years* before the supposed first invention and nam-ing of the twelve constellations of the zodiac by the ancient Babylonians and Greeks.[33] For anyone with a strong commitment to this orthodox theory it is therefore unsettling to read Edge's persuasive analysis of the similarities between the (supposedly Graeco-Babylonian) zodiacal constellation of Taurus the Bull – as computer simulations show it would have looked in the pre-dawn on the summer solstice in 15,000 BC – and the Great Lascaux Bull:

> Take a moment to consider the six prominent black dots just above the Lascaux Bull's shoulders. It is through the comparison of these dots with the Pleiades that we can discover that the relationship between the Lascaux Bull and the constellation of Taurus is profound. Not only are these six dots the same number as the easily visible Pleiades, but they are arranged in a pattern closely resembling the spatial relationships among the members of the Pleiades, and they have approximately the same relationship as the Pleiades to the head and face of the related bull. So striking is the resemblance of this Ice Age bull to the traditional picture of Taurus that if the Lascaux Bull had been discovered in a medieval manuscript rather than on a cave ceiling the image would immediately have been recognized as Taurus.[34]

SKYBEARERS

Edge's conclusions concerning a very ancient identity for Taurus have recently been extended to other zodiacal constellations by Alexander Gurshtein, Professor of Astronomy at Mesa State College, Colorado. In a paper published in *Scientific American* in May 1997, Gurshtein argues that 'the identities of the most famous constellations may be far more ancient than previously thought'.[35]

His case is broadly based on the fact that 'of the 12 signs of the zodiac, just four have any real significance at any particular time. These are the constellations in which the sun rises at the spring and autumn equinoxes and at the two solstices of midwinter and midsummer'[36] (see boxed section overleaf, 'Some Astronomical Basics'). Gurshtein's point is that today the four 'skybearing' constellations are Pisces, Virgo, Gemini and Sagittarius (with Pisces at the spring equinox). 'In Babylonian times,' however:

> it would have been quite different. The wobble of the earth on its axis ['precession'] shunts the point at which the sun appears at the equinoxes and the solstices around the sky at the rate of one degree every 72 years. So around 700 BC [the time that the Graeco-Babylonian zodiac was supposedly invented], the most important constellations would have been Aries [at the spring equinox], Libra, Cancer and Capricorn. That might explain why the Babylonians incorporated these four signs in their maps, but why did they bother to name the other signs? The answer may lie in pushing even further back in the past. For all twelve of the zodiac signs to have been 'significant' by the time of the Babylonians, the naming process would have had to begin back in 4400 BC when the key constellations would have returned to those of today. In other words, Pisces, Virgo, Gemini and Sagittarius may be the most ancient signs of the zodiac, having been named during the Stone Age.[37]

SOME ASTRONOMICAL BASICS

The sun always rises generally 'in the east', but does not always rise at the same point on the eastern horizon. Instead, like the needle of some giant gauge, it slowly sweeps along the horizon during the solar year, passing a little further on each successive dawn from the observer's right (south of east) to the observer's left (north of east) and then back from left to right again, and so on *ad infinitum*. A complete cycle from extreme south of east to extreme north of east, and then back to extreme south of east, requires a full year to unfold.

There are four natural 'high-points' in the year, towards which enormous numbers of ancient and sacred ceremonies were once directed all over the world. In the modern Western calendar these points are:

- the December Solstice (21 December, midwinter and the shortest day in the northern hemisphere), when the sun reaches the most southerly extent of its annual range;
- the March Equinox (21 March, the beginning of spring in the northern hemisphere), when the sun rises precisely due east and night and day are of equal length;
- the June Solstice (21 June, midsummer and the longest day in the northern hemisphere), when the sun reaches the most northerly extent of its annual range;
- the September Equinox (21 September, the beginning of autumn in the northern hemisphere), when the sun rises precisely due east (as it did before in March) and night and day are of equal length (as they were before in March).

Each 'high-point' really represents a location in the earth's annual orbit of the sun. Tilted on its axis at an angle of approximately 23.5 degrees from the vertical, our planet's north pole points most directly *away* from the sun on 21 December and most directly *towards* the sun on 21 June, and lies effectively broadside on to the sun on 21 March and 21 September.

The earth orbits the sun at a distance of 146 million kilometres in a fixed plane that lies enclosed by limitless expanses of empty space and distant stars. Since we live on a spherical planet, human beings naturally perceive this cosmic environment as a vast sphere towards the edges of which we look outwards from the centre. About two thousand visible stars decorate it, more plentiful in some quadrants than in others, but distributed everywhere.

A belt of these stars, which ancient astronomers organized into the twelve constellations of the zodiac, lies in the plane of the earth's orbit. If we envisage that orbit as a circular railway-line then we might imagine this belt of constellations as a series of stations distributed around it.

As the earth proceeds along its orbit, it is obvious that the sun will at various times lie between it and every one of these 'stations' – i.e. every one of the twelve constellations of the zodiac. The effect, for the observer on earth, is that the sun at dawn is seen to rise against a background of star-images that slowly rotate during the course of the solar year, spending roughly a month 'in' each 'house' of the zodiac.

On a longer scale of time, there is the astronomical phenomenon known as 'precession'. It is a very slow wobble of the axis of the earth and its effect for earth-bound observers is to cause an equally slow cyclical slippage of the belt of the zodiac against the rising point of the sun. The result, in any particular epoch, is that the four key constellations will only mark the equinoxes and the solstices *temporarily*. This precessional slippage, which operates at the rate of one degree every 72 years, means that each constellation houses the sun at each point for an average of 2160 years. All twelve of the constellations therefore cycle past all of the four key points of the year in a total of just under 25,920 years.

OPPOSITE:

1   *Position of the rising sun throughout the year.*
2   *The spin axis of the Earth.*
3   *The position of the Earth in its orbit around the sun at the four pivotal points of the year.*
4   *The rising position and path of the sun at the four pivotal points of the year, which dictates the length of the hours of daylight (northern hemisphere only).*
5   *The 'background' constellation against which the sun is seen to rise changes each month due to the orbit of the Earth around the sun.*
6   *Sun rising in Leo.*
7   *One month later it is seen to rise in Virgo.*
8   *Due to the precessional cycle the background of stars against which the sun rises on any given date moves anticlockwise 1° every 72 years.*

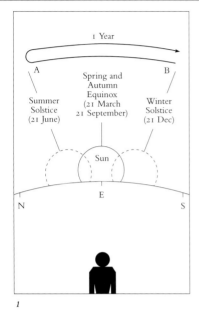

1

1 Year

A   B

Summer
Solstice
(21 June)

Spring and
Autumn
Equinox
(21 March
21 September)

Winter
Solstice
(21 Dec)

Sun

N   E   S

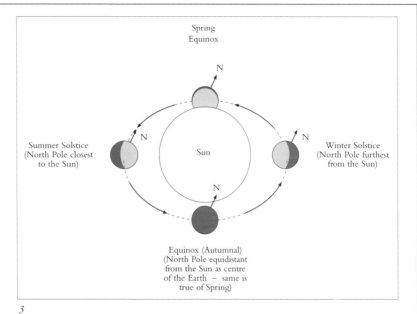

3

Spring
Equinox

N

Summer Solstice
(North Pole closest
to the Sun)

Sun

N

N

Winter Solstice
(North Pole furthest
from the Sun)

N

Equinox (Autumnal)
(North Pole equidistant
from the Sun as centre
of the Earth – same is
true of Spring)

2

Pole of the
Ecliptic

N

23.5°

Equator

Earth

Spin axis of
the Earth

Plane of Earth's orbit
around
the Sun (Ecliptic)

S

Summer Solstice

S

E   W

Autumn Equinox

S

E   W

Winter Solstice

S

E   W

Spring Equinox

S

E   W

4

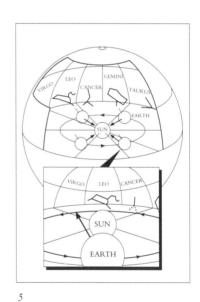

VIRGO   LEO   CANCER   GEMINI

TAURUS

EARTH

SUN

VIRGO   LEO   CANCER

SUN

EARTH

5

6

Cancer

Leo

Ecliptic

Sun

Earth

7

Leo

Libra   Virgo

Ecliptic

Sun

Earth

8

30°

Virgo   Leo   Cancer

Sun   Ecliptic

Earth

1° every 72 years

Gemini

Cancer

Leo

Sun   Ecliptic

Earth

Computer simulations of the effects on stellar positions of the 'precessional' wobble of the earth's axis show Gurshtein's proposed date of 4400 BC to be conservative. If his logic is correct this must be regarded as *the very latest date* at which the naming process could have begun, because the constellation of Gemini had then all but ceased to 'house the sun' on the spring equinox and was being remorselessly replaced, as a result of precession, by its neighbour Taurus. In the same period, at the summer solstice, Virgo was being replaced as the ruling constellation by Leo, Sagittarius was being replaced at the autumn equinox by Scorpio, and Pisces was being replaced at the winter solstice by Aquarius. By contrast, if we go back 1600 years earlier, to 6000 BC, we find that the sun has just begun its precessional journey through Gemini on the spring equinox – and thus through Virgo on the summer solstice, Sagittarius on the autumn equinox and Pisces on the winter solstice.[38]

The full implication of Gurshtein's argument is that these constellations could have been recognized – very much as we see them today – as early as 6000 BC. If he is right, then the zodiac is not an invention of the Greeks and the Babylonians. They must have received it as a legacy from a far earlier source which, theoretically, could have influenced many other ancient cultures as well.

CITIES OF THE ZODIAC

Long before Edge and Gurshtein, Stansbury Hagar had been puzzled by the Western zodiac's striking similarity to one in use by the pre-Colombian Maya of Central America. Indeed, he judged the two zodiacs to be 'so similar as fully to justify the statement that they were probably derived from the same source, even though we do not know where or how the communication occurred'.[39]

Principal monuments of Uxmal soaring above the jungle, reflecting the late evening sun. Left to right: the 'Nunnery Quadrangle' – the terrestrial counterpart of the constellation of Virgo according to Stansbury Hagar; the 'House of Turtles' (Cancer); the Pyramid of the Magician (Scorpio); and the 'House of the Governor' (Gemini).

The 'House of Pigeons', Uxmal,
correlated by Hagar with the zodiacal
constellation of Taurus.

The central Ball Court, representing
the constellation of Leo in the
sky–ground scheme of Uxmal.

'House of the Governor', detail from
the façade.

Following up his earlier work on 'celestial imitation' at Teotihuacan, Hagar
discovered a zodiacal pattern in the layout of Mayan temples showing that:

> Many and perhaps all of the Maya cities were planned to reflect on earth the
> supposed design of the heavens ... In four places – Uxmal, Chichen Itza,
> Yaxchilan and Palenque – an almost complete zodiacal sequence can be
> recognized and sequences of varying extent are seen in many other places,
> including the temples and stele of Copan and Quirigua, the oldest-known
> Maya ruins ... [40]

Hagar's most detailed work in this field was done at the Mayan city of Uxmal,
some 200 kilometres west of Chichen Itza, and set out in a scholarly paper en-
titled *The Zodiacal Temples of Uxmal* published in 1921.[41] As our diagram shows,
he identifies a temple near the south-western boundary of the site as the terrestrial model of the constellation of Aries, the so-called 'House of Pigeons' with its

The zodiacal temple of Uxmal.

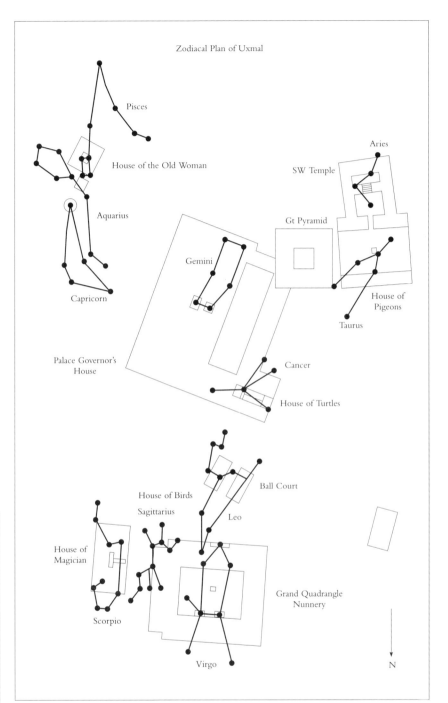

Zodiacal Plan of Uxmal

Pisces

House of the Old Woman

Aquarius

Capricorn

Palace Governor's
House

Gemini

Gt Pyramid

SW Temple

Aries

House of
Pigeons

Taurus

Cancer

House of Turtles

House of Birds

Sagittarius

House of
Magician

Scorpio

Ball Court

Leo

Grand Quadrangle
Nunnery

Virgo

N

*Twin pumas back-to-back at Uxmal,
set along an alignment from the 'House
of the Governor' to the summer
solstice sunset. Compare to the ancient
Egyptian symbol of the Akeru – two
lions back-to-back, also associated with
the rising and the setting of the sun at
the solstices. See stela of Sphinx, photo
page 93.*

distinctive roof-comb as the counterpart of Taurus, and the 'House of the
Governor' as the counterpart of Gemini (because the 'twin serpents [that] extend
over the entire façade occupy the position of Gemini in Maya zodiacs'[42]). In addition the 'House of the Turtles' represents Cancer, the central Ball Court is Leo,
the so-called 'Nunnery Quadrangle' is Virgo, the 'House of Priests' is Libra, the
'House of Birds' is Sagittarius, the distinctive 'Pyramid of the Magician' is the
counterpart of Scorpio, and Capricorn, Aquarius and Pisces are matched by three
temples near the south-eastern boundary of the site.[43]

Pyramid of the Magician, part of a great cosmic scheme to mirror heaven on earth.

'This great cosmic scheme,' Hagar concludes:

> was based upon the belief that everything in this world is the shadow or reflection of the perfect reality that exists in the celestial realms. Imitation of the observed celestial plan therefore brings down upon earth some of the celestial perfection. The sacred celestially planned city partakes of that perfection and casts upon its inhabitants the benificent influences from the stars.[44]

THE STELLAR CITY OF ORION

In addition to the zodiacal template identified by Hagar, there is one Mayan city which is known to be modelled upon a *non*-zodiacal constellation. This city is Utatlan, the late capital (AD 1150–1524) of the Quiche Maya of the Postclassic Guatemalan highlands. The Quiche were the authors of the book of wisdom known as the *Popol Vuh* – a book that contains repeated enigmatic references to the 'ground-earth' and the 'sky-earth' as reflections, or imitations, of one another.[45] It therefore feels somehow appropriate, as the archaeologists José Fernandez, Robert Cormack and others have established, that 'the settlement core of Utatlan' was designed 'according to the celestial scheme reflected by the shape of the constellation of Orion'.[46]

Fernandez was also able to prove that all of Utatlan's major temples 'were oriented to the heliacal setting points of stars in Orion',[47] and argued that the Milky Way, alongside which Orion stands, 'was thought of as a celestial path connecting the firmament's navel with the centre of the underworld'.[48]

Tablet of the Foliated Cross, Palenque, circa 700 AD. The figure on the left is Chan Bahlum, the newly crowned king. He stands on top of 'the first true mountain' — the primeval mound which rises out of the waters of the 'Primordial Sea'. The figure on the right is the deceased Lord Pacal, the former ruler of Palenque in spirit form. Between them is a symbolic device representing the Milky Way – which both the Mayans and the ancient Egyptians saw as a central feature of the netherworld.

Surely of relevance in this regard is a fact well known to scholars, namely that Orion was extensively involved in Mayan rebirth beliefs – which describe the constellation as nothing less than the location of 'the place of creation'.[49] Its three prominent belt stars are generally depicted in the *Popol Vuh* either as three stones set in the 'hearth of creation'[50] or as the 'turtle of rebirth'[51] and there are scenes in the Madrid Codex in which the tree of the Maize God, symbolizing the cosmic axis, rises in resurrection out of the turtle's back.[52] As we shall see in Part III, almost identical imagery occurs in the reliefs of Angkor Wat in Cambodia where the god Vishnu in his tortoise (or turtle) incarnation is shown supporting on his back the world axis represented by Mount Mandera whilst it 'churns the Milky Ocean' to generate the elixir of immortality.[53] Meanwhile in Egypt, as we show in Part II, the constellation of Orion was envisaged as the sky-image of Osiris, the god of rebirth, and the three stars of Orion's belt provided the celestial prototype according to which the three Great Pyramids of Giza were laid out on the ground – again on a site that was thought of as 'the place of creation'.

In the tomb of Senmut in Upper Egypt (see photograph, page 99), the constellation of Orion, clearly identified by its three belt stars, is placed in close juxtaposition to the figures of two turtles of the kind associated by the Maya both with Orion and with rebirth.

And on a mural in the Temple of the Foliated Cross at the Mayan site of Palenque in Mexico's Chiapas province we see the Milky Way represented by a maize tree rising from 'the place of Creation near Orion'.[54] The Milky Way is flanked by two figures – the spirit of Lord Pacal, the deceased former ruler of Palenque, and his son and successor Chan-Bahlum, who are shown in psychic communication with one another. As the father ascends to the heavens, the son is transformed from 'the status of heir-apparent into king'.[55] At the same time it is understood that the deeds and rituals performed by the son are essential if the father's hoped-for rebirth amongst the stars is to be achieved. Indeed, one of the central teachings of this mural is that the father is in some way *engendered* by his son – a teaching that has been described by David Freidel, Linda Schele and Joy Parker as 'the great central mystery of Maya religion'.[56]

It is extremely curious, as we shall see in Part II, that an identical mystery lies at the heart of the ancient Egyptian rebirth cult – where Osiris plays the role of the transfigured father and Horus is the engendering son. In Egypt, as amongst the Maya, the stellar context involves Orion and the Milky Way. In Egypt as in Mexico a journey through the netherworld must be undertaken by the deceased. In Egypt as in Mexico religious teachings assert that life is our opportunity to prepare for this journey – an opportunity that should under no circumstances be wasted.

No precedents

Such correspondences lead us to suspect that the lineaments of an ancient rebirth ritual – wrapped up in sophisticated astronomical observations and descended from a worldwide cosmological system that also left its legacy in Egypt and in south-east Asia – lie scattered and fragmented throughout the land of Mexico. This system, which taught the duality and interpenetration of ground and sky, earth and heaven – matter and spirit – urged the initiate to shed attachments to the sense-world (as he might drop broken flutes) and to ascend upwards, through self-sacrifice and the quest for knowledge, to the celestial realms.

There can be no doubt that all the early Central American civilizations professed such ideas. But the problem is that scholars have no theories as to their source. They are simply there, fully formed at the beginning of the Maya, fully formed at the beginning of Teotihuacan. *Everything* is unprecedented, not only the spiritual and cosmological dimensions but even much more mundane matters – for example Teotihuacan's extensive and impressive city plan. As Professor Michael Coe has observed: 'Perhaps the strangest fact regarding this great city plan is that there is absolutely no precedent for it anywhere in the New World.'[57]

The Maya, too, are in every way unprecedented – by virtue of their profound

Olmec head, La Venta.

Olmec head, La Venta.

Figure from La Venta known as 'the Walker', seeming to show a bearded man with Caucasian features.

RIGHT: *The Great Sphinx of Giza.*

spiritual beliefs, by virtue of the breathtaking accomplishments of their architecture, and by virtue of their immense competence as astronomers. However, their amazingly precise calendar, described by the historian of science Otto Neugebauer as 'one of the most fertile inventions of humanity',[58] is even harder to explain.

This is the calendar that offers the richest detail concerning the vast scheme of world ages in which our epoch is regarded as the Fifth Sun. It is a work of immense complexity, incorporating a more accurate calculation of the length of the solar year than the modern Gregorian calendar, an exact calculation of the period of the moon's orbit around the earth, and an exact calculation for the synodical revolution of Venus.[59] Unlike other apparent inventions of the Maya, however, the calendar is not strictly speaking 'unprecedented'. On the contrary, there is now firm evidence from inscriptions, accepted by scholars, that precisely the same system was in use amongst the Olmecs – the so-called 'mother culture' of Central America.[60]

The only problem, however, once again, is that we have no really solid information on the origin of the 'Olmecs'. Even their name is artificial, having been given to them by archaeologists, who freely admit that 'the proto-Olmec phase remains an enigma ... it is not really known in what time, or in what place, Olmec culture took on its very distinctive form'.[61]

Rather alarmingly, all that we have as a bridge between ourselves and these distant people are a few hundred remarkable stone artefacts that they made – and then deliberately concealed, to await the passage of the centuries. These artefacts are utterly extraordinary.

SERPENT ASCENDING

It was late in the day and we stood in the shadow of a gigantic 'Olmec head' from La Venta, a 3500-year-old site, dominated by a strange conical pyramid with fluted sides, that was now largely destroyed by industrial development.

RIGHT AND BELOW: *Images of bearded Caucasians from Monte Alban.*

Image of bearded Caucasian intertwined with serpent, La Venta.

The head, which had been rescued from La Venta along with dozens of other equally unusual pieces, was carved in one piece out of solid grey granite. It was the head of an old man, a pronouncedly African head, that stood 2.5 metres tall and 2 metres wide and weighed more than 20 tonnes. The mouth and lips were full, protruding, the nose was wide and flat, the eyes deeply etched, gazing directly ahead. The sculptor had given the figure a distinctive head-dress, marked with panels in relief to resemble a striped cloth that fell down over the ears to below the chin. The comparisons with the head of the Great Sphinx of Egypt were obvious.

Orthodox historians do not accept the presence of any Africans in the New World prior to the time of Columbus and have tried to sidestep the implications of the obviously African features of the 3000-year-old Olmec heads – 16 of which have so far been found.[62] It may at least be taken as a sign that there is no racism

Altar, La Venta.

Seated figure, La Venta. The style of the head-dress is similar to the nemes *head-dress worn by the Pharaohs of ancient Egypt.*

Stela of the Bearded Man, La Venta. The central figure with a hooked nose and ceremonial false beard was christened 'Uncle Sam' by Matthew Sterling, the archaeologist who discovered it in the 1940s.

Side panel of the so-called 'Altar of Infant Sacrifice', La Venta. Note the crowns of the two adult figures. The one of the left closely resembles the deshret *crown of Lower Egypt. The crown on the right is in the form of a step-pyramid – a symbol that occurs repeatedly in the Americas and in Egypt.*

in archaeology that there are also supposed to have been no Caucasians in the New World before Columbus! Scholars have therefore predictably raised quibbles about the Quetzalcoatl myth of the tall bearded white man and have sought to dismiss any suggestions that it might be reflected in the numerous reliefs of Caucasian faces that have been excavated in some of the oldest archaeological sites of Mexico. In the Olmec area several were found in the same strata as the African heads and sometimes side by side with them, but images of Caucasians have also been excavated as far afield as Monte Alban in the south-west, a site dated to between 1000 and 600 BC.[63]

In 1996 and 1997, as we reported in the last chapter, the discovery of Caucasian bones more than 9000 years old in the Americas seems, quite suddenly, to have validated the Quetzalcoatl myth. It is therefore now legitimate to ask how long it may be before another lucky turn of the archaeologist's spade will uncover the bones of individuals who could have served as the prototypes for the great Olmec heads.

'Man in Serpent', La Venta. The serpent is distinctively plumed, making this possibly the earliest surviving example of the symbol of Quetzalcoatl in the Americas. Feathered serpents are also found in the sacred art of ancient Egypt, great sky-serpents dominate the religious iconography of the temples of Angkor in Cambodia and Hindu scriptures teach that the Supreme God Vishnu 'slept on the lap' of the serpent Sesha before awakening to create our present universe. In all these cultures the serpent served as a symbol of ineffable cosmic forces and as a metaphor for rebirth and spiritual renewal.

We show some of these heads in photographs, as well as other curious Olmec sculptures and reliefs. There is no explanation for them, no precedent. As the archaeologist L. A. Parsons remarks: 'Antecedents for Olmec stone sculpture have not been found . . . I now believe they will never be found in Mesoamerica.'[64]

These orphaned works of art include the earliest-ever representation of the plumed serpent, Quetzalcoatl, shown encircling the figure of a man, calling to mind the fact that in Nahuatl symbolism 'the plumed serpent is . . . the sign of the heavenly origin of man.'[65] More significantly, the presence of the plumed serpent amongst the Olmec remains tells us that Quetzalcoatl's cult of rebirth and spiritual renewal, complete with all its symbols, was being practised in Central America at least as early as 1500 BC – the supposed genesis date of Olmec civilization.

The architecture of the cult always involved pyramids – again even as early as La Venta – which were always seen as 'portals to the otherworld'[66] and connected to 'the ordering of earth and sky'.[67] Around 2500 BC, as we shall see in Part II, a cult with identical views flourished in Egypt at a place then called Rostau, 'the gateway to the otherworld' – in the sacred domain of the Great Pyramid.

EGYPT

CHAPTER 3

SANCTUARIES
OF THE COSMOS

PREVIOUS PAGE: *Sunrise behind the temple of Luxor.*

ABOVE: *The shadow of the Great Pyramid soon after sunrise on the spring equinox.*

BELOW: *Boat found buried beside the Great Pyramid has the design of a sea-going vessel.*

WE ARRIVED AT THE top of the Great Pyramid of Egypt soon after dawn on 21 March 1996, the spring equinox. To our west the monument's shadow stretched out far into the desert, sharply defined as though it had been cast by the gnomon of a giant sundial. At the base of the pyramid to our east we could see a series of empty rhomboidal trenches carved out of bedrock – these were 'boat pits' or 'boat graves', which at one time had contained ceremonial wooden boats. On the south side of the base an ugly modern structure known as the 'boat museum' reared up towards us, containing an intact boat, 43.5 metres long, that had been found beside the pyramid. Off to our south-east, hemmed in by two archaic megalithic temples, crouched the Great Sphinx, its forepaws extended towards the horizon, the emblem of enigmatic antiquity.

We could feel a cool breeze blowing steadily towards us from the delta-country to the north. We looked up and saw the blue sky, dotted with early morning clouds, curving above us like a lens. Beneath our feet the man-made mountain tapered jaggedly upwards while the earth stretched away in all directions from its broad square base and rolled around the far-off circle of the horizon – the ever-turning earth, eternally revolving eastwards upon its own axis.

SCALE MODEL

Known to the ancient Egyptians as Rostau, 'the Gateway to the Otherworld', there is something about Giza that draws the mind to contemplation of the cosmos. Stand at the foot of the pyramids at night, position yourself so that a prominent star lies vertically above the apex of any one of them and within ten minutes you will observe the effects of the rotation of the earth – for that star will no longer rest right above the pyramid but will have been shifted perceptibly to the west by the earth's spin. Similarly if you were to stand on the summit of the Great Pyramid at dusk and were to position a colleague at the base of the monument, the curvature of the earth means that he would see the sun 'set' first while it would still be in full view to you.

Historians sometimes debate whether the ancients ever worked out that the earth is a sphere. Generally the conclusion is that they did not, at least until the

PREVIOUS PAGES: *The Pyramids and the Great Sphinx of Giza, photographed around 4 p.m., one week before the winter solstice, December 1997. The Sphinx faces perfectly due east. The pyramids are oriented precisely north, south, east and west, with each face aligned to a cardinal direction.*

time of the Greeks. Is it therefore a pure fluke (see diagram) that the height of the Great Pyramid multipled by 43,200 yields a figure very close to modern measurements of the polar radius of the earth and that its base perimeter multiplied by 43,200 yields a figure very close to the equatorial circumference of the earth? Scholars accept that the ratio exists but deny that it has any significance, asking, 'Are we seriously expected to recognize this as a "message" rather than a number arrived at quite accidentally?'[1]

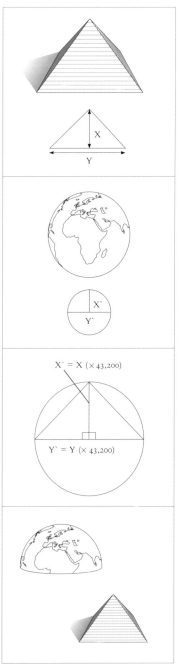

The Great Pyramid as a model of the northern hemisphere.

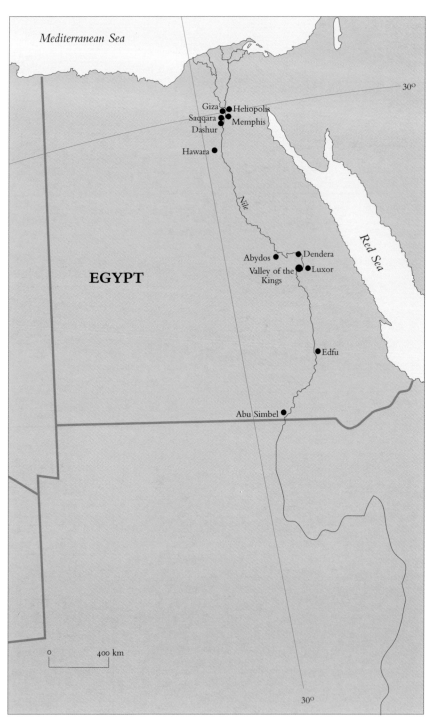

Perhaps the builders of the Great Pyramid were determined that it must be very massive so that it would resist the passage of time and the destructive hands of ignorant people. In relation to the northern hemisphere of the earth a ratio anywhere between 1:40,000 and 1:50,000 would have satisfied such a 'design requirement'. Perhaps the builders deliberately opted for the figure of 1:43,200 because it falls within these parameters and is at the same time a significant *number – one that should supposedly be recognizable to any astronomically competent civilization. Perhaps it is a form of communication – an attempt to use the universal language of precession to send a message across the ages?*

What makes us suspect a message, or at any rate some attempt at communication, is the peculiar fact that the number 43,200 is not random but one of a series that is generated mathematically by the 'precession' of the axis of the earth. We have already shown in Chapter 2 that this slow cyclical wobble displaces the positions of all stars at the rate of one degree every 72 years (and thus by 360 degrees in 25,920 years). The best-known effect is the apparent gradual rotation of the 12 constellations of the zodiac in relation to the rising point of the sun on the spring equinox, allowing each constellation to 'house' the sun on that special day for a period of 2160 years.

The heartbeat of the mechanism is the number 72 – the number of years required for one degree of precession. Six hundred such heartbeats amount to 43,200, the figure used in the pyramid–earth ratio.

A LOST SCIENCE?

Is the Great Pyramid a mathematical model of the northern hemisphere of the earth deliberately built on a scale derived from the earth's own characteristic motion, i.e. the precession of its axis?

Most mainstream scholars do not want to 'waste their time' considering the astronomical and geodetic characteristics of the Pyramid – or of other ancient monuments. This may be in part because the orthodox theory of the past conditions its supporters to regard our remote ancestors as *too stupid* to have made accurate astronomical and geodetic observations. Furthermore, as the author John Michel has pointed out, it is only human nature that established academics should seek to 'protect their territory', stubbornly preferring their 'own picture of primitive antiquity to any facts that might disturb it'.[2]

The astronomical effects of precession, and the rate at which they proceed, were supposedly discovered by the Greek astronomer Hipparchus at around 100 BC – having never previously been noted by mankind. This opinion, though still passed off as fact in encyclopedias and text-books, seems hardly tenable in the light of the growing body of evidence, some reported in the last chapter, concerning Stone Age observations of the zodiacal constellations.

Such 'historically anomalous' evidence was first seriously considered by Dr Hertha von Dechend of Frankfurt University and the late Giorgio de Santillana, Professor of the History of Science at the Massachussets Institute of Technology. In their immense study *Hamlet's Mill*, published in 1969, they argue that a body of scientific astronomical knowledge was in existence in the world at least '6000 years before Virgil' (i.e. at around 6000 BC) and that this lore used precise, idiosyncratic and widely disseminated mythological conventions to describe celestial events that have been proved by astronomical calculations to have occurred in the skies in the epoch of 6000 BC.[3] The evident maturity of these conventions even at that early date troubled Santillana and von Dechend, who at length, rather hesitatingly, ascribed the origins of the astronomical lore to 'some almost unbelievable ancestor civilization' that 'first dared to understand the world as created according to number, measure and weight'.[4] The two academics further stated their view that this lost culture had succeeded in diffusing a profound influence to later, historical cultures as far afield as Egypt and India, Greece and Mexico. Somehow, before anything normally thought of as history had begun, a 'ruthless metaphysics' had evolved – a 'prodigiously vast' cosmological theory 'that dilated the mind beyond the bearable, although without destroying man's role in the cosmos'.[5]

Amongst the most persuasive proofs concerning the existence of this lost spiritual and astronomical culture is the fact that accurate values for precession, in the form of specific numbers, can be found in the most ancient traditions of mankind. The values given and the symbols used are so consistent that the authors are forced to conclude that they have stumbled upon the vestiges of a lost science – a coded wisdom tradition with its own technical language – a 'great, worldwide archaic construction', upon the remains of which 'the dust of centuries had settled when the Greeks came upon the scene'.[6]

It takes 36 years for what astronomers call the 'vernal point' (i.e. the sun's 'address' against the background of the stars on the spring equinox) to move half a degree around the band of the zodiac and 72 years to complete a one-degree shift. Since the sun's light entirely obscures the stars during the day, such observations can only have been made by ancient astronomers in the hour before dawn, looking towards the eastern horizon, when it is possible to distinguish the stars that will 'house' the sunrise. In observational terms the one-degree shift brought about by precession over 72 years – an entire human lifetime – is barely perceptible, being roughly equivalent to the width of a forefinger held up towards

There is something about Giza that draws the mind to contemplation of the cosmos. The site was known to the ancient Egyptians as Rostau – literally the 'Gateway to the Otherworld'.

'The visible accomplishments of ancient cultures – to mention only the pyramids or metallurgy – should be a cogent reason for concluding that serious and intelligent men were at work behind the stage, men who were bound to have used a technical language...' Giorgio de Santillana and Hertha von Dechend, Hamlet's Mill, *page 58.*

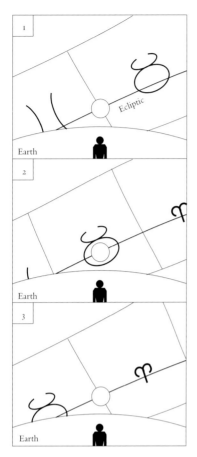

Due to precession the sun's 'address' at the spring equinox (known as the vernal point) changes its zodiacal sign every 2,160 years.

the horizon. A 30-degree shift – through one zodiacal constellation – would, by contrast, be impossible to miss but its progression could only be properly recorded and noted by many generations of observers (30 degrees at 72 years per degree equals 2160 years). A 60-degree shift, i.e. through two zodiacal constellations, takes 4320 years (2160 × 2 = 4320) which is why a 360-degree shift (all 12 zodiacal constellations) requires the grand total of 25,920 years.

These are the basic ingredients of a number code, let us call it the 'precessional code', which Santillana and von Dechend have shown – in hundreds of pages of meticulously documented evidence – to be present in ancient myths and sacred architecture all around the world.[7] In common with other esoteric numerological systems, the code is one in which it is permissible to shift decimal points to left or right at will and to make use of almost any conceivable combinations, permutations, multiplications, divisions and fractions of certain *essential* numbers (all of which relate, very precisely, to the rate of the precession of the equinoxes).

The 'ruling' number in the code is 72. To this was frequently added 36, making 108, and it was permissible to divide 108 by 2 to get 54 – which could then be multiplied by 10 and expressed as 540 (or as 54,000, or as 540,000, or as 5,400,000, etc.). Also highly significant is 2160 (i.e. the number of years required for the vernal point to transit one complete zodiacal constellation). This could be divided by 10 to give 216, or multiplied by 10 and factors of 10 to give 216,000 or 2,160,000, etc. The number 2160 was also sometimes multiplied by 2 to give 4320 – or 43,200, or 432,000, or 4,320,000, and so on.

OTHER FEATURES

Santillana and von Dechend's evidence provides a possible explanation for the 1:43,200 scale in the pyramid–earth ratio. Moreover, as one might expect if they are really architectural expressions of a 'precessional code', the monuments of Giza incorporate many other precessional, astronomical and geodetic features.

For example the precessional number 216 is found, very subtly concealed, in the heart of the Great Pyramid, in a triangle that is formed by three of the basic dimensions of the so-called 'King's Chamber'. This austere and uninscribed red granite room, in which no Pharaoh was ever found entombed, is a 2:1 rectangle exactly 20 Egyptian royal cubits in length and 10 royal cubits in width (10.46 metres × 5.23 metres). As our diagram shows, the chamber also 'contains' a right-angled triangle with its short dimension (15 cubits) drawn diagonally across the west wall from the lower south-west corner to the upper north-west corner, its median dimension (20 cubits) drawn along the entire length of the floor on the south side of the chamber, and its long dimension (25 cubits) drawn from the upper north-west corner of the chamber to the lower south-east corner.[8]

These side-lengths of 15 cubits, 20 cubits and 25 cubits can be expressed as the ratio 3:4:5 (because if we allocate the value '3' to the length of 15 cubits then 20 cubits must naturally have the value '4' and 25 cubits must have the value '5'). All right-angled triangles with side lengths in this special 3:4:5 ratio are called 'Pythagorean' – after Pythagoras, the Greek philosopher, mathematician and religious teacher of the sixth century BC who was supposedly the first to discover that

View into the King's Chamber from its east side looking west towards the granite 'sarcophagus'. There is no evidence that the Chamber was ever used for the burial of a Pharaoh. We shall suggest in later chapters that the sarcophagus was part of the physical apparatus of a sophisticated rebirth ritual – a 'virtual reality game' of the afterlife journey of the soul. The builders of the King's Chamber endowed it with harmonious mathematical and geometrical proportions. Even its location was carefully selected. The floor of the Chamber is precisely positioned at a level where the vertical section of the Pyramid is halved, where the area of the monument's horizontal section is exactly half that of the base, where the diagonal from corner to corner is equal to the length of the base, and where the width of the face is equal to half the diagonal of the base.

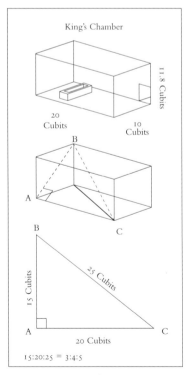

Right-angled triangle contained within the measurements of the King's Chamber.

they share a unique characteristic. This is that the square of the short side (3 units × 3 units = 9 units) added to the square of the median side (4 units × 4 units = 16 units) together result in a figure equal to the square of the long side (5 units × 5 units = 25, i.e. the sum of 9 plus 16).[9] The real 'secret magic' of the triangle, however, as the Icelandic mathematician Einar Palsson has pointed out,[10] is only revealed when the numbers are cubed.

Then we get:

$$3 \times 3 \times 3 = 27$$
$$4 \times 4 \times 4 = 64$$
$$5 \times 5 \times 5 = 125$$

The total of 27 plus 64 plus 125 is the precessional number 216 and we do not think that it found its way into the dimensions of the King's Chamber by accident.

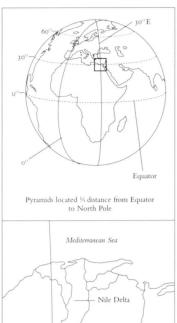

Pyramids located ⅓ distance from Equator to North Pole

EYGPT

Longitude and latitude of the Pyramids – located at one-third of the distance from the Equator to the North Pole.

View of the three Giza pyramids from the south looking north, with the Pyramid of Menkaura in the foreground. All the pyramids are accurately aligned to the cardinal directions, but the highest standards are found in the Great Pyramid, which targets true north more precisely than the meridian building of the Greenwich Observatory in London. The pyramids could only have been aligned to such tolerances by master astronomers taking sightings on stars. Moreover, as we shall see, the very pattern made by the three pyramids on the ground appears to have been dictated by the sky pattern of the three belt stars of the constellation of Orion.

Likewise, the many geodetic and astronomical characteristics of the Giza monuments also have the feeling of clever and conscious deliberation on the part of the builders, rather than of the random workings of chance.

The pyramids, for example, straddle the 30th parallel – latitude 30 degrees north – and thus lie precisely one-third of the way between the equator and the north pole of our planet.[11] With a similar dedication to precision, their north faces are aligned due north towards the pole, their east faces due east, their south faces due south, and their west faces due west. Indeed, the Great Pyramid is so precisely

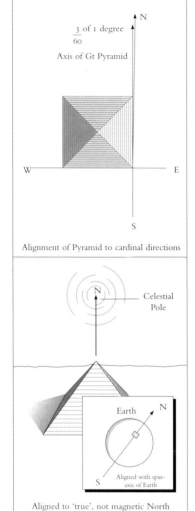

Alignment of Pyramid to cardinal directions

Aligned to 'true', not magnetic North

The alignment of the Great Pyramid to the cardinal directions dictated by the spin axis of the earth.

aligned to the cardinal directions that its meridian axis (see diagram) stands away from true north–south by the minute error of just three-sixtieths of a single degree.[12]

It is worth emphasizing that we do not speak here of compass directions, which orient to the magnetic north pole. 'True north–south' means, very specifically, the *geographical* north and south poles of the planet – the actual points around which its axis spins. The normal way of establishing the direction of true north is through observations of stars lying at or near the 'celestial north pole' (i.e. the point in the northern heavens, presently marked by the 'north star' Polaris, directly in line with the infinitely extended axis of the earth). In addition, sightings of stars in the southern heavens are also used by astronomers to establish an accurate north–south line – for such stars lie due south of the observer as they 'culminate' – i.e. reach their highest altitude – each night at the meridian. The fine accuracy displayed by the Great Pyramid suggests that both northern and southern sightings must have been used in its setting-out. What is certain, moreover, is that sightings of this precision could only have been undertaken by master astronomers. Is it really so unreasonable to suppose that adepts like these might have known about the obscure phenomenon of precession – and learned how to predict its effects?

ASTRONOMICAL OBSERVATORIES

On the edge of the upper Egyptian city of Luxor, on the east bank of the sacred river Nile 26 degrees north of the equator, the immense central hall of the temple of Amen-Ra at Karnak defines a narrow, dead-straight axis for about a kilometre along a precise bearing – 26 degrees south of east to 26 degrees north of west. This *tour de force* of monumental architecture, built up during the reigns of many Pharaohs in the second millennium BC,[13] is considered by Egyptologists to have served purely ceremonial functions. Nevertheless, as we stood at the

The temple of Amen-Ra at Karnak, viewed from the south-east across the sacred lake. Karnak is suggestively described in an inscription on the obelisk of Hatshepsut (to the right of the frame) as 'the horizon of heaven on earth'.

Karnak axis at dawn, photographed from the interior of the temple two weeks before the winter solstice.

Karnak axis at dawn, 7 December 1995, two weeks before the winter solstice, photographed from the western entrance gateway. Compare page 57.

western end of the axis and gazed east on a succession of dawns around the winter solstice in December 1995, we observed that the sun's rising point each morning edged ever more closely into direct alignment with our angle of view. Three weeks before the solstice its disk did not become visible in the 'sights' of the axis until it had climbed to an altitude of almost 15 degrees above the horizon; by 21 December, however – the shortest day – the axis was brought to bear on it far lower in the sky, allowing it to 'crown' a distant ceremonial gateway.

It was the opinion of the nineteenth-century British astronomer Sir J. Norman Lockyer that none of this is accidental since *all* the principal temples of ancient Egypt, 'whatever view may be entertained with regard to their worship or the ceremonial in them, were undoubtedly constructed, among other reasons, as astronomical observatories – the first observatories that we know of in the world'.[14] In his classic study *The Dawn of Astronomy* (1894), Lockyer lavished particular attention on the temple of Amen-Ra and argued that its axis had been deliberately oriented towards the solstices (winter solstice sunrise in the south-east and summer solstice sunset in the north-west) and that it was therefore 'fair to say that many thousand years ago the Egyptians were perfectly familiar with the solstices, and . . . more or less fully with the yearly path of the sun'.[15]

A FLASH IN THE SANCTUARY

With this practical astronomical observation Lockyer planted the seeds of a long-drawn-out controversy. He did so because the yearly 'path' of the sun, from its

Karnak axis at dawn, 21 December 1997, the winter solstice.

Sunrise on the winter solstice silhouettes the Colossi of Memnon on the west bank of the Nile opposite the temple of Luxor.

rising point furthest south of east at the winter solstice to its setting point furthest north of west at the summer solstice is not fixed and immutable but rather undergoes slow, almost imperceptible changes drawn out over very long periods of time. These changes, *which should not be confused with the astronomical cycle of precession*, correspond to actual changes in the tilt of the earth's axis in relation to the plane of its orbit round the sun – a phenomenon known as the 'obliquity cycle', which has a duration of more than 40,000 years.[16] Because these changes can be modelled mathematically, it is theoretically possible, if you trust the accuracy of the ancient builders, to work out from the degree to which a temple is *misaligned*, the date at which it must originally have been surveyed (i.e. the date at which it would have pointed accurately at a given solstitial sunrise or sunset). In the case of Karnak this date has been variously calculated at 11,700 BC, 3700 BC, and anywhere between 2000 and 1000 BC.

The second date, 3700 BC, is Lockyer's own, based on observations made in 1891.[17] Thirty years later, in 1921, the astronomer F. S. Richards – using more refined observations and formulae – came up with the earlier date of 11,700 BC.[18] This was dismissed by even the least orthodox of researchers, and by Richards himself, as being 'ridiculously remote'.[19] More recently Karnak's axis has been resurveyed by Professor Gerald Hawkins of the Smithsonian Institution who took his sightings from a roof chapel above the sanctuary and argued that the alignments here indicate a setting-out date of somewhere between 2000 and 1000 BC.[20]

Hawkins' date accords with orthodox Egyptological chronology. However, the contortions that he went through to achieve it, taking sightings from well above

ground-level, should arouse scepticism. As Norman Lockyer observes, the axis of Karnak is best understood as an immense 'instrument' – in some ways like a telescope – designed to focus light and 'to carry it to the ... extremity of the temple, into the sanctuary, so that once a year ... the light passed without interruption along the whole length of the temple'.[21] Lockyer is adamant that the sighting of this event was registered at ground-level, 'at the absolute moment of sunrise'[22] and that the effect of this would have been a 'flash' in the sanctuary which would have remained visible for perhaps as long as a 'couple of minutes'.[23]

The argument and the observations sound reasonable enough. Yet if Lockyer is correct, then the date that the obliquity cycle indicates for his 'flash' in the sanctuary reverts to 11,700 BC, as F. S. Richards calculated in 1921.

DENDERA AND ITS ZODIACS

One hundred miles north of Karnak stands the Ptolemaic temple of Dendera, dedicated to Hathor, the goddess of wisdom and love, who was associated with the night sky. Built during the first century BC, this beautiful edifice is almost three times as old as the oldest Gothic cathedral. It is in an amazing state of preservation, with deeply worn inner stairways leading up through cool, dark corridors in the walls to a roof that is still intact.

The entire temple is covered with richly coloured reliefs which depict a brilliant symbolic array of astronomical figures, led by the twelve familiar constellations of the zodiac and by the star-god: 'Osiris, Lord of Doubles ... Orion who treads his Two Lands, who navigates in front of the stars of the sky'.[24]

Entering the temple by way of its massive vestibule of 24 columns, we looked up to observe the so-called 'Square Zodiac of Dendera' with the constellation of

The obliquity cycle.

BELOW: *The Temple of Dendera under a modern sky.*

OPPOSITE: *Entrance to the hypostyle hall, Dendera, viewed from the vestibule.*

ABOVE: *Columns and ceiling of the vestibule, Dendera.*

Detail from the Square Zodiac: constellation of Sagittarius, top right.

Leo leading off northwards on the western side of the ceiling. In a roof chapel above, the same zodiacal figures appear in circular configuration devised by someone with a good knowledge of precession.[25]

THE CYCLE OF THE AGES

The rising point of the sun on the spring equinox was traditionally regarded by the ancients as the 'governing moment of the year'[26] and the character of each astrological age was determined by the zodiacal constellation that 'housed' the sunrise on that special day.

We presently live 'at the dawning of the Age of Aquarius' with the constellation of the water-bearer poised to succeed Pisces as the 'house of the sun' on the spring equinox. This also means, when the New Age is born, that sunrise on the autumn equinox (21 September) will take place in the house of Leo (when it is now in Virgo), that the winter solstice (21 December) will be housed by Scorpio (now Sagittarius), and that the summer solstice (21 June) will be housed by Taurus (presently Gemini).

In other words, each succeeding astrological Age has its own distinctive co-ordinates made up of two *pairs* of zodiacal constellations standing opposite one another at the cardinal moments of the year. Taken together, they form an

The Circular Zodiac, Dendera.

*Dendera zodiac, depicting the
constellations at the cardinal points of
the year 4000 BC.*

amazing cosmic mechanism, dancing around the equinoxes and the solstices, in
which Aquarius always pairs Leo, Scorpio always pairs Taurus, Pisces always pairs
Virgo and Sagittarius always pairs Gemini. The other four zodiacal signs
(Aries–Libra and Capricorn–Cancer) are also strictly paired so that, when the
mechanism shifts, *everything shifts*.

These twinned pairs of constellations were the principal focus of Professor
Alexander Gurshtein's research – which has pushed back the antiquity of the
zodiac at least as far as 6000 BC.[27] They were frequently personified in ancient
myths as the 'keepers' or 'bearers' of the sky,[28] and are depicted on the circular
zodiac of Dendera as female figures with their arms upstretched.

What is strange, although the temple was built during the first century BC at
the beginning of the Age of Pisces, is that the positioning of these figures does
not reflect a 'Piscean' sky. Nor – which might be understandable – does it depict
the sky of the previous Age, when Aries housed the sun on the spring equinox.
Instead it depicts a sky in the 'Aquarian configuration' – a sky whose four principal bearers stand in opposing pairs beneath the constellations of Leo–Aquarius
and Taurus–Scorpio.

Computer simulations (see diagrams) suggest that this configuration was meant
to represent the characteristic sky of the epoch from approximately 4380 BC to
2200 BC when the constellation of Taurus stood at the spring equinox and Leo
marked the summer solstice.[29] Although this epoch was already fabulously ancient
in the first century BC, we know that Dendera was habitually linked by Egyptian
priests to far earlier times. One inscription in the temple even informs us that the
original building plans were a legacy of the 'early primeval age'. They were found,
it seems, 'in old delineations written upon leather of animal skin of the time of
the Followers of Horus'.[30]

Panel from the Square Zodiac,
constellation of Leo, top right.

ABOVE: *Detail from the Square*
Zodiac, constellation of Scorpio, top
right.

RIGHT: *Roof chapel, Dendera.*
The human-headed birds represent an
element of the soul – the ba *– whose*
characteristic was free and unfettered
movement in the afterlife.

The god Osiris-Orion, centre, as depicted at Dendera.

EDFU

Throughout the 3000 years of their recorded history, the ancient Egyptians honoured a tradition which asserted that no site was sacred unless it had been built upon the foundations of an earlier sacred site. It is a tradition which is richly expressed at the great temple of Horus – the solar deity whose mythical parents were the star-gods Isis (Sirius) and Osiris (Orion) – which stands on the west bank of the Nile at Edfu in Upper Egypt. In the marvellously preserved form in which it survives today, this temple is not old, at any rate not by ancient Egyptian standards, since the building of the central structures did not begin until 237 BC and continued sporadically until 57 BC.[31] Nevertheless, archaeologists take note of the fact that vestiges of far more ancient engineering works are still in evidence at Edfu. The inner and outer enclosure walls, for example, date from the Old Kingdom (2575–2134 BC), and a later wall running outside the outer one dates from the First Intermediate Period (2134–2040 BC). There are remains of other structures that have been dated to the Second Intermediate Period (1640–1532 BC), and to the New Kingdom (1550–1070 BC).[32]

In short, archaeology tells us that Edfu was continuously maintained and developed as a sacred site over a period of well over 2000 years – from at least the third millennium BC until around the time of Christ. This evidence confirms the essential accuracy of a vast 'library' of written information that has come down to us in the form of acres of hieroglyphs carved on the towering limestone walls of the temple itself. These 'Edfu Building Texts' repeatedly describe the temple as *a copy*

OPPOSITE: *Hypostyle hall, Edfu.*

RIGHT: *Shrine behind santuary, Edfu.*

Horus in his manifestation as a falcon, Edfu.

of an earlier, pristine original, and speak of the various stages of building and rebuilding that preceded its present form.[33] Where the texts vary from the archaeological record is in the time-frame they envisage – a time-frame that steps outside all known history and returns us to a forgotten age thousands of years before the first Pharaoh of the First Dynasty sat on the throne of Egypt. As the late Dr Eve Reymond of Liverpool University has shown, it was believed that:

> the constitution of the historical temple was determined by a pre-existing entity of a mythical nature ... The temple is, in a strict sense, the concretizing of its Ancestor ... 'made like unto that which was made in its plans of the beginning'.[34]

The Ark of Horus, Edfu.

The texts speak of the sanctuary of the historical temple at Edfu as the god's 'genuine Great Seat of the First Occasion'[35] and refer again and again to ancient books and writings which apparently were used to guide the construction of the temple.[36] These documents, it seems, had been handed down from the legendary epoch known to the ancient Egyptians as the 'First Occasion' (also referred to as the 'First Time' – 'Zep Tepi' – the 'early primeval age', the 'time of Osiris', the 'time of Horus', etc.).[37] It was an epoch, very far away in the past, in which a group of divine beings known sometimes as 'the Seven Sages' and sometimes as 'the builder gods' were believed to have settled in Egypt and to have established 'sacred mounds' at various points along the Nile. These mounds were to serve as the foundations, and to define the orientation, of temples to be built in the future.[38] More specifically, and the Edfu Texts are very clear on this, it was intended that the development of these sites should bring about nothing less than 'the resurrection of the former world of the gods'[39] – a world that had been utterly destroyed. We are told that this lost domain, the 'Homeland of the Primeval Ones', was 'an island which, in part, was covered with reeds and stood in darkness in the midst of the primeval water . . .'[40] We are told that 'the creation of the world began on this island, and that it was here that 'the earliest mansions of the gods were founded.'[41] At a certain point during the primeval age, however, this blessed 'former world' was overwhelmed, suddenly and totally, by a great flood, the majority of its 'divine inhabitants' were drowned and the 'mansions of the gods were inundated'.[42]

NETHERWORLD

The notion that a temple – or a former world – might be reborn seems strange to us today, since our civilization is accustomed to think of time in a linear rather than in a cyclical fashion. But in ancient Egypt the image of time as an eternal serpent, endlessly swallowing its own tail, conditioned all thought about the past, present and future. For this reason it was not difficult for people to believe that every living, conscious soul, and every characteristic 'epoch' of the earth, would return again and again into existence. Indeed, temples themselves were considered to be *living beings*,[43] all of which were descended from a common ancestor – 'a temple that once really existed', as Reymond comments, 'in the dim past of predynastic Egypt'.[44] She adds:

> The Edfu tradition, and so perhaps the tradition of many other temples, evidently looked on this far-distant temple as the work of the gods themselves in which the creation of the Earth was completed.[45]

It is entirely consistent with the cyclical time-frame of the Edfu Texts that the 'far-distant' temple to which Reymond refers should itself have been seen as a copy of an *even earlier* archetype. When the gods began to build it, we are told, they modelled it upon a place 'that was believed to have existed *before* the world was created'. This place was called the Duat-N-Ba, literally the 'Netherworld of the Soul'.[46] Its location, which was in the sky, as we shall demonstrate more fully in the next chapter, is hinted at by a curious detail concerning the orientation of the temple of Edfu that has come down to us in the Building Texts. This inscription states that the temple was not aligned to any of the annual rising or setting points of the sun but that its 'orientation lay from Orion in the south to the Great Bear in the north'.[47] A related inscription confirms the general picture by telling us that the temple was built according to a plan 'which fell from heaven'.[48]

Edfu hieroglyphs. The temple was believed to have been built according to a plan which 'fell from heaven'.

IN THE HALL OF THE DOUBLE TRUTH

ON THE WEST BANK of the Nile, opposite Luxor and Karnak, stands the strange and beautiful temple of Deir el Medina – which, like Edfu and Dendera, is a product of the final days of the once-remarkable civilization of ancient Egypt. Dedicated in the third century BC to Maat, the Egyptian goddess of cosmic equilibrium, its walls are inscribed with hieroglyphic texts expressing archaic religious and spiritual ideas.

The temple is built around an axis oriented south-east to north-west. We entered it through a gate in the south-eastern wall leading to a courtyard dominated by four elaborately adorned columns with floral capitals. Beyond these we passed into a central hall at the end of which we came to three doorways leading into three separate enclosed shrines. The southernmost of these shrines, dark and unprepossessing though it seemed at first, proved to contain a finely crafted and almost complete scene of what scholars describe as the Psychostasia, or Weighing-of-the-Heart (derived from the Greek *psyche* = soul, i.e. heart, and *stasis* = balance).[1]

We took time to examine this scene, which consists of a chapter from the ancient Egyptian *Book of the Dead*, *Reu Nu Pert Em Hru*, literally the 'Book of Coming Forth By Day', one of a large corpus of funerary texts copied and recopied at all periods of Egyptian history, that concerned themselves with 'the freedom granted to spirit forms which survived death to come and go as they pleased'.[2]

Standing in the doorway of the shrine our eyes were drawn to the wall on our left containing an elegant relief of Ptolemy IV Philopator (ruled 221–205 BC), the Macedonian Greek Pharaoh on whose orders this Temple of Maat was built.[3] Represented as a deceased soul, dressed in sandals and a simple linen kilt, this scene shows him being ushered into a spacious hall at the head of which, in partially mummified form, sits Osiris, the high god of death and resurrection, identified in the ancient Egyptian sky-religion with the great southern constellation of Orion.[4]

The place to which Ptolemy has been brought is sometimes referred to as the Judgement Hall of Osiris, and sometimes as the Hall of the Double Maati – which translates as 'the Hall of the Two Truths' or possibly 'the Hall of Double

Detail from the 'weighing of the soul', Deir el Medina. Top, some of the Assessors who hear the 42 Negative Confessions; left, ibis-headed Thoth, god of wisdom, records the verdict; centre, Ammit, the Eater of the Dead, the agency of the soul's extinction; right, Osiris, Judge of the Dead and agency of the soul's resurrection.

Justice'.[5] It is not a place to which the soul was believed to have come immediately after death. Indeed, it could only be reached by those who were spiritually 'equipped' to complete a long and hazardous post-mortem journey through the first five of the twelve divisions of the Duat – the fearful parallel dimension, shadowy and terrifying, filled with fiends and nightmares, that was believed by the ancient Egyptians to separate the land of the living from the kingdom of the blessed dead.[6] The reader will recall that it was this same Duat, referred to as the Duat-N-Ba (the 'netherworld of the soul'), that was said to have provided the model for the mysterious 'primeval temple' spoken of in the Edfu Building Texts.

Ptolemy stands in the posture of salutation, left hand clenched across his right breast, right hand raised. On either side of him is a figure of Maat (hence 'Double Maati') – a tall and beautiful goddess, sensual and full-breasted, wearing a headdress topped by her characteristic ostrich-feather plume (the hieroglyph for 'Truth'). The figure behind Ptolemy is empty-handed, and seems to be guiding him into the hall; the figure facing him holds in her right hand a long staff and in her left hand the hieroglyph *ankh*, the 'cross' or 'key' of life – the symbol of eternity.[7]

In a double row at the side of the Hall 42 dispassionate figures crouch in the manner of scribes pouring over papyrus, each wearing the feather of Maat. These are the 42 Judges or Assessors of the Dead, before each of whom the deceased must be able to declare himself innocent of a particular wrong – the 42 so-called 'Negative Confessions'. For example:

No. 4   'I have not stolen';

No. 5   'I have not slain man or woman';

No. 6   'I have not uttered falsehood';

No. 19 'I have not defiled the wife of a man';

No. 38 'I have not cursed the God'.[8]

Having completed this stage of his examination, Ptolemy now finds himself confronted by an immense pair of scales beneath the arms of which are to be seen representations of Anubis, the jackal-headed guide of souls, and Horus the falcon-headed son of Osiris. One pan of the scales contains an object, shaped like a small urn, symbolizing the heart of the deceased, 'considered to be the seat of intelligence and thus the instigator of man's actions and his conscience'.[9] In the other pan stands the feather of Maat, symbolizing once again . . . Truth.

On this encounter of the heart with Truth everything hinges.

For at this moment an irrevocable Judgement will be passed which will offer the prospect of eternal life to the soul that triumphs, and eternal annihilation to the soul that fails. Beyond the scales is depicted the agency of the soul's extinction: a monstrous hybrid, part crocodile, part lion, part hippopotamus, who is known as Ammit, the 'Devourer', the 'Eater of the Dead'. And beyond Ammit,

Complete scene of the weighing of the soul, Deir el Medina.

The goddess Nepthys, benefactor and protector of the dead. Tomb of Seti I, Valley of the Kings. Nepthys was the mother of Anubis.

Hypostyle hall, Karnak: Thoth, the god of wisdom (ibis-headed, left) writes the name of Pharaoh Seti I (centre) on the tree of life. In later times Thoth became known to the Greeks as Hermes Trismegistus. He was the keeper of the knowledge that opened the door to immortality.

seated in majesty on his throne at the extreme right of the scene, our eyes are drawn again towards the mummified figure of the star-god Osiris, the agency of the soul's resurrection.

Horus and Anubis test the scales, and proceed to measure the weight.[10] Meanwhile, to the immediate right of the scales, between the deceased and the snarling, slavering jaws of Ammit, we observe the tall ibis-headed figure of Thoth, the 'personification of the mind of god . . . the all-pervading and directing power of heaven and earth . . . the inventor of astronomy and astrology, the science of numbers and mathematics, geometry and land surveying'.[11] Mysteriously referred to in archaic inscriptions as 'three times great, great',[12] Thoth was the ancient Egyptian god of wisdom, 'the recorder of souls', who – from Ptolemaic times onwards – would also come to be known to the Greeks under the name of Hermes Trismegistus ('Hermes the Thrice Great').[13] In the Judgement Scene he is shown as a powerful man dressed in a short tunic, wearing his characteristic avian head-mask. In his left hand he holds up a palette and in his right a fine reed pen.

Heart and feather stand poised in equilibrium, as they must if the soul is to be admitted to the afterlife kingdom of Osiris.

Horus confirms the balance.

Anubis announces the verdict.

Thoth records . . .

THOTH AND MAAT

The scales of Maat.

The deities Thoth and Maat are present in the earliest surviving scriptures of mankind – the ancient Egyptian Pyramid Texts of the third millennium BC – and continue to play pivotal spiritual and cosmic roles throughout the entire 3000-year span of Pharaonic history. Standing either side of Atum-Ra, the sun-god, as he sails the celestial ocean in his 'boat of millions of years', they are portrayed in the *Book of the Dead* as eternal presences or principles whose function is to guide and balance the motion of the universe: 'Thoth ... Lord ... self-created, to whom none hath given birth ... he who reckons in heaven, the counter of the stars, the enumerator of the earth and of what is therein, and the measurer of the earth'.[14] Elsewhere we read: 'The land of Manu [the West] receiveth thee [Ra, the sun-god] with satisfaction, and the goddess Maat embraceth thee both at morn and at eve ... the god Thoth and the goddess Maat have written down thy daily course for you every day.'[15]

The word *maat* has many meanings in addition to 'truth' – for example, 'that which is straight', and, in the physical and moral sense, 'right, real, genuine, upright, righteous, just, steadfast, unalterable', etc. *Khebest maat* is 'real lapis-lazuli' as opposed to blue paste. *Shes maat* means 'ceaselessly and regularly'. *Em un maat* indicates that a thing is really so. The man who is good and honest is *maat*. And the truth, *maat*, 'is great and mighty and it hath never been broken since the time of Osiris'.[16] It is perhaps not surprising that in some versions of the Psychostasia the goddess Maat, with her arms outstretched, takes the form of the scales themselves.[17]

The feather and the heart, the two objects weighed in these scales, combine to convey a potent symbolic message. The former, as we have seen, is the type and symbol of the goddess herself, whilst it cannot be an accident that the latter, resembling a small vase with two handles, is not only used as the ancient Egyptian hieroglyph for 'heart' but also forms the 'determinitive' (defining sign) of the word *tekh*, 'a weight'.[18] From this etymology – *tekh* through *tehuti* – some scholars derive the origins of the name Thoth, a derivation which the Egyptians themselves appear to have favoured.[19] Let us also note in passing that the towering granite obelisks found in temples along the Nile were called *tekhen* by the ancient Egyptians – 'a word of unknown origin' according to Martina D'Alton of New York's Metropolitan Museum of Art.[20]

As we shall see in later chapters, obelisks played a special role in the quest for immortality that was pursued for millennia by Egypt's high initiates. From the remotest times this quest was intimately associated with the cult of Thoth, whose will and power were believed to keep the forces of heaven and earth in equilibrium: 'it was his great skill in celestial mechanics,' observed Sir E. A. Wallis Budge, 'which made proper use of the laws (*maat*) upon which the foundation and maintenance of the universe rested'.[21]

After an exhaustive analysis of funerary texts from all periods of ancient Egyptian history, Budge also comments on the manner in which Thoth is ubiquitously portrayed as possessing 'unlimited power' in the afterlife realm of the Duat.[22] It is this power that is symbolized by his role as recording angel in the

The goddess Maat, personification of truth, justice and cosmic harmony.

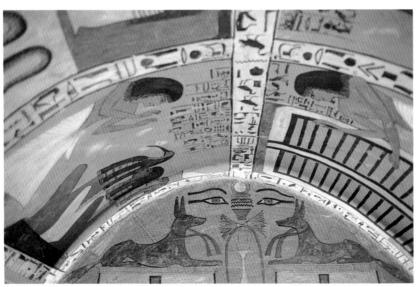

Judgement Scene. According to the *Book of What is in the Duat* (numerous representations of which survive in the tombs of Pharaohs from the Eighteenth Dynasty onwards): 'The examination of the words takes place, and he [Thoth] strikes down wickedness – he who has a just heart, he who bears the words in the scales – in the divine place of the examination of the mystery of mysteries of the spirits.'[23]

But what exactly is meant by 'wickedness' and what is the real nature of the mystery that is examined in the Judgement Hall of Osiris?

THE BOOKS OF THOTH

At stake in the Judgement Scene is something more than moral character. This is clear because questions pertaining to moral behaviour are addressed at quite an early stage in the proceedings by the soul of the deceased. This is the function of the 42 Negative Confessions. It follows, therefore, that the 'weighing' of the heart must be an evaluation of something else – a measuring of some other quality or character or 'truth' that the individual has been given the opportunity to add to during the course of his or her life. It is even possible that this may be the source of the Judgement Hall's 'Double Truth' – the concept that it is a place where two distinct and different levels of assessment must be undergone. This would explain why, as one eminent authority has observed:

> the testing of the soul in the Balance in the Hall of Osiris is not described as the judging or 'weighing of actions' [which the 42 Negative Confessions certainly are] but as *utcha metet*, the 'weighing of words'.[24]

Additional light is shed on this curious formula when we remember that Thoth was regarded by the ancient Egyptians as a god who could teach 'not only words of power but the manner in which to utter them'.[25] Knowledge of these 'words' was believed to be essential if the deceased was to hope to complete his afterlife quest through all twelve of the 'Divisions' of the Duat:

> The words … must be learned from Thoth, and without knowledge of them, and of the proper manner in which they should be said, the deceased could never make his way through the Duat. The formulae of Thoth opened the secret pylons for him, and provided him with the necessary meat, and drink, and apparel, and repelled baleful fiends and evil spirits, and

Detail from the Book of Gates, *tomb of Rameses VI, Valley of the Kings. Like the* Book of What is in the Duat, *the* Book of Gates *depicts a journey through the Duat. The journey is by boat. In it, protected by the coils of a cosmic serpent, the sun god Ra stands flanked by the figures of 'Mind' (fore) and 'Magic' (aft). In the* Book of Gates *the Judgement Hall of Osiris occupies the Sixth Division of the Duat.*

they gave him the power to know the secret or hidden names of the monsters of the Duat, and to utter them in such a way that they became his friends and helped him on his journey . . .[26]

It was believed that *Reu Nu Pert Em Hru*, the *Book of the Dead* – 'a sort of Baedeker for the transmigration of the soul'[27] – was a composition of Thoth and that certain chapters of it had been written 'with his own fingers'.[28] In addition numerous passages from the ancient texts have survived in which we learn that the wisdom god was also seen as the author of certain other 'books'[29] – books which anyone who sought the prize of immortality should attempt to discover during his lifetime: 'I am endowed with glory, I am endowed with strength, I am filled with might, I am supplied with the books of Thoth and I have brought them to enable me to pass through . . .'[30]

What the texts imply is that only he or she who has sought and found the books of Thoth can attain eternity. 'How long have I to live' the deceased asks in some versions of the Judgement Scene. If all is well at the 'weighing of words' Thoth replies by offering the coveted prize: 'Thou art for millions of years, a period of life of millions of years . . .'[31]

A QUEST FOR KNOWLEDGE

According to Clemens Alexandrinus (*Stromata VI*) there were 42 books of Thoth, a number that provides a curious sense of balance with the 'first truth' – the 'weighing of actions' – examined by means of the 42 Negative Confessions. These books of the 'second truth' – the 'weighing of words' – were thought to be divided into seven categories[32] covering, amongst other subjects, cosmography and geography, the construction of temples, the history of the world, the worship of the gods, medical matters, the hidden meaning of hieroglyphics, and treatises on astrology and astronomy including 'the ordering of the fixed stars, the positions of the sun, moon and planets, the conjunctions and phases of the sun and the moon, and the times when stars rise'.[33]

The tradition of the books of Thoth persisted well into the Christian era, associated with Graeco-Egyptian temples such as Deir el Medina, Dendera, Edfu and the Temple of Isis at Philae where the ancient Egyptian hieroglyphs continued to be used and understood until as late as the fourth century AD.[34] It is therefore hardly surprising that Clemens (AD 150–215) should have been aware of this tradition which was, indeed, set down afresh in writing in his adopted city of Alexandria at about this time. These writings, the so-called *Corpus Hermeticum*,[35] repeatedly describe Thoth (the 'Hermes Trismegistus' of the Greeks) as 'he who won knowledge of all'.[36] He:

saw all things, and seeing understood, and understanding had the power both to disclose and to give explanation. For what he knew he graved on stone; yet though he graved them onto stone he hid them mostly, keeping sure silence [so] that every younger age of cosmic time might seek for them . . . [37]

A quest, then, appears to have been envisaged for these stone tablets, or 'books', of Thoth/Hermes. Indeed the *Corpus Hermeticum* leaves us in no doubt about this matter, telling us that the wisdom god used magic to postpone for as long as possible the rediscovery of his treasures of knowledge:

> Ye holy books ... which have been anointed with the drug of imperishability ... remain ye undecaying through all ages, and be ye unseen and undiscovered by all men who shall go to and fro on the plains of this land, until the time when Heaven, grown old, shall beget organisms worthy of you.[38]

Walter Scott, the translator of this passage into English, appends the following explanatory note concerning the term 'organisms': 'Literally "composite things"; that is, men composed of soul and body. After long ages there will be born men that are worthy to read the books of Hermes.'[39]

A SERPENT WHICH CANNOT DIE ...

The urge to read them must be very old because it can be traced back deep into ancient Egyptian times, long before the compilation of the *Corpus Hermeticum*. For example, a papyrus of the Ptolemaic period preserves the story of a certain Setnau-Khaem-Uast, a son of Rameses II (ruled 1290–1224 BC), who sought for a 'book written by Thoth himself'.[40] Information had come Setnau's way, as a result of diligent research, that this book – which was said to contain a spell capable of granting immortality – lay concealed in an antique tomb in the Memphite necropolis (an extensive burial area stretching for some 35 kilometres along the west bank of the Nile from Meidum to Giza):

> Setnau went there with his brother and passed three days and nights seeking for the tomb ... and on the third day they found it. Setnau recited some words over it, and the earth opened and they went down to the place where the book was. When the two brothers came into the tomb they found it to be brilliantly lit up by the light which came forth from the book.[41]

Another papyrus, this time from the Middle Kingdom (the Westcar Papyrus, circa 1650 BC), preserves an even older story from the time of Khufu (ruled 2551–2528 BC), the supposed builder of the Great Pyramid of Giza. The papyrus speaks of a 'building called "Inventory"', located at the sacred city of Heliopolis (18 kilometres north-east of Giza), in which was stored 'a chest of flint' containing a mysterious object that Khufu is reported to have 'spent much time searching for'. The context suggests it could have been a document of some kind because it recorded the 'number of the secret chambers of the sanctuary of Thoth'.[42]

It is generally agreed that the Westcar Papyrus reports – or at any rate touches upon – real events. According to Professor I. E. S. Edwards it contains a 'kernel of truth' and 'was certainly a copy of an older document'.[43] Edwards further points out that Heliopolis, the site of the 'Inventory Building', had been a centre of astronomical and astrological science in Egypt since times immemorial and that the title of the high priest of that city was 'Chief of the Astronomers'.[44]

The Egyptologist F. W. Green expresses the opinion that the 'Inventory Building' could well have been a 'chart room' at Heliopolis 'or perhaps a "drawing room" where plans were made and stored'.[45] Similarly, Sir Alan H. Gardner argues that 'the room in question must have been an archive' and that Khufu 'was seeking for details concerning the secret chambers of the primeval sanctuary of Thoth'.[46]

The central image of the Westcar Papyrus of some great secret of Thoth lying sealed away in a box is repeated in another text which tells how the wisdom god had deposited one of his books 'in an iron box in the middle of the Nile at Coptos'[47] (an ancient site some kilometres to the north of Luxor):

> The iron box is in a bronze box, the bronze box is in a box of palm-tree wood, the palm-tree wood box is in a box of ebony and ivory, the ebony and ivory box is in a silver box, the silver box is in a gold box ... The box wherein is the book is surrounded by swarms of serpents and scorpions and reptiles of all kinds, and round it is coiled a serpent which cannot die.[48]

Last but not least amongst many similar sources that we could cite, there is a Coffin Text, circa 1900 BC, that speaks of the journey of the soul towards immortality. 'I open the chest of Thoth', states the deceased, 'I break the seal ... I open what the boxes of the god contain, I lift out the documents ...'[49]

So there is a sense in all of this that what is weighed in the Judgement Scene at the 'weighing of words' must in some way have to do with the possession of *knowledge* by the deceased, the kind of knowledge that can be inscribed on to tablets of stone or written down in books and 'documents'.

THE WORD

Like so many of the other funerary and rebirth texts of ancient Egypt, the Coffin Texts are manuals to guide the afterlife journey of the soul – the terrifying quest in the dark valley of the Duat that culminates with the Judgement Scene. The texts are so called because they were inscribed inside coffins, presumably so as to be easily accessible to the dead. They date from the First Intermediate Period (2134–2040 BC) and were particularly favoured during the Twelfth Dynasty (1991–1783 BC).[50] In the early spells we read:

> The young god [the deceased entering the afterlife kingdom of Osiris having found immortality] is born of the beautiful West, having come here from the land of the living; he has got rid of the dust which was on him, he has filled his body with magic, he has quenched his thirst with it ... *he has mastered the land through what he knew.*[51]

In a later spell an almost identical formula occurs:

> See, Your Majesty has come, you have acquired all power, and nothing has been left behind by you ... You have filled your body with magic, you have quenched your thirst with it ... you have mastered the land with what you know like those to whom you have gone down.[52]

And later still we read of the triumph of the 'equipped spirit' and may begin to guess as to what it is with which he is 'equipped':

I have passed over the paths of Osiris; they are in the limit of the sky. As for him who knows this spell for going down into them, he himself is a god, in the suite of Thoth; he will go down to any sky to which he wishes to go down to. But as for him who does not know this spell for passing over these paths, he shall be taken into the infliction of the dead which is ordained, as one who is nonexistent.[53]

CELESTIAL CO-ORDINATES

There can be no dispute that the equipped spirit was thought to master the land of the Duat with '*what he knew*'. But what exactly was this knowledge? The suggestion in the texts that it was used to 'go down to any sky' hints very strongly that astronomy might have been involved. This accords with what has been learnt concerning the astronomical interests of the priests of Heliopolis. It also makes sense of an important characteristic of the Duat to which few modern Egyptologists have paid attention: the afterlife region was *not* at any time conceived of by the ancient Egyptians as an 'underworld' in the conventional Judaeo-Christian sense. On the contrary, as Dr R. O. Faulkner of the British Museum

Detail from the Book of What is in the Duat, *tomb of Thutmosis III, Valley of the Kings. Astronomically, the Duat was located in the sky between the constellations of Orion and Leo, but it was also a parallel universe which was always depicted as a maze of narrow corridors and passageways and rising galleries and chambers, populated by monsters. Compare to the passageway system of the Great Pyramid, facing page.*

RIGHT: *The Grand Gallery of the Great Pyramid.* BELOW: *The descending corridor. Scholars have as yet failed to consider the possibility that the Pyramids and perhaps even the Sphinx of Giza could have been built as three-dimensional models of the 'inner world' of the Duat – places of preparation in which initiates may have been selected to immerse themselves, perhaps in total darkness, perhaps for days, in order to gain fore-knowledge of the afterlife realm. Yet there is nothing inherently improbable about such a proposition. We already know that the various ancient Egyptian 'books of the dead' provide textual explanations and visual images of the Duat with the explicit purpose of preparing the deceased for the after-life journey. To create a large-scale three-dimensional 'model' of the Duat – a sort of simulated Netherworld – would be no more than an extension of this practice.*

long ago observed, it is better described as a 'netherworld' since it was 'part of the visible sky'.[54]

In fact the Duat had very specific celestial co-ordinates. The first systematic attempt to chart these co-ordinates was undertaken in the 1940s by the Egyptologist Selim Hassan. Through a painstaking study of a mass of funerary and rebirth texts he established that the Duat had been conceived of by the ancient

RIGHT: *The Subterranean Chamber of
the Great Pyramid.* BELOW: *Scenes
from the* Book of What is in the
Duat, *tomb of Thutmosis III.*

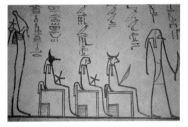

Egyptians as having been 'localized in the eastern part of the sky' when the
bright star Sirius – identified with the goddess Isis[55] – and the stars of the Orion
constellation – Osiris – were visible there in the pre-dawn. This was clear, he
reasoned, from passages in the oldest texts which tell us: 'Orion has been
enveloped by the Duat while he who lives on the Horizon purifies himself. So
this [Sirius] has been enveloped by the Duat while he who lives on the Horizon
purifies himself.'[56] Hassan understood that such passages must have been based on
observational astronomy:

> as the sun rises and purifies himself in the Horizon, the stars Orion and
> Sothis are enveloped by the Duat. This is a true observation of nature, and
> it really appears as though the stars are swallowed up each morning by the
> increasing glow of the dawn. Perhaps the determinative of the word Duat,
> the star within a circle, illustrates this idea of the enveloping of a star.[57]

More recently the author Robert Bauval has been able to pin down the location
of the Duat in time and space still further with a crucial observation that Hassan
missed. Because of the earth's orbit, the background stars against which the sun is
seen to rise each morning *very slowly change* throughout the course of the solar
year. This means that the sun does not rise in concert with Orion and Sirius on
every dawn, but only at certain and specific dawns (when the sun lies roughly
between the earth and these stars). Furthermore, because of another characteristic motion of the earth, the *season* in which the 'swallowing up' of Orion and
Sirius takes place also very slowly changes. This motion is precession, which
retards the moment of the sun's arrival at any given stellar 'address' at the rate of
one degree every 72 years.

Precessional calculations for 2500–2300 BC – when the oldest surviving funerary texts from ancient Egypt were supposedly compiled – indicate that in that

The sky region of the Duat on the summer solstice circa *2500* BC *– also showing the trajectory of Orion until its culmination at the meridian.*

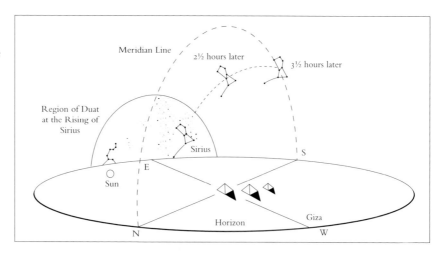

epoch the Duat could only have been regarded as being 'active' (i.e. with Orion and Sirius rising just ahead of the sun) at around the summer solstice – the longest day of the year.[58] At this time, and at no other season, would it have been believed to open its gates to the assembled souls of the dead. At one gate stood the constellation of Leo. At the other, divided from Leo by the glowing river of the Milky Way, stood Sirius, Orion and the constellation of Taurus. In 2500 BC this sacred portal in the heavens was said to 'open' at the summer solstice because the sun rose in it at that time of the year. Today, because of the effects of precession, the sun 'swallows up' Orion and Sirius at the autumnal equinox. In 10,500 BC phenomenon could only have been witnessed on the spring equinox.

Is it possible that the initiate's skill at 'going down to any sky' could be a reference to an ability to make precessional calculations – i.e. to harness intellect to imagination and to visualize the skies of former and future epochs?

Was it such knowledge that was believed to be sufficiently powerful to counterbalance the feather of Maat on the scales of Judgement and to triumph over nonexistence?

> This is the word which is in darkness. As for any spirit who knows it he will live among the living ... he will never perish ... he will never die.[59]

SUPERSTITION, OR SCIENCE?

Undeniably powerful and even disturbing, the ideas conveyed in the funerary and rebirth texts of ancient Egypt have been described by Dr Stephen Quirke, Curator of the Department of Egyptian Antiquities at the British Museum, as belonging to an:

> everlasting world ... in which the endeavour to outlast eternity reaches its most self-conscious. [They] spell out the precise phrasing by which a dead person could be made into an eternally rejuvenated being. Today we call these ancient texts 'funerary literature', but this technical term does them little justice: these are texts to transfigure the dead, to make human beings into immortal gods.[60]

Book of What is In the Duat: *the spiritualization of the deceased, flanked by two Eyes of Horus amidst a landscape of stars and winged serpents.*

The 'sarcophagus' in the King's Chamber of the Great Pyramid. If the Pyramid was built as a model of the Duat then the sarcophagus may have been used as part of the initiate's preparation for his own inevitable death and afterlife journey and hoped-for rebirth.

The ancient Egyptians themselves often called the texts *sakhu*, Quirke reports:

> meaning recitations that would turn a person after death into an *akh*, 'a transfigured spirit'. The only alternative was to die and remain *mut*, 'dead'. These opposites of *akh* and *mut* are roughly equivalent to the European contrast between the blessed and the damned. As in the European tradition, paradise is envisaged in terms of light, and the word *akh* itself is one of a group in which the idea of light and radiance is paramount, such as the Egyptian for 'horizon', *akhet*, the home of light. Faced with the alternative, the Egyptians concentrated all their resources into securing this eternal radiance.[61]

In other words, although Quirke does recognize the lofty goal expressed in the texts – to transform human beings into immortal gods – he believes that it was sought for reasons that are largely *psychological*. Quite simply, he argues, the ancient Egyptians found the alternatives to eternal life – nonexistence, annihilation – too horrible to contemplate and therefore created an elaborate fantasy world which they imagined that their souls could enter and in which, if suitably 'equipped', they hoped that they might win the prize of immortality.

In line with Quirke's view, it has become customary amongst Egyptologists today to disparage the texts as little more than wishful thinking – 'a strange accretion of spells and mumbo-jumbo ... a reflection of humanity's earliest supreme revolt against the darkness and silence from which none returns'.[62] Some scholars have even gone so far as to insist that:

> In spite of their meticulous attention to detail in practical matters, the Egyptians of the Pyramid Age never evolved a clear and precise conception of the After-Life ... The impression made on the modern mind is that of a people searching in the dark for a key to truth and, having found not one but many keys resembling the pattern of the lock, retaining all lest perchance the appropriate one should be discarded.[63]

Similarly Dr Margaret Murray observes that 'the horror of death is very marked in the religious texts of the Egyptians ... Knowing that death is inevitable [the Egyptian] tried to prepare for it by a knowledge of the magic which would enable him to come back to the land and home he loved so well ...'[64]

It is the fundamental proposition of *Heaven's Mirror* that matters are by no means so simple and that the Egyptian scriptures contain extraordinary material with an importance vastly deeper and darker than mere mumbo-jumbo, and far, far older than scholars have imagined.

OPPOSITE: *Seti I Temple, Abydos: Isis, goddess of magic, offers ankh, the gift of eternal life, to the soul of Pharaoh Seti I. Behind the symbolism, and the ethereal beauty of the reliefs, the sense of a lofty and ancient purpose animates the sacred art of Egypt.*

CHAPTER 5

HIDDEN CIRCLES

The Pharaohs of ancient Egypt sought to prepare their souls for an afterlife journey through an astronomical netherworld. Detail of the starry ceiling in the tomb chamber of the Sixth Dynasty Pharaoh Teti at Saqqara.

OPPOSITE: *'Earth is this king's detestation; this king is bound for the sky.'*

THE EDFU BUILDING TEXTS state that they are records of 'the words of the Sages' that had long previously been written down in a book by Thoth, the ancient Egyptian god of wisdom.[1] We have seen that an identical claim is made in the 'Hermetic' texts, compiled in the Egyptian city of Alexandria in the second and third centuries AD. It is also made in all the funerary and rebirth texts of ancient Egypt including the *Book of the Dead*, the Coffin Texts, the Pyramid Texts, the *Book of Gates* and the *Book of What is in the Duat*.

The Pyramid Texts, the earliest of these writings to have survived, were already so old in 2300 BC, when they were copied on to the tomb walls of Fifth and Sixth Dynasty pyramids at Saqqara, that some of the scribes could not understand them.[2] Weirdly, however, though modern scholars are adamant that there is no connection, they convey the same essential message as the Hermetic Texts, i.e. that immortality is a 'pearl of great price'[3] which can only be won through right thought and right deeds – for 'souls have at stake in this life their hope of eternity in the life to come'.[4]

What must we do to ensure that we do not lose that stake?

The Hermetic Texts tell us that it is 'a man's duty not to acquiesce in his merely human state, but rather, in the strength of his contemplation of things divine, to scorn and despise that mortal part which has been attached to him because it was needful that he should keep and tend the lower world'.[5] In exactly the same manner, in the Pyramid Texts, the initiate strives to turn his back on earthly matters and to focus his intellect on the heavens: 'Earth is this King's detestation . . . This King is bound for the sky,' he states at one point.[6] 'Grant that I may seize the sky and take possession of the horizon,' he asks at another.[7] 'A stairway to the sky is set up for me,' he affirms later, 'that I may ascend on it to the sky.'[8]

Elsewhere in the Pyramid Texts we can read intensely Hermetic maxims such as 'The spirit is bound for the sky, the corpse is bound for the earth.'[9] We also frequently come across exhortations to the initiate such as the following: 'Arise, remove your earth, shake off your dust, raise yourself, that you may travel in company with the spirits . . . Cross the sky . . . Make your abode among the imperishable stars.'[10]

It is surely a very similar cluster of ideas that we find expressed much later in

The second shrine of Tutankhamun, detail top, and full tableau above. The connection that the ancient Egyptians saw between intellect, insight and astronomy, and the link to the afterlife, could hardly be made more explicit.

Egyptian history in the second shrine of the young Eighteenth Dynasty Pharaoh Tutankhamun (ruled 1333 to 1323 BC). Here, carved in gold, we see a row of human figures, gazing upwards, each connected through the forehead to a star or to a celestial orb by means of a series of rays. The atmosphere is one of meditation, intense concentration, stillness. It is hard to think of a more appropriate image of the initiate's endeavour to 'ascend' – to contemplate the mysteries of the cosmos and to win an understanding of its deepest secrets: 'The celestial portal to the horizon is open to you and the gods are joyful at meeting you. They take you to the sky with your soul ...'[11]

COPYING THE SKY

It was believed that the initiate's ability to 'ascend to the sky', in other words to make visionary journeys amongst the stars of heaven, would be enhanced if he were to divorce himself from the concerns and 'attachments' of material exist-

Astronomical ceiling of the tomb of Seti I, Valley of the Kings. Scholars agree that the crocodile riding on the back of the hippopotamus to the right of the scene represents the constellation known today as Draco and that the 'thigh' on which the hippo rests its hand represents the stars of the Big Dipper (Ursa Major). It is tempting, although contrary to the scholarly consensus, to associate the lion, outlined by stars, with the constellation of Leo, the standing man facing it with Orion, and the bull above the standing man with Taurus. On this interpretation Seti I's tomb shows us the entrance to the 'land of the Duat' overlooked by the constellation of Draco.

ence. It was also believed that his contemplations would be richer if they were to be performed in an ambience that would make him acutely aware of his place in the cosmos. This was essential because in Egypt, as in ancient Central America, 'all the world which lies below' was viewed as having been:

> set in order and filled with contents by the things which are placed above; for the things below have not the power to set in order the world above. The weaker mysteries, then, must yield to the stronger ... the system of things on high is stronger than the things below ... and there is nothing that has not come down from above.[12]

The Hermetic Texts insistently describe Egypt as 'an image of heaven',[13] as 'the temple of the whole world',[14] and sometimes as 'the sanctuary of the Cosmos'[15] – the precious land where 'all the operations of the powers which rule and work in heaven have been transferred to earth below'.[16] Similarly, in the *Book of What is in the Duat* we learn that the essential requirement for those seeking the life eternal was that they should build on the ground perfect copies 'of the hidden circle of the Duat in the body of Nut [the sky]':[17]

> Whosoever shall make an exact copy of these forms, and shall know it, shall be a spirit well-equipped both in heaven and in earth, unfailingly, and regularly and eternally.[18]

> Whosoever shall make a copy thereof, and shall know it upon earth, it shall act as a magical protector for him both in heaven and upon earth.[19]

Ceiling of the burial chamber, Tomb of Rameses VI, Valley of the Kings. The goddess Nut is represented twice along the entire length of the ceiling, symbolizing the morning and evening skies.

Is it not obvious that people thinking in this way would have been almost compelled to express their ideas not only in texts but also in architecture? How they might have done so is hinted at in the Edfu Texts which tell us that when Thoth arrived at the foundation-ground of the original mythical temple – of which the historical Edfu temple was a later copy – he recited the following words over it:

> I cause its long dimension to be good, its breadth to be exact, all its measurements to be according to the norm, all its sanctuaries to be in the place where they should be, *and its halls to resemble the sky*.[20]

Scholars think it is entirely possible that this mysterious edifice 'resembling the sky' did once exist. According to Dr Eve Reymond it is almost certain that it would have stood 'near to Memphis, which the Egyptians looked on as the homeland of the Egyptian temple'.[21]

Relief from the temple of Hathor at Dendera showing standing man grasping the hieroglyph 'sky' decorated with stars. Is it possible that ancient Egyptian temples and pyramids could have been part of a deliberate and long-term programme to copy sky to ground – literally to build heaven on earth?

NEW DISCOVERIES

Memphis (originally 'Men-nefer', meaning 'established and beautiful'),[22] was the first historical capital of ancient Egypt – dating back to at least 3000 BC. Its ruins lie some 24 kilometres south of Cairo, near the modern village of Mit Rahina, but they are few and unimpressive. Of far greater interest, and far more striking, are the extensive remains of the so-called 'Memphite necropolis' that lies to the west of Memphis itself. This huge burial ground, overlooking the Nile and spilling westwards into the surrounding desert, runs 35 kilometres from north to south. As well as hundreds of shaft burials and gigantic *mastaba* tombs, it includes sites on which many pyramids stand – for example Abu Roash, Zawiyet el-Aryan, Abusir, Saqqara and Dashur. By universal consensus, however, the most remarkable monuments of the Memphite necropolis are found at Giza, just 10 kilometres south of Cairo, where stand the three great Pyramids and the Great Sphinx of Egypt.

Located so close to Memphis, is it possible that Giza – together with nearby Heliopolis (a great academy of priestly learning as we shall see) – could have been

The so-called 'Bent' (left) and 'Red' Pyramids at Dashur. Both appear to have been built by Sneferu (2575–2551 BC) the first Pharaoh of the Fourth Dynasty and the father of Khufu, the supposed builder of the Great Pyramid. If Pyramids are tombs and tombs only then why did Sneferu need two?

The three great Pyramids of Giza, viewed from the south-east. To the right, the Great Pyramid of Khufu, centre the Pyramid of Khafra, left the smaller Pyramid of Menkaura.

the 'predynastic religious centre' that Reymond felt must have existed? More specifically, is it possible that the great Pyramids and Sphinx of Giza could in some way 'resemble the sky' – like the original mythical temple that is spoken of so clearly in the Edfu Texts?

Remarkable scientific discoveries made at Giza in the 1990s have deepened the mystery surrounding these monuments, revealing that they do indeed reflect a celestial plan and hinting at vast antiquity. The discoveries have received wide attention, having been reported in three bestselling books (*The Orion Mystery*, *Fingerprints of the Gods*, *Keeper of Genesis*). They have also been covered in many newspapers and magazines and in a series of international television documentaries watched by millions (*The Mystery of the Sphinx*, *Genesis in Stone*, etc.). For readers who are not familiar with these new discoveries, and to refresh the memories of those who are, we set out below a short summary of what is now known.

The Great Sphinx and its megalithic temples. Geological evidence suggests that these structures may be thousands of years older than Egyptologists have realized.

1 The geology of the Sphinx

Thanks to the work of Egyptologist John Anthony West and Dr Robert Schoch, Professor of Geology at Boston University, we now know that there are serious scientific objections to the presently accepted dating of the Great Sphinx of Egypt. We will not repeat here the evidence – extensively outlined elsewhere[23] – which casts fatal doubt on the orthodox Egyptological theory attributing the monument to the Fourth Dynasty Pharaoh Khafre (ruled 2520–2494 BC). It is sufficient to state, as at least one senior member of the profession has been honest enough to admit, that 'there is not a single ancient inscription which connects the Sphinx to Khafre'.[24]

The monument, in other words, is anonymous stone. It is carved in one piece out of the bedrock of the Giza plateau and, although later patched up with repair blocks (in both ancient and modern times) is still recognizable in essence as a gigantic monolith. As such it is not susceptible to carbon-dating, which can measure the age of organic materials only. Indeed just as there is not a single ancient inscription concerning the Sphinx so also there is not a single test presently in existence which can tell us accurately *when* the monument was carved. Theoretically that could have happened at any time after the limestone of the Giza plateau was originally laid down by former oceans tens of millions of years ago.

Professor Robert Schoch's work helps to narrow the search by establishing a *minimum age* for the Sphinx. But Schoch's findings are fiercely controversial because they set that minimum age very high, at 7000 years or older – i.e. deep in predynastic times, at least 2500 years earlier than the date accepted by Egyptologists. The Boston geology professor is, however, unrepentant:

> I've been told over and over again that the peoples of Egypt, as far as we know, did not have either the technology or the social organization to cut out the core body of the Sphinx in pre-dynastic times . . . However, I don't see it as being my problem as a geologist. I'm not seeking to shift the burden, but it's really up to Egyptologists and archaeologists to figure out who carved it. If my *findings* are in conflict with their *theory* about the rise of civilization then maybe its time for them to re-evaluate that theory. I'm not saying that the Sphinx was built by Atlanteans, or people from Mars, or extra-terrestrials. I'm just following the science where it leads me, and it leads me to conclude that the Sphinx was built much earlier than previously thought.[25]

The granite stela between the paws of the Sphinx is not contemporary with the monument but was inscribed to commemorate a restoration campaign undertaken by Pharaoh Thutmosis IV (ruled 1401–1391 BC). The stela states enigmatically that the Sphinx marks 'the Splendid Place of the First Time'.

Schoch, a world expert on the weathering of limestone, bases this conclusion on a careful study of the erosion of the Sphinx. He believes (and hundreds of other geologists have endorsed his view)[26] that the great monument could not possibly have been carved as late as 2500 BC. This is so because it bears the unmistakable marks of 'precipitation-induced weathering', deep vertical fissures and undulating, horizontal coves that could only have been caused by thousands of years of heavy rain – rain that must have fallen on the Sphinx *after* it was carved.

The problem is that in 2500 BC Egypt was as bone dry as it is today, getting less than an inch of rain a year. Palaeoclimatologists, however, are able to tell us, very accurately, when the weather was wetter. Their conclusion is that the last time

The 'Queen's Chamber' of the Great Pyramid. To the left, on the east side of the room, is a curious corbelled niche of unknown function. To the right, on the south wall of the room, is the entrance to the Great Pyramid's mysterious southern shaft, which targeted the star Sirius in the Pyramid Age. In 1993 the German robotics engineer Rudolf Gantenbrink explored this shaft with a mechanized robot camera. After making a journey of 65 metres up the steeply inclined shaft, the camera came to what appears to be a small portcullis 'door' with two metal handles. Could there be an intact chamber beyond this little door at the top end of the shaft – perhaps even as large as the Queen's Chamber at the bottom end of the shaft?

sufficient rain fell in the eastern Sahara to have caused the characteristic weathering of the Sphinx was between 7000 and 5000 BC.[27]

It is because of this evidence, choosing the most recent and conservative date possible, that Professor Schoch arrives at his minimum age for the Sphinx of 7000 years. His colleague John Anthony West, on the other hand, thinks that the monument may be much older (a possibility by no means ruled out by the weathering studies). 'My conjecture,' he told us:

> is that the whole riddle is linked in some way to those legendary civilizations spoken of in all the mythologies of the world. You know – that there were great catastrophes, that a few people survived and went wandering around the earth and that a bit of knowledge was preserved here, a bit there . . . My hunch is that the Sphinx is linked to all that. If I were asked to place a bet I'd say that it predates the end of the last Ice Age and is probably older than 10,000 BC, perhaps even older than 15,000 BC. My conviction – actually it's more than a conviction – is that it's vastly old.[28]

2   The astronomy of the Pyramids and the Sphinx

The geological work upsets orthodox Egyptological chronology but cannot substitute a precise alternative date, leaving the floor open anywhere between 15,000 BC, if John West is right, and 5000 BC if Robert Schoch's estimate is preferred. The astronomical studies of Robert Bauval have provided researchers with an indispensible additional technique by means of which the true genesis date of the Giza monuments can be assessed.

The first breakthrough was published in *The Orion Mystery* in which Bauval showed that the four narrow 'air shafts' of the Great Pyramid (features that are unique to this monument) might more properly be called 'star shafts' since they would have aligned, circa 2500 BC, with four prominent stars to which the ancient Egyptians accorded immense ritual importance. The shafts are 'meridional', i.e. two of them point due north and two of them point due south, and would have targeted these stars at their nightly transit of the north and south meridians of the sky. In the epoch of 2500 BC the stars 'acquired' in this way by the two northern

The stellar alignments of the four 'star shafts' of the Great Pyramid circa *2500 BC.*

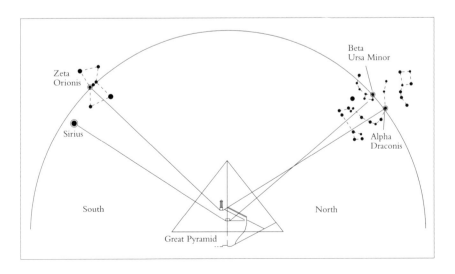

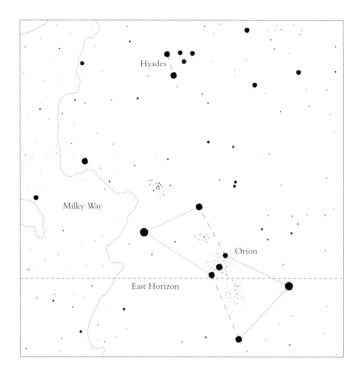

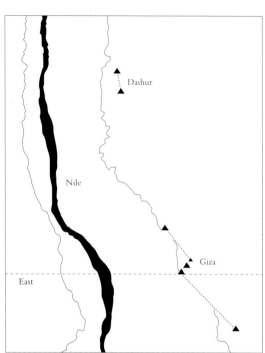

LEFT: *The constellation of Orion.*
RIGHT: *The layout of the pyramids of Giza duplicating the position of the constellation of Orion circa 10,500 BC, with the Nile corresponding to the Milky Way. Note also the correlation between the Dashur pyramids and the Hyades.*

shafts were Kochab (Beta Ursa Minor), in the constellation of the Little Bear, and Thuban (Alpha Draconis) in the constellation of Draco – the Dragon or Serpent. At the same period the southern shafts targeted Sirius (the bright star in the constellation of Canis Major that the ancient Egyptians identified as the celestial counterpart of their goddess Isis) and Al Nitak (Zeta Orionis) the brightest of the three stars of Orion's belt – which, as we have seen, 'the ancient Egyptians identified with Osiris, their high god of resurrection and rebirth and the legendary bringer of civilization to the Nile Valley in the remote epoch referred to as "Zep Tepi", the "First Time"'.[29]

Bauval's investigations into the orientations of the shafts were an extension of earlier work by the Egyptian architect Alexander Badawy and the American astronomer Virginia Trimble and, in themselves, do not challenge the orthodox dating of the Pyramids to 2500 BC. Bauval, however, did not stop here but went on to draw attention to the curious *pattern* of the three great Pyramids on the ground – his 'Orion correlation theory', a vital breakthrough in the study of ancient Egypt:

An overhead view [of the Giza plateau] shows that the Great Pyramid and the second pyramid stretch out along a diagonal running 45 degrees to the south and west of the former's eastern face. The third pyramid . . . is offset somewhat to the east of this line. The resulting pattern mimics the sky where the three stars of Orion's belt also stretch out along a 'faulty' diagonal. The first two stars (Al Nitak and Al Nilam) are in direct alignment like the first and second pyramids, and the third star (Mintaka) lies offset somewhat to the east of the axis formed by the other two.

The visual correlation, once observed, is obvious and striking on its own. Additional confirmation of its symbolic significance . . . is provided by the

'What is a solstice or an equinox? It stands for the capacity of coherence, deduction, imaginative intention and reconstruction with which we could hardly credit our forefathers.' Giorgio de Santillana, Preface to Hamlet's Mill.

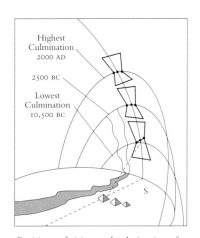

Positions of rising and culmination of Orion through the ages. The pattern of the stars in 10,500 BC marks a beginning, or 'First Time', of the cycle, reproduced in the layout of the three Great Pyramids of Giza.

Milky Way, which the ancient Egyptians regarded as a kind of celestial Nile and which was spoken of in archaic funerary texts as the Winding Waterway. In the heavenly vault the belt stars of Orion lie to the west of the Milky Way, as though overlooking its banks; on the ground the Pyramids stand perched above the west bank of the Nile.

Faced by such symmetry, and by such a complex pattern of interlocking architectural and religious ideas, it is hard to resist the conclusion that the Pyramids of Giza represent a successful attempt to build Orion's belt on the ground.[30]

After all the Orion constellation was seen by the ancient Egyptians as the celestial image of Osiris, Lord of the Duat. Is it not therefore rather suggestive of an attempt to make the Pyramids 'resemble the sky' that they turn out to have been built in the pattern of the belt stars of Orion/Osiris? Perhaps this is what is meant by a Coffin Text which states: 'I am a builder and I have knowledge . . . I am the semblance of Osiris . . . I am the image of Osiris.'[31]

But an image of Osiris when? Since Osiris was Orion, Bauval reasoned that it might be possible to find an answer to this question in the stars – specifically in the slow changes in stellar positions brought about by the precession of the earth's axis.

Several astronomical computer programs such as Skyglobe and Redshift allow researchers to simulate the effects of precession on all the stars in the sky and to view those stars from any point on the earth's surface. What Bauval discovered about the constellation of Orion, viewed from Giza, is that its three belt stars appear to slide up and down the meridian during the precessional cycle – 13,000 years 'up' (i.e. *gaining* altitude above the horizon at meridian transit) and 13,000 years 'down' (i.e. *losing* altitude above the horizon at meridian transit). The lowest point in the cycle last occurred at around 10,500 BC and its highest point will next occur between AD 2000 and 2500.

Bauval also realized that it is not only the altitude of the three belt stars that is affected by the precessional cycle: their orientation in relation to the meridian simultaneously undergoes constant changes, shifting almost imperceptibly, century by century, in a clockwise direction. Using Skyglobe to 'wind the stars back', and compare what he saw in the skies to the pattern of the three great Pyramids on the ground, he discovered that there was only one epoch in which heaven and earth locked together perfectly. This was the epoch of 10,500 BC, the lowest point, or beginning – effectively the 'first time' – of the current precessional cycle of the constellation of Orion. It is in that epoch, and only in that epoch, that the pattern of the Pyramids on the ground would have exactly replicated the pattern of the three belt stars.

Of course, it could be a fluke that the Orion correlation – although evident and obvious in all epochs – is only perfect in the astronomical 'first time' of 10,500 BC. But if so then it must also be a fluke that Osiris/Orion is repeatedly referred to in ancient Egyptian scriptures as the god of the 'First Time'.

Likewise it must be a fluke (but how many can there be?) that a second dramatic sky–ground correlation occurs at Giza in exactly the same epoch. Reported

ABOVE: *The Sphinx gazes east to where the constellation of Leo reclines on the horizon, an hour before sunrise, spring equinox 10,500 BC.*

BELOW: *The Sphinx aligned to Leo and the Pyramids aligned to the stars of Orion's belt at sunrise, spring equinox 10,500 BC.*

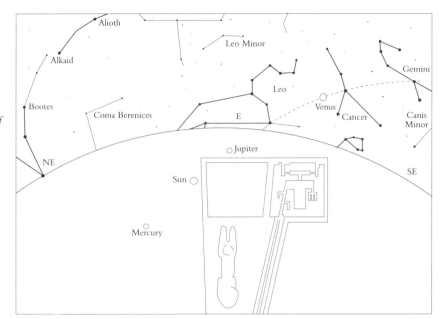

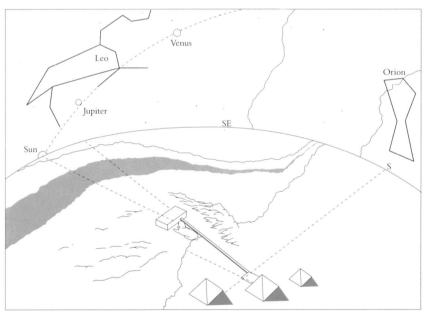

Sphinx, who attracts all,
Eyes closed with serenity at dusk,
Eyes open projecting strength at dawn.
What secrets do you hold?
What knowledge do you bear?

in *Keeper of Genesis*, this correlation involves the Sphinx, which is aligned with painstaking accuracy to gaze due east – the direction of sunrise on the spring equinox. Computer simulations show that in 10,500 BC the constellation of Leo housed the sun on the spring equinox – i.e. an hour before dawn in that epoch Leo would have reclined due east along the horizon in the place where the sun would soon rise. This means that the lion-bodied Sphinx, with its due-east orientation, would have gazed directly on that morning at the one constellation in the sky that might reasonably be regarded as its own celestial counterpart.

The sense of sky and ground matching one another deepens an hour later. As Leo rises higher and at the precise moment that the top of the solar disk breaks over the horizon due east, perfectly in line with the gaze of the Sphinx, the computer shows us that the three stars of Orion's belt would have been positioned due

Ceiling of the tomb of Senmut, west bank, Luxor. Senmut lived in the fifteenth century BC. He was renowned for his wisdom and astronomical prowess, and as the architect of the female Pharaoh Hatshepsut – near whose mortuary temple he is buried. In an inscription he claims to have 'penetrated all the writings of the divine prophets' and to have been 'ignorant of nothing that has happened since the beginning of time'. His tomb ceiling features the constellation of Orion (represented as the figure of Osiris standing in a boat with staff in hand, looking over his shoulder). The figure is prominently associated with the 'trinity' of the belt stars, suggesting that these three stars may have been used by the ancient Egyptians as a kind of 'short-hand' symbol for the whole constellation.

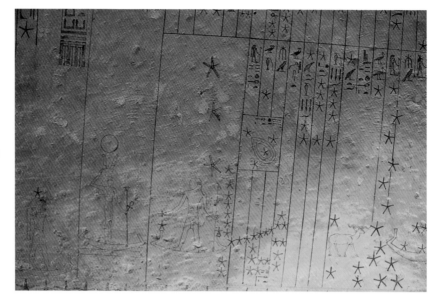

south, 'culminating' at the meridian in a pattern identical to the ground-plan of the Giza Pyramids:

> The question reduces to this: is it a coincidence, or more than a coincidence, that the Giza necropolis as it has reached us today out of the darkness of antiquity is still dominated by a huge equinoctial lion statue at the east of its 'horizon' and by three gigantic pyramids disposed about its meridian in the distinctive manner of the three stars of Orion's belt in 10,500 BC? And is it also a coincidence that the monuments in this amazing astronomical theme park manage to *work together* – almost as though geared like the cog-wheels of a clock – to *tell the same time*.[32]

Obviously, we do not think coincidence was involved. Leo and Orion were both constellations that were highly significant to the ancient Egyptians, standing at the gates of the Duat sky region through which the souls of the dead were believed to travel before they could attain immortal life. For a people whose religious teachers instructed them to build perfect copies on the ground of 'the hidden circle of the Duat' it therefore seems to us that the Egyptians could not possibly have failed to make the connections between the lion statue and the pyramids 'below' and the constellations of Leo and Orion 'above'.

One unsolved mystery is the remote *date* transcribed by the astronomy of the monuments – the date of 10,500 BC that accords so well with the geology of the Sphinx. Is it a coincidence? Or are we to understand that the monuments were actually built in 10,500 BC? Alternatively, why should they not have been built piecemeal – one at a time over thousands of years – according to a site plan based on observations of the sky recorded in 10,500 BC? Or again, perhaps their builders did not need such records. Perhaps they were people who could calculate precession as accurately as we can with computers today. Perhaps, as so many of the ancient texts hint, they were adepts who knew the 'spells' for 'going down to any sky they chose to go down to' and who looked in those skies for the 'path' which would allow the soul to triumph over the 'infliction of the dead'.

CHAPTER 6

MYSTERY TEACHERS OF HEAVEN

Tomb chamber, Pyramid of Unas, Fifth Dynasty, Saqqara. Under a ceiling of stars, the chamber is inscribed with the hieroglyphs of the Pyramid Texts, the oldest surviving scriptures of mankind.

FOR THOSE EQUIPPED WITH knowledge of precession, a comparison of the ground-pattern of the three Giza Pyramids to the sky-pattern of the three stars of Orion's belt draws attention to the remote epoch of 10,500 BC.

By contrast, the Great Pyramid's four star-shafts seem to tell another story. We find it hard to believe that the builders of such accurate pointers would not have realized that the angles selected for each shaft, once linked to precession, could also be used to calculate an epoch – indeed the precise date on which they would all have been perfectly 'locked on' to their four target stars: Kochab, Thuban, Sirius and Zeta Orionis. As we have seen, that date was not 10,500 BC but 2500 BC. So it is almost as though the monuments, which physically simulate the stars of the earlier epoch – but which also use angled shafts to pinpoint the location of the same stars in the later epoch – could have been deliberately designed to force contemplation of the long, slow changes in the sky that are brought about by precession. When we remember the ancient funerary and rebirth texts, which state that the fundamental attribute of the 'equipped spirit' was his ability to 'go down to any sky', it seems perfectly legitimate to ask whether the Pyramids might not have been conceived of by their makers as vast laboratories intended to 'equip' the human spirit for immortality with esoteric *knowledge* of the cycles of the stars?

A passage in the Pyramid Texts adds force to this question with what could be interpreted as a reference to the Great Pyramid itself – a monument which has a flattened summit platform because its top few courses and its apex are missing:

> O Height which is not sharpened, gate of the sky . . . grasp the King by his hand and take the King to the sky, that he might not die on earth among men.[1]

In a sense this is not controversial. Egyptologists agree that pyramids were regarded as 'immortality devices' designed to transmit the souls of the Pharaohs buried in them directly to the heavens.[2] What they do not consider, however, is the possibility that the Great Pyramid could have been used as such a device by Pharaohs who were *not* buried in it. This is the possibility raised in the above text which is inscribed inside the tomb chamber of the Fifth Dynasty Pharaoh Unas – who ruled in Egypt almost two hundred years after the Great Pyramid was supposedly sealed up as the tomb of the Fourth Dynasty Pharaoh Khufu.

OPPOSITE: *Detail of the texts, Pyramid of Unas.*

Pyramid of Unas, texts and stars.

MYSTERIOUS AND INEXPLICABLE

Egyptologists regard the 'Old Kingdom obsession with immortality in the sky'[3] – and the apparent belief that pyramids had a part to play in obtaining it – as a mere bagatelle, a 'childish' form of sympathetic magic driven by a 'naïve' desire to live for ever. This view accords with the widespread historical theory of the stupidity of the ancients according to which it is perfectly reasonable to propose that the Giza Pyramids, weighing in total around 15 million tonnes and precisely aligned to the cardinal directions of sky and ground, could in fact have been built as 'tombs and tombs only' by megalomaniac Pharaohs whose only desire was to project their own egos into eternity.

That interpretation is possible. But there is very little evidence for it in the specific case of the Giza Pyramids. Their supposed 'tomb-chambers' were empty when they were first opened by Arab adventurers in the ninth century AD and no trace of any Pharaonic burial has ever been found inside them. In addition they carry no inscriptions – not a word – to tell us why they were built or how they were used.[4] Accordingly, the Egyptological claim that they are the tombs of the Fourth Dynasty Pharaohs Khufu, Khafre and Menkaure built over the 80-year period between 2551 and 2472 BC *cannot be considered as established fact* but is merely a theory.

Such theories, happily, are not our only guide to the real nature of the religious rituals once practised in the vicinity of these monuments. Much more reliable is the great quantity of primary information that is found in the ancient Egyptian funerary and rebirth texts. Amongst these, as we have seen, the oldest surviving are the Pyramid Texts – so called because they were inscribed inside the Fifth and Sixth Dynasty pyramids at Saqqara between the twenty-fourth and twenty-second centuries BC.

Where and when did these texts really originate?

The dates of inscription are not in dispute, but all scholars agree that the Pyramid Texts contain a mass of internal evidence suggesting that they were already ancient in the Fifth and Sixth Dynasties. Indeed, they show unmistakable signs of having been copied from much earlier source documents that have not come down to us. As the Egyptologist James Henry Breasted points out, they contain:

> a great company of archaic words which have lived a long and active life in a world now completely lost and forgotten. Hoary with age like exhausted runners, they totter into sight above our earliest horizon for a brief period, barely surviving in these ancient texts, then to disappear forever, and hence are never met with again. They vaguely disclose to us a vanished world of thought and speech, the last of the unnumbered aeons through which pre-historic man has passed till he finally comes within hailing distance of us as he enters the historic age.[5]

Is it not mysterious and almost inexplicable that at that moment, when prehistoric man first came 'within hailing distance of us', there already existed in Egypt an organized establishment with the skills and manpower 1) to build the gigantic and scientifically aligned monuments of the Memphite necropolis including (whatever their true function) the Great Sphinx and the Pyramids of Giza, and 2) to promulgate a body of ideas as complex and as evolved as those found in the Pyramid Texts?

Although professing to see no mystery, or anything to explain in any of this, Egyptologists have concluded that such an establishment did exist. They have identified it as the revered predynastic religious academy at Heliopolis – and point out that 'the entire ritual of the Egyptian temple was Heliopolitan in origin'.[6]

What purpose did that 'ritual' serve? What ideas might it have conveyed? And why should these ideas have involved pyramids and stars, the rising place of the sun, god-kings and the quest for immortality?

Cartouche (royal name) of Unas.

Detail of texts, Pyramid of Teti.

THE MANIFESTATION OF ATUM

Ancient Heliopolis now lies entirely covered by the suburb of Mattariya on the eastern side of Cairo. Of its glories nothing remains except a single obelisk erected on the site by the Twelfth Dynasty Pharaoh Senuseret I (1971–1926 BC). All that archaeology knows about the city for sure is that it already must have been very old in Sunuseret's time – indeed that it flourished 'as far back as the middle of the third millennium BC and will doubtless have been vastly older.'[7] Professor I. E. S Edwards asserts that even in predynastic times it was 'the most important city in Egypt'.[8] Peter Tompkins describes it as 'the capital of a predynastic state'.[9] Sir J. Norman Lockyer, the British astronomer, calculated that it must have been founded long 'before 4000 BC'.[10] And the Greek traveller Diodorus Siculus, who visited it in the first century BC, was told by its inhabitants that they accounted it 'older than any other city in Egypt'.[11]

The story of Heliopolis inhabits the borderland between myth and history. It cannot be told without telling also of the Nun and of the Primeval Mound, of the Bennu bird and of the Benben stone.

The earliest surviving versions, which naturally are found in the Pyramid Texts, open on the eve of the 'First Time', the golden age, which scholars suppose to be mythical, said to have existed before history began, 'before anger came into being . . . before strife came into being . . . before tumult came into being.'[12]

The 'First Time' is so called because it is the beginning of the present reality, the present incarnation of the earth – the first tick of the cosmic clock that beats out the divinely generated cycle of our existence. On such grave matters the texts of the Egyptians are more than usually enigmatic and obscure; nevertheless, as though seen through a veil, one sometimes glimpses the vertiginous chasms of other potential 'realities' lying all around and interpenetrating our own – each one of them an idea in the mind of God waiting to materialize into the world of forms.

The mythical Benben stone was frequently depicted in hieroglyphs by the form of a stepped pyramid and also as a true pyramidion.

Before the first tick of the First Time, we are told that Atum, the god before all gods, the 'Complete One', the All-Father, lay motionless in the waters of the primordial void – 'the infinity, the nothingness, the nowhere and the dark'[13] –

The Benben, or pyramidion, from the largely ruined pyramid of the Twelfth Dynasty Pharaoh Ammenenhat III (1894–1797 BC), which stands at Dashur. The Benben is cut from highly polished black granite. An inscribed winged sun-disk stretches over two Eyes of Horus above the hieroglyph neferu *('perfection') repeated three times.*

which the ancient Egyptians called the Nun. Though inert, the god was filled with magic, energized by the power to 'author his own forms',[14] able to realize endless possibilities of creation and transformation. Alone, naked, suspended in a dark ocean, enclosed by meaninglessness, he was not extinguished. As the texts put it, he 'possessed his limbs',[15] maintaining full control over the force of his own will, the strength of his intellect, and the wisdom of his countless existences.

At a moment preordained, he stirred and surged up through the viscid waters of the Nun, sending out ahead of him a spell of fabrication, giving shape and texture to the wraiths of his imagination, causing *something* to emerge out of nothing.

This triumphant emergence of existence out of nonexistence was a piece of divine magic which the ancient Egyptian scriptures used three parallel and inter-linked symbols to express. One was the Primeval Mound – the first land to burst forth from the waters of the Nun – the second was the sacred Benben stone, the third was the Bennu bird. All three symbols were grouped together in the Egyptian mind at Heliopolis, within one sacred enclosure – Het Benben, the 'Temple of the Benben', which was also sometimes referred to as the 'Mansion of the Phoenix'.[16]

THE MOUND AND THE STONE

The ancient Egyptians called Heliopolis 'Innu', 'the pillar',[17] and regarded it as the first part of the land-world to have forced itself free of the Nun.[18] This is made clear in a text that describes the condition of Atum before the creation of the world – a time, he recalls, 'when I was still alone in the waters, in a state of inert-ness, before I found anywhere to stand or sit, before Heliopolis had been found-ed that I might be therein'.[19]

The god materialized as the Primeval Mound: 'O Atum!' we read in Utterance 600 of the Pyramid Texts. 'When you came into being you rose as a High Hill, you shone as the Benben stone in the Temple of the Phoenix.'[20]

The place that the Egyptians regarded as the site of the original Primeval Mound, at the exact geographical centre of Heliopolis, was a raised area called the 'High Sand'.[21] However, it was not believed to be the the first piece of solid matter created by Atum. This was the Benben stone, a 'stone fallen from heaven' – some scholars have suggested that it might have been a meteorite[22] – originating from a drop of the seed of Atum which fell into the primeval ocean of the Nun.[23] As Henri Frankfort points out, this mysterious object was once actually on display at Heliopolis, enshrined on the 'High Sand' in the 'Mansion of the Phoenix',[24] where we discover from the *Book of What is in the Duat* that it was visited by initiates but protected from the eyes of the ignorant:

> These are they who are outside Het Benben ... They see Ra [the sun god] with their eyes and they enter into his secret images ... I protect my hidden things which are in Het Benben.[25]

The original Benben stone, said to have carried the secrets of 'the hiddenness of the Duat',[26] was lost from Heliopolis in antiquity and where it might be today is unknown. A large number of detailed textual references, paintings and reliefs, however, leave us in no doubt that it was pyramid-shaped, 'variously shown', scholars confirm, 'as a rounded object, a pyramidion, or a stepped object'.[27] In the Pyramid Texts, observes Frankfort, its determinative sign 'shows a tapering, somewhat conical shape which became stylized for use in architecture as a small pyramid, the pyramidion; covered with gold foil it was held aloft by the long shaft of the obelisk'.[28]

Pyramid at Giza, seeming to rise out of the morning mist as a potent image of the Benben stone – said to have fallen from heaven.

RIGHT: *Obelisk of Thutmosis I, Karnak, 21.3 metres high and weighing an estimated 143 tonnes. An inscription tells us that the pyramidion was originally covered with electrum – an alloy of gold and silver.*

The Benben provided the model, and was in fact the name used by the ancient Egyptians, for the capstones (pyramidions) of all pyramids and for the tips (but not the shafts) of all obelisks. In Heliopolis the original obelisk was the megalithic pillar, Innu, after which the city received its Egyptian name, which stood in the courtyard of the temple Het Benben – a temple that was almost certainly open to the skies.[29] In Frankfort's opinion, this erect stone pillar, tipped by the pyramidion of the Benben – 'a drop of the seed of Atum' – looks very much as though it might originally have been 'a phallic symbol of Heliopolis, the pillar city'.[30]

What strengthens this analysis is the fact that in the ancient Egyptian language the root word *bn* and its duplication *bnbn* are 'connected with various outflows, including those of a sexual nature ...'[31] In a passage on the creation of the Benben we read that Atum is the god 'who begat [*bnn*] a place [*bw*] in the primeval ocean, when seed [*bnn.t*] flowed out [*bnbn*] the First Time [i.e. at the beginning of the present creation] ... It flowed out [*bnbn*] under him as is usual, in its name "seed" [*bnn.t*]'.[32]

THE FLIGHT OF THE PHOENIX

So the Benben had sexual, reproductive connotations, in quite a direct physical sense. What, then, of the fabulous Bennu bird, a name also deriving from the root *bnn*?

The model for the phoenix of the later Greeks,[33] the Bennu was another manifestation of Atum, this time in the form of a grey heron that was said to have appeared at the moment of creation, perched atop a pillar on the Primeval Mound. It is important to note, as the Egyptologist R. T. Rundle Clark has pointed out, that the rising of the mound and the appearance of the phoenix were not viewed as consecutive events but rather as 'parallel statements, two aspects of the supreme creative moment'.[34]

In the texts that moment is epitomized as the victory of light and the spirit over darkness and death and specifically as 'that breath of life which emerged from the throat of the Bennu bird, in whom Atum appeared in the primeval nought'.[35] In Rundle Clark's eloquent evocation of this scene:

> One has to imagine a perch extending out of the waters of the Abyss. On it rests a grey heron, the herald of all things to come. It opens its beak and breaks the silence of the primeval night with the call of life and destiny.[36]

This avatar of Atum, the self-begotten and deathless Bennu, crosses the dimensions between spirit and matter, symbolizing the eternal transmigration of the

The ancient Egyptian phoenix, the Bennu bird, was depicted as a grey heron and symbolized the eternal transmigration of the soul.

This late carving on the south exterior wall of the sanctuary at Karnak shows Philip Arrhideus, the half-brother and successor of Alexander the Great, being crowned as an Egyptian Pharaoh. The function of king-making in Egypt was traditionally the privilege of a mysterious brotherhood known as the Followers of Horus, whose presence was thought of as already very ancient as far back as the epoch of the Pyramid Texts.

soul. All the high initiates in the temples of ancient Egypt sought what it represented. Indeed, they hoped, metaphorically, to become it. 'I am the Bennu bird,' declares the deceased at one point in the *Book of the Dead*, 'and upon earth shall come forth again.'[37]

It should be obvious that an evolved doctrine of reincarnation (for which there is indeed a mass of evidence in the Egyptian texts[38]) must underlie this aspiration. In such a doctrine, observed the German philosopher Hegel in a little-known essay on the phoenix:

Spirit – consuming the envelope of its existence – does not merely pass into another envelope, nor rise rejuvenescent from the ashes of its previous form; it comes forth exalted, glorified, a purer spirit. It certainly makes war upon itself – consumes its own existence; but in this very destruction it works up that existence into a new form, and each successive phase becomes in its turn a material, working on which it exalts itself to a new grade.[39]

THE FOLLOWERS OF HORUS

'When a message comes from heaven,' we are informed in the ancient Egyptian Leyden Papyrus, 'it is heard in Heliopolis.'[40] But by whom would it have been heard?

Historical records indicate that at any one time as many as 12,000 priests may have been employed in the service of the Temple of the Benben[41] – that mysterious pyramid-shaped object sometimes spoken of as a 'stone fallen from heaven'.[42] The *Book of the Dead* informs us: 'Behold, the starry sky is in Heliopolis.'[43] The Heliopolitan High Priest, as we have seen, was called the 'Chief of the Astronomers'.[44] And associated with Heliopolis are monuments such as the great Pyramids and the Sphinx incorporating precise astronomical alignments.

Despite such fairly obvious clues Egyptologists continue to maintain that the priests of Heliopolis were 'half-savage' and that their religion consisted of nothing more than empty rituals and primitive superstitions dressed up in some sort of mumbo-jumbo sky 'cult'.[45] By contrast, at the opposite extreme, a few dissenting voices suggest that 'the temple at Heliopolis, although it was represented to the uninitiated as a place of religious worship, was in reality an astronomical observatory designed and equipped by scientists for scientific purposes'.[46]

In this relief from his temple at Abydos, Pharaoh Seti I shows his young son Rameses II a list of all the kings of Egypt back to the time of Menes, the first Pharaoh of the First Dynasty. On the opposite wall the list continues into prehistory, denominating the names of the Followers of Horus – and before them the Gods – who had ruled in Egypt in promordial times. The Followers of Horus were remembered in tradition as the 'founders of Heliopolis' and as the 'mystery teachers of heaven'.

ABOVE: *Horus performing the rituals engendering the resurrection of his father Osiris to eternal life in the heavens (temple of Seti I, Abydos).*

BELOW: *The mysterious Djed pillar on a column adjacent to the above scene, thought to symbolize the 'world tree' or cosmic axis. Compare to the Tablet of the Foliated Cross, Palenque (see page 36).*

The complex and sophisticated astronomical characteristics of the Pyramids and the Great Sphinx of Giza, so close to Heliopolis and known to have been under the control of the Heliopolitan priesthood, obviously offer strong support for the observatory theory. We see no reason, however, to agree with the dissenters that religion was merely a sort of 'cover story' for the practice of a 'secret science'. On the contrary we think it is possible, using rigorous methods of investigation including but not limited to astronomy, that what the scientists of Giza/Heliopolis really sought was an entirely religious and spiritual outcome – nothing less ambitious than the phoenix-like 'renewal' and transfiguration of the human soul.

If the ancient texts are to be believed, this quest for life after death was imported into Egypt in the 'First Time' – many thousands of years before the First Dynasty of Pharaohs. From the beginning it was associated with a shadowy group of semi-divine beings called the 'Shemsu Hor' – the 'Followers of Horus'. Remembered by tradition as the 'founders of Heliopolis'[47] and as 'the mystery teachers of Heaven',[48] their emblem was the falcon-headed god Horus, the divine child of Osiris and Isis whose celestial counterparts were Orion and Sirius.

Horus stood for many things, most importantly the sun. In this capacity he merged at Heliopolis with Atum to form the composite sun-god Ra, whose symbol was a man with the head of a falcon, surmounted by the solar disk and cobra (the *uraeus* serpent). In later times it was this worship of Atum and Horus in the form of Ra – the sun – that led the Greeks to call ancient Innu 'Heliopolis', the 'City of the Sun'.

Isis, right, suckling the infant Horus (temple of Dendera).

The story of Horus is the story of the Osirian resurrection. It is the story of the initiate's quest to pass beyond the trials of death, to 'gather his bones together' and to burst forth again triumphantly into life. It tells how the god Osiris ruled in Egypt, in the First Time, over a kingdom established according to the rules of cosmic justice. Murdered in his prime by his envious brother Set – who was said by tradition to have had 72 co-conspirators[49] – Osiris was restored briefly to physical life by the sorcery of his sister Isis, who then took the form of a kite and hovered over his phallus, receiving his seed. In this way was conceived Horus, who grew to manhood and took his revenge upon Set, subjugating him and restoring his father's earthly kingdom. Magically his actions also served to restore his father to spiritual life in the heavens,[50] where Osiris underwent resurrection as Lord of the Duat and would preside for all eternity over the judgement of the souls of the dead.

This archetypal myth, in many ways so similar to the Central American traditions of Quetzalcoatl, was the basis for the rule of the Pharaohs, the god-kings of ancient Egypt. Each whilst he lived was known as the 'Horus King' but aspired, on his death, to ascend the heavens and to join Osiris – indeed, literally, to become 'an Osiris'. Each Pharaoh, in other words, directly identified himself in life with the god Horus and in death with the god Osiris – and also at all times (however confusing it may seem to the modern mind) with Ra, the sun, of whom the Pyramid Texts state: 'Horus has caused that you enclose for yourself all the gods within your embrace.'[51]

It is fair to assume that such doctrines would have been taught by a group called the 'Followers of Horus' who lodged themselves at Heliopolis, site of the Benben stone, perch of the Bennu bird – the first sacred domain of ancient Egypt where scholars agree that the oldest surviving recensions of the Pyramid Texts must have been compiled.[52]

Identical in all respects other than name to the 'Sages' and the 'Builder Gods' spoken of in the Edfu Texts,[53] the 'Followers of Horus' were said to have carried with them a 'knowledge' of the 'divine origins' of Egypt[54] and of the divine purpose of this land, 'which once was holy and wherein alone, in reward for her devotion, the gods deigned to sojourn upon earth'.[55]

In addition it has recently been realized that within the term 'Shemsu Hor' the word 'Shemsu' – 'Followers' – is not properly understood if thought of in the sense of companions or disciples but 'literally means *Followers of Horus* in the precise sense of "those who follow the path of Horus", that is the "Horian way", also called the solar way, or *paths of Ra*'.[56] It is perhaps because of their celebrated knowledge of this special 'path' in the heavens, and because they were teachers who could transmit this knowledge to others, that the Pyramid Texts tell the initiate: 'The Followers of Horus will cleanse you, they will recite for you the spell of Him who Ascends.'[57]

The Pharaoh making an offering to Osiris, temple of Seti I, Abydos. Each Pharaoh saw himself as the Horus King - literally the embodiment of Horus upon earth – and hoped on his death to join Osiris in the heavens.

PATH OF REBIRTH

Modern astronomers speak frequently of the 'path of Ra', which they call the 'path of the sun' or, more technically, the 'ecliptic'. It is defined as the 'extension

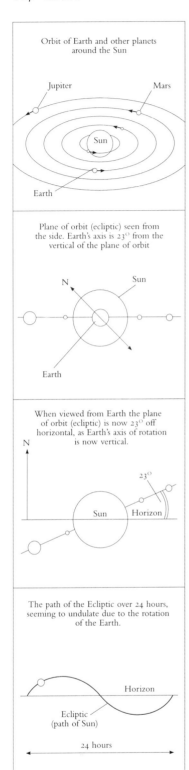

Orbit of Earth and other planets around the Sun

Jupiter

Mars

Sun

Earth

Plane of orbit (ecliptic) seen from the side. Earth's axis is 23° from the vertical of the plane of orbit

N

Sun

Earth

When viewed from Earth the plane of orbit (ecliptic) is now 23° off horizontal, as Earth's axis of rotation is now vertical.

N

23°

Sun   Horizon

The path of the Ecliptic over 24 hours, seeming to undulate due to the rotation of the Earth.

Horizon

Ecliptic (path of Sun)

24 hours

The ecliptic.

into the celestial sphere of the earth's plane of revolution about the sun . . . As far as terrestrial observers are concerned, this circle traces out the annual *motion of the sun on the sky* relative to the background of distant stars.'[58] To this the *Penguin Dictionary of Astronomy* adds:

> From the point of view of the observer on Earth, the relative orbital motion of the Earth and Sun makes it look as if the Sun were travelling around the earth once a year. The *Sun's path* on the celestial sphere traces out the ecliptic plane, and is often marked as the ecliptic on celestial charts.[59]

There exists, in other words, scientifically recognized and labelled, a real 'path of the sun', a circular path amongst the stars completing a full cycle in approximately 365 and one quarter days – the solar year. The possibility has therefore recently been considered that by styling themselves 'followers of the paths of Ra', i.e. of the 'path of the sun', the Shemsu Hor could have given us a clue about their true interests – a long-term tracking of astronomical events along the ecliptic.[60] Had they kept careful records for sufficiently extended periods, such 'astronomer-priests' would not have failed to observe the effects of precession, in particular the gradual rotation of the twelve zodiacal constellations against the background of which the sun rises at dawn on the spring equinox.[61]

The numinous power of this great cosmic cycle as a symbol of rebirth and renewal after long periods of apparent extinction, and thus of life after death, seems to be repeatedly hinted at in many ancient texts. In the Hermetica we read:

> The lapse of terrestrial time is marked by changing states of the atmosphere, and the variations of heat and cold; while that of celestial time is marked by the return of the heavenly bodies to their former positions as they move in their periodic revolutions. The Kosmos is that in which time is contained; and it is by the progress and movement of time that life is maintained in the Kosmos. The process of time is regulated by a fixed order; and time in its ordered course renews all things in the Kosmos by alteration.[62]

The same text continues: 'God is self-contained, and self-derived . . . He moves not in time but in eternity . . . Into eternity all movements of time go back, and from eternity all movements of time take their beginning.'[63]

Finally we are told that god and eternity contain 'a Kosmos which is imperceptible to sense. This sensible Kosmos [i.e. the material universe that we see all around us] has been made in the image of that other Kosmos, *and reproduces eternity in a copy.*'[64]

The powerful Hermetic idea of material, terrestrial things being copies of celestial originals is obviously linked to the ancient Egyptian conviction that temples should 'resemble the sky'. Moreover, in both cases, exactly as at the pyramid-city of Teotihuacan in Mexico, the work of creating a 'resemblance' – the deliberate mimicking on the ground of the patterns, cycles and mysteries of the cosmos – was clearly understood as being part of a serious and intelligent endeavour to turn men into immortal gods.

PART III

CAMBODIA

CHAPTER 7

DRACO

THE NOTION OF A LAND that is the 'image of heaven' on which are built cosmic temples with 'halls that resemble the sky' is not confined to ancient Egypt and ancient Mexico. Precisely the same idea also took root in south-east Asia, in the Cambodian Hindu and Buddhist cities of Angkor Wat and Angkor Thom, a thousand years after the collapse of the civilization of the Pharaohs.

No connection is recognized by historians, who are adamant that neither Egypt nor Mexico could have had any influence – either directly or through diffusion – on the Angkor temples. As the French archaeologist George Coedes has written, these monuments are only to be understood as the products 'of a Hindu civilization transplanted to Indo-China ... As soon as one looks behind the external forms for the motivating inspiration, one finds an Indian idea.'[1]

Though original works of art, with their own unique characteristics, it is absolutely true, and conspicuous, that the monuments of Angkor Wat and Angkor Thom are in many ways 'Indianized'. It is true that Sanscrit, the classical religious language of the Indian subcontinent, was the exclusive medium in which religious writings were set down at Angkor – whilst the indigenous Khmer language was used only for secular inscriptions. And it is also true that Hinduism (with particular emphasis on the cults of Vishnu and Shiva), as well as Buddhism of the Mahayana school, have left an obvious and unmistakable imprint on the sacred architecture and symbolism of Angkor.

Nevertheless, a major problem does exist. Quite simply, as scholars admit, 'nothing is yet known about the prehistory and protohistory of Indo-China'.[2] Since it is precisely out of this 'prehistory and protohistory' that the temples of Angkor loom towards us, with almost no advance warning, we should perhaps not be so sure that we have understood all the influences that shaped them.

PREVIOUS PAGE AND ABOVE: Apsaras *(celestial dancers) decorate many of the temples at Angkor.*

OPPOSITE: *Entrance causeway and central towers of Angkor Wat at sunrise, two days before the autumn equinox.*

SEVENTY-TWO DEGREES EAST: THE GOD HORUS LIVES

The name 'Angkor', although supposedly a corruption of the Sanscrit word *nagara*, 'town',[3] has a very precise meaning in the ancient Egyptian language – 'the god Horus lives'.[4] Other acceptable translations of 'Ankh-Hor' or 'Ankhhor' are 'May Horus Live', 'Horus Lives' and 'Life to the Horus'.[5]

A Buddhist monk burns incense in the central sanctuary of Angkor Wat – still a place of living religion to this day.

In Egyptian traditions the Followers of Horus, legendary founders of the sacred city Heliopolis, are depicted as the carriers of a profound astronomical *gnosis*. Scholars have always dismissed this attribute as a myth. Yet it is a fact that the monuments and hieroglyphic writings associated with Heliopolis – the great Pyramids of Giza, the Great Sphinx and the Pyramid Texts – are all intensely astronomical in nature. In particular, these key monuments and scriptures appear to encode scientifically exact observations of the precession of the equinoxes which proceeds at the rate of one degree every 72 years. It is not impossible that an entire geodetic system may have been built up around this number and expressed in the ratio between the dimensions of the Great Pyramid and the dimensions of the earth – 1:43,200, as we saw in Part II, i.e. 600 times 72.

No historian can say precisely where or when the mathematical and geodetic convention arose of dividing spheres and circles into 360 degrees. It is in accord with this convention, however, that geographers and map-makers now divide the sphere of our planet into 360 degrees of longitude (vertical segments running from pole to pole) each of which, at the equator, has a width of approximately 112 kilometres. Since times immemorial, a similar convention has also been extended outwards by astronomers to the 'celestial sphere' of the sun and the moon, the planets and the stars. An obvious convenience is that a sphere of 360 degrees can be neatly sliced up into a variety of harmonious combinations, e.g.

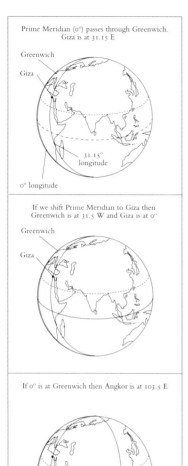

Prime Meridian (0°) passes through Greenwich.
Giza is at 31.15 E

If we shift Prime Meridian to Giza then
Greenwich is at 31.5 W and Giza is at 0°

If 0° is at Greenwich then Angkor is at 103.5 E

But if 0°= Giza then Angkor is at 72° E

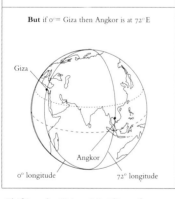

*Shifting the Prime Meridian of
longitude from Greenwich to Giza,
to reveal traces of a possible ancient
world grid.*

two hemispheres of 180 degrees, four quadrants of 90 degrees, etc. The sphere can also be equally divided into five segments each precisely 72 degrees in width.

Today any schoolroom globe will show the net of vertical longitude and horizontal latitude lines with which geographers have gridded the earth. Latitude can be calculated with precision in the northern hemisphere by measuring the angle of the pole star above the horizon (other simple techniques are also available). Yet historical civilizations were unable to measure longitude accurately until the eighteenth century, when reliable marine chronometers were invented.

This is not the place to consider the implications of large numbers of ancient maps which have been found, drawn up before the eighteenth century (and based on copies of even older source maps), which display almost perfect relative longitudes.[6] As has been shown elsewhere, such maps could be the legacy of a race of seafarers in remote prehistory – a lost advanced civilization that had measured the globe and surrounded it with a net of geodetic co-ordinates more than 12,000 years ago.[7]

The exact siting of longitude lines in the 'net' that we use today is a political matter. It is therefore by a convention that has only been universally accepted for the last century or so that the first line of the net – zero degrees longitude, the so-called 'prime meridian' – is marked by the Royal Greenwich Observatory in London. The Pyramids of Giza lie 31.15 degrees of longitude east of the Greenwich Meridian, ancient Heliopolis lies 31.20 degrees east and the temples of Angkor lie 103.50 degrees east. The distance between the sacred domain of

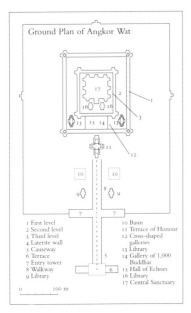

Ground Plan of Angkor Wat

1 First level
2 Second level
3 Third level
4 Laterite wall
5 Causeway
6 Terrace
7 Entry tower
8 Walkway
9 Library
10 Basin
11 Terrace of Honour
12 Cross-shaped galleries
13 Library
14 Gallery of 1,000 Buddhas
15 Hall of Echoes
16 Library
17 Central Sanctuary

0   100 m

Plan of Angkor Wat.

Giza-Heliopolis, ruled by sages who followed the astronomical 'path' of Horus, and the sacred domain of Angkor – which, as 'Ankhhor', means literally 'the god Horus lives' – therefore reduces in round numbers to the geodetically significant figure of 72 degrees of longitude (103 degrees minus 31 degrees = 72 degrees).

PYRAMID

There have been times, in the past few decades, when Angkor has seemed like the heart of darkness – for it lies in the midst of a dark forest in a land where dark deeds have been done.

Approaching the temples from the south, we drove past the ossuary where the skulls of Khmer Rouge victims lie in a chest-high heap and then continued due north for three miles on a straight pot-holed road, sometimes passing shadowy figures at the edge of the enclosing trees. It was mid-November 1996, an hour before dawn. To our left, dominated by the constellation of Orion, the sky-region that the ancient Egyptians called the Duat had sunk low down in the west and would soon set. To our north, directly ahead of us, the dragon or 'snake'[8] constellation of Draco was rising. To our right, in the east, the sky seemed compressed by the pressure of the sun below the horizon.

Now the road veered sharply left, diverted due west by a wide moat surrounding an imposing rectangular island. On the island, across the shining water, rising above a formidable perimeter wall, we could just discern the distant outlines of a broad, powerful pyramid surmounted by five lofty towers.

The complex we had come to was Angkor Wat, one of the largest stone buildings ever created, yet in itself only part of a fantastic archipelago of tombs, temples and great geometrical 'cities' straddling an area of almost 300 square kilometres in the flood plains of the Mekong river.[9]

From ground level it is difficult to see clearly how the many monuments of Angkor relate to one another. But if you raise yourself above them in your mind's eye and look down on them from far above – as we propose to do in the rest of this chapter – the sense of a grand plan gradually begins to force itself upon you.

Angkor Wat consists of a series of five internested rectangular enclosures. The short dimensions are aligned with high precision to true north–south, showing 'no deviation whatever' according to modern surveys.[10] The long dimensions are

Apsara, Angkor, *and modern Khmer girl photographed at the temple of Ta Sohm.*

The 'pyramid' of Angkor Wat, with its five lofty towers.

Angkor Wat: the megalithic causeway viewed from inside the western entrance gate looking towards the temple.

oriented, equally precisely, to an axis that has been deliberately 'diverted 0.75 degrees south of east and north of west'.[11]

The first and outermost of the five rectangles that we find ourselves looking down on from the air is the moat. Measured along its outer edge it runs 1300 metres from north to south and 1500 metres from east to west.[12] Its 'ditch', 190 metres wide,[13] has walls made from closely fitted blocks of red sandstone set out with such precision that the accumulated surveying error around the entire 5.6 kilometres of the perimeter amounts to *barely a centimetre.*[14]

Angkor Wat's principal entrance is on the west side where a megalithic causeway 347 metres long and 9.4 metres wide[15] bears due east across the moat and then passes under a massive gate let into the walls of the second of the five rectangles. This second enclosure measures 1025 × 800 metres.[16] The causeway continues eastward through it, past lawns and subsidiary structures and a large reflecting pool, until it rises on to a cruciform terrace leading into the lowest gallery of the temple itself. This is the third of the five internested rectangles visible from the air and precision engineering and surveying are again in evidence – with the northern and southern walls, for example, being of identical lengths, *exactly* 202.14 metres.[17]

Ascending to the fourth rectangle, the fourth level of Angkor Wat's gigantic central pyramid, the same precision can be observed. The northern and southern walls measure respectively 114.24 and 114.22 metres. At the fifth and last enclosure, the top level of the pyramid – which reaches a height of 65 metres above the

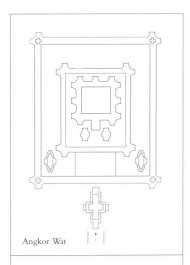

Angkor Wat

Mandala

Angkor Wat as a mandalic structure.

Detail of Naga balustrade, Angkor Thom.

entrance causeway – the northern wall is 47.75 metres in length and the southern wall 47.79 metres.[18] According to a study published in the journal *Science* these minute differences, 'less than 0.01 per cent', demonstrate an 'astounding degree of accuracy' on the part of the ancient builders.[19]

MANDALAS OF THE MIND

In imagination we are flying, hovering directly above the fantastically carved central tower of Angkor Wat – the summit of a bizarre pyramid-mountain, decorated with Gothic turrets, rising out of a landscape of sacred geometry.

Academic authorities recognize that the pattern is a 'mandala', not painted on paper or cloth, as is more usual, or traced in coloured sands, but made in water and stone – a work of beauty and science: 'a symbolic diagram used in the performance of sacred rites and as an instrument of meditation'.[20] As it is most frequently employed by Buddhist monks, the mandala is:

> a representation of the universe, a consecrated area that serves as . . . a collection point of universal forces. Man, by mentally 'entering' the mandala and 'proceeding' towards its centre, is by analogy guided through the cosmic processes of disintegration and reintegration.[21]

Through representing, 'copying' and symbolizing the universe, 'physical' mandalas, whether painted on paper or built in stone, have been described as 'convenient means for promoting true mandalas in people's minds'.[22] In Buddhist practice they are initiatory instruments designed to condition certain mental processes which can assist neophytes on the straight and narrow path towards illumined *gnosis* – the state of 'enlightenment', 'realization', or 'awakening', that will allow them to attain 'knowledge of the truth'.[23]

In our imaginations, let us now continue our flight above Angkor Wat, move to the western side of the enclosure, cross the moat and then look due north.

Below us, less than 2 kilometres dead ahead but running across our entire field

Plan of Angkor.

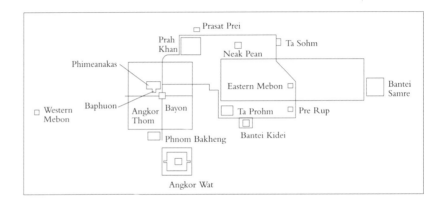

of view because it is almost 5 kilometres in length, is the southern side of another immense moat. Inside the moat is a square island on which towers a 12-metre-high perimeter wall, also square, with each of its four sides identical in length to all the others – 4 kilometres per side, a total extent of 16 kilometres. Into the wall are let five gates approached by five causeways lined with rows of powerful stone figures pulling on the body of a huge Naga serpent, a mythical hooded cobra, also made out of stone.

Although on a far larger scale than Angkor Wat, is this geometrical enclosure, divided by water and stone, also a mandala intended to promote 'mandalas of the mind'?

Its name is Angkor *Thom*, meaning Angkor 'the Great' (whilst Angkor *Wat* means Angkor 'the Temple'), and it contains three notable temples of its own – the Phimeanakas, the Baphuon and the Bayon – which we shall examine more closely in later chapters. Each of these structures takes the form of a pyramid with the Phimeanakas ('the Palace of Heaven')[24] lying to the northwest of the centre of the enclosure and the Baphoun ('the tower of bronze')[25] lying 200 metres due south of it. The third temple, the Bayon ('Father of Yantra')[26], is sited with scientific precision at the exact geometrical centre of Angkor Thom.

Bizarre, striking, vast and surreal, the Bayon is surmounted by a forest of towers that almost conceals its step-pyramid shape. Its name, however, reveals much – since a 'yantra' is a specialized form of mandala which 'provides an advanced focus for meditation'.[27] As is the case with Angkor Wat it is therefore generally agreed that at least one of the functions of the Bayon and its surroundings, however strange and incomprehensible they seem at first sight, must have been to serve as symbolic diagrams of the universe into which initiates would enter to equip their spirits with some form of esoteric, 'cosmic' knowledge.

THE DRACO–ANGKOR CORRELATION

OPPOSITE: *The Bayon, at the heart of Angkor Thom, was conceived by its builders as a symbolic diagram of the universe. It is surmounted by 54 towers, each with four gigantic sculpted faces – 216 faces in all.*

In our mind's eye let us now spiral ever higher into the atmosphere above the sacred domain so that we can look down on its full extent of almost 300 square kilometres. The large rectangle of Angkor Wat and the four times larger square of Angkor Thom lie on the western side of the landscape stretching away beneath

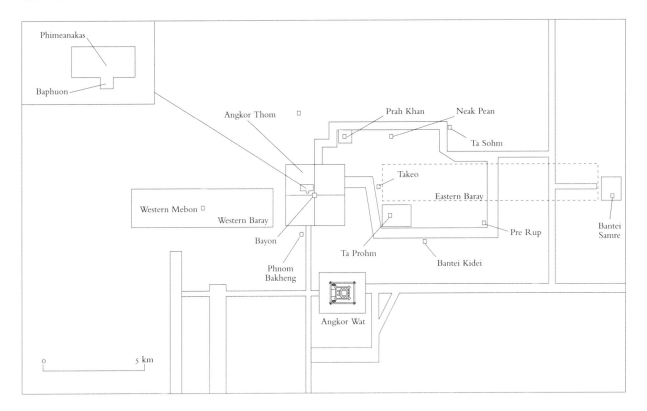

Angkor from the air.

us. In a band extending 25 kilometres to the east and 15 to the north, surrounded by jungle but just visible from our perspective, we can see the ruins of many other temples built by the same Khmer god-kings who made the larger monuments. Small or large, these temples all repeat in their own design the geometrical enclosures of the classic mandala or yantra.

Are they linked in some way? Or is it even possible that the entire *group* could have been conceived of as a mandala in its own right? A mandala on a very large scale copying some very large feature of the cosmos?

An enigmatic phrase occurs in one of the surviving triumphal inscriptions of Jayavarman VII, the Khmer king who built Angkor Thom and the Bayon in the twelfth century AD. The inscription is on a stela, excavated from the royal palace,[28] and asserts with no preamble, context or explanation, that 'the Land of Kambu' (Cambodia) is 'similar to the sky'.[29]

For anyone versed in ancient Egyptian sky–ground dualism, this curious but powerful statement by a god-king raises an obvious question: could it be a reference to the practice of building architectural 'scale models' or 'copies' on the ground of particular stars or constellations in the sky?

In 1996 John Grigsby, a 25-year-old Ph.D. student working for us to compile a database of facts about Angkor, made a brilliant and original discovery. Just as the three great Pyramids of Giza in Egypt model the belt stars of the southern constellation of Orion, so too do the principal monuments of Angkor model the sinuous coils of the northern constellation of Draco.[30]

There can be no doubt that a correlation exists: the correspondences between the principal stars of Draco and at least fifteen of the main pyramid-temples of Angkor are too close to be called anything else (see diagram). Furthermore, these correspondences extend to a number of neighbouring constellations in the same general sky-region. The only matter of doubt, therefore, is whether such complex and detailed similarities could have come about purely by chance or whether they are deliberate. But, as Grigsby points out:

> If this is a fluke then it's an amazing one. Not only do the stars of Draco seem to sit over the temples of Angkor when both images are aligned north. Also the distances between stars as represented by the distances between monuments are roughly accurate and indeed are very accurate once it is realized that this represents the result of a difficult process, achieved without the use of detailed photographs of the constellation, but with maps made manually. There is allowance for human error in the transference of the constellation on to a map, and then the transference of the fallible map on to difficult terrain over hundreds of square kilometres with no method of checking the progress of the site from the air.

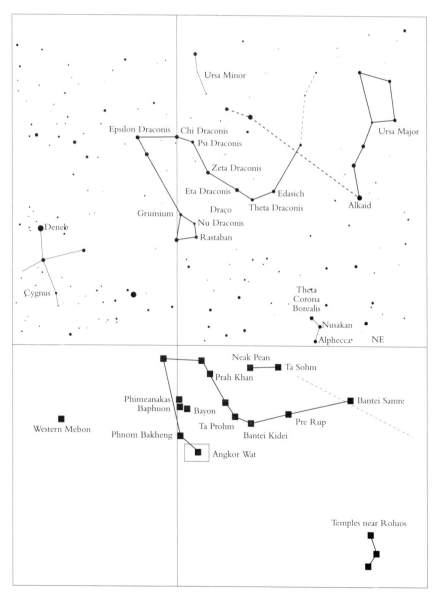

The Angkor–Draco correlation.

All this considered, it seems even more possible that the stars of Draco were indeed the template for the site of Angkor. Indeed, not only do the stars of Draco seem to be depicted, but also the nearby stars of Alkaid and Kochab, which form a straight line with Thuban in the sky – which also happens 'coincidentally' on the ground – and Deneb in the constellation of Cygnus, which finds its matching temple in the Western Mebon. Furthermore, the temples were built over some 250 years and there is evidence for reusing old sites, such as the Bayon, the Baphuon and the Phimeanakas. It could be possible, then, that the positions of the temples were mapped out at the start of the enterprise.[31]

But *when* was the start of the enterprise?

In our search for an answer to this question we were to stumble upon an extra-ordinary mystery.

AD 1150

We began with Grigsby's basic Draco–Angkor correlation and with the same computer program, Skyglobe 3.6, that has revealed the astronomical 'template' of the three great Pyramids and the Great Sphinx of Giza. The chief advantage of this program is that it calculates the effects of precession on stellar positions and generates accurate simulations which enable researchers to view the stars exactly as they would have looked over any point on the earth's surface, in any chosen epoch, down to a precise month, day, hour and minute.

Grigsby had not considered the date at which his correlation might have occurred, simply noting that it was general and obvious in all epochs. It seemed to us, however, that if the correlation was significant then it should be much more precisely datable. We reasoned that if the layout of the temples on the ground was indeed the result of a deliberate architectural plan to 'copy' the stars of Draco, then precessional calculations should enable us to discover *exactly which sky* the temples copy – in other words the sky of which precise epoch.

Indisputable archaeological and inscription evidence proves that the temples of Angkor were built by named and known Khmer monarchs, almost all of whom reigned during the four centuries between AD 802 and AD 1220. We therefore supposed, if the correlation was deliberate rather than accidental, that it would prove to be based upon the skies of these four centuries – which, because precessional changes in such a short period are barely noticeable, can effectively be regarded as the same sky from the beginning to the end of the four hundred years.

We picked the date of AD 1150 to begin our search for an exact match between the pattern made by the temples of Angkor on the ground and the pattern made by the stars of Draco in the sky. This was the death-date of Suryavarman II, 'Protected by the Sun', the Khmer god-king who built Angkor Wat as his funerary temple.[32] And since Angkor Wat is undoubtedly the largest and most elaborate single edifice in the entire Angkor scheme, indeed the 'ruling temple', we chose to pay special attention to its firm and uncompromising east–west

The sun rising over the central tower of Angkor Wat at dawn on the spring equinox.

orientation which marks it out as 'equinoctial' (in the same sense that the Sphinx is 'equinoctial' – i.e. aligning with, targeting and in some way announcing the precise directions of sunrise and sunset on the spring equinox). The purposeful offsetting of Angkor Wat's axis by 0.75 degrees south of east and 0.75 degrees north of west is part of this scheme interacting with the overall layout of the temple to give observers a 'three-day warning' of the equinox. This effect has been well described in the journal *Science*:

> On the day of the spring equinox, an observer standing on the southern edge of the first projection of the causeway (just in front of the western entrance gate) can see the sun rise directly over the top of the central tower of Angkor Wat. Three days later the sun can be seen rising exactly over the top of the central tower from the center of the causeway, just in front of the western entrance gate ... This precise observation of the sun at the spring equinox is extremely important.[33]

Because we were aware of this analysis, we reasoned that the right time to look at the sky above Angkor should be at dawn on the spring equinox in AD 1150. We thought that such an exercise would be a good test of the 'Draco–Angkor correlation'. If it was there in the heavens at that precise and highly significant moment then it would provide further support for Grigsby's view that it is 'too much of a coincidence that so many of the features of Angkor have a counterpart in Draco'.[34] On the other hand, if the correlation did not occur over Angkor on the spring equinox in AD 1150 then it would look somewhat less likely to have been the result of a deliberate plan.

Draco is a northern constellation, circumpolar at higher latitudes. It therefore does not travel very much east or west during its nightly journey but seems to rotate, slowly, around the north pole of the sky. From this it follows, and should be obvious, that to look at Draco the observer must look north. It follows, too, if the temples of Angkor are indeed meant to 'copy' Draco on the ground, that they too should be viewed looking north. Ideally what should arise to prove the correlation is a moment at which an observer stationed due south of Angkor at dawn would be able to look due north and 'see' (in his imagination, of course) the great geometrical enclosures and temples on the ground and, in the northern sky immediately overhead, the constellation of Draco the snake-dragon sprawled out across the meridian. Clearly, such powerful 'mental imaging' would have been enhanced by the confident knowledge that experienced astronomers have of the precise positions of all the stars in the sky, *whether those stars are visible or not*, at midday or at midnight, at dusk or at dawn. In other words, although all the stars in the sky are 'swallowed up' by the rising light of the sun at least half an hour before the actual moment of sunrise, we were prepared to accept that adept astronomers with the levels of skill shown in the alignments of Angkor Wat would have had no more difficulty than modern astronomers in computing the precise position of Draco in the sky at the exact moment of sunrise on any dawn.

Skyglobe took us back to AD 1150, to the spring equinox, and to 6.23 a.m.,

BELOW LEFT: *'Wrong correlation, wrong time'. The view north from Angkor at sunrise on the spring equinox AD 1150. The constellation of Draco is at the highest culmination and is clearly visible, but it is in the wrong pattern.*

BELOW RIGHT: *'Right correlation, wrong time'. The view north from Angkor on the spring equinox AD 1150. The constellation of Draco is at the lower culmination, in the right pattern, but is not visible as it is below the horizon.*

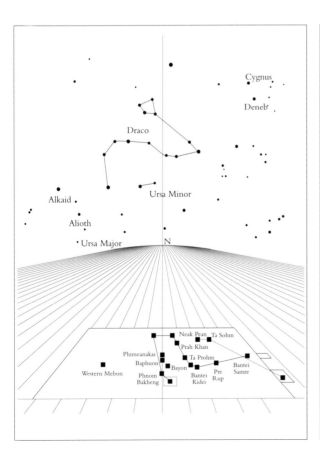

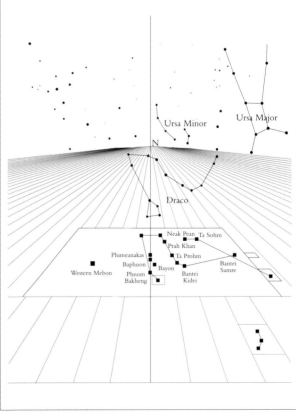

when the half-risen disk of the sun would have bisected the horizon perfectly due east. There, sprawled across the meridian at the point in its nightly journey known by astronomers as its 'upper culmination'[35] was Draco – but not at all as we had hoped. When compared with the vast pattern traced out on the ground by the Angkor temples the constellation appeared, mockingly, to be exactly 'upside down' (see diagrams), i.e. rotated through 180 degrees.

We set Skyglobe to scroll through a complete 24-hour cycle of the sky looking for a time when Draco would be the 'same way up' as the temples. That moment came exactly 12 hours later, at 6.23 p.m., the precise time of the constellation's 'lower culmination'.[36]

Discouragingly, however, there was still no correlation. Although now in the 'right' configuration, Draco would have been buried far below the horizon at 6.23 in the evening – completely out of the sky–ground arena.

Still wanting to give the correlation another chance we scrolled through the entire year of AD 1150, and then through the entire epoch from the ninth to the thirteenth centuries AD, to see if there was any time at which Draco would have been above the horizon at lower culmination.

In a way we were almost surprised to find that there was none. Yet the reason was simple: the constellation's altitude was so low in the twelfth century AD that its lower culminations *always* took place beneath the horizon.

10,500 BC

In other words, in the period during which we know that the Angkor temples were built there was not a single occasion – let alone a spring equinox – on which the whole of Draco would have lain above the northern horizon at lower culmination.

Our first instinct was to accept this as evidence that the correlation was very probably coincidental. Coincidence or not, however, we found it difficult to ignore the fact that the temples do appear, stubbornly, consistently, and with a high degree of accuracy, to outline Draco's principal stars at lower culmination. It is a fact too that all these stars *were present* in the sky over Angkor – and indeed on the meridian – at sunrise on the spring equinox in AD 1150. The problem, however, is that they were then at upper culmination, and thus 'upside down'.

How serious is this problem? And does it really amount to an argument in favour of the correlation being coincidental? Isn't it, in its own way, quite extraordinary that all the stars and matching temples are present but that the sky 'template' appears to have been rotated by precisely 180 degrees away from the 'model' on the ground?

Precession is an engine that 'rotates the sky' – very slowly over its 25,920 years cycle – and that equally slowly changes the altitude at which stars cross the meridian. Is it possible that a computer search might reveal an epoch when Draco was at a higher altitude and when, therefore, it might have appeared *above* (rather than below) the northern horizon in the pattern that is matched by the temples?

We were reminded of Giza, where the sky–ground correlations were imperfect in 2500 BC when the Pyramids and the Sphinx are supposed to have been built but which computer models show to have been in precise alignment at sunrise on the spring equinox in 10,500 BC.[37] We were reminded, too, of the way in which the Pyramids and the Sphinx appear almost to have been purposefully designed, like mandalas of the mind, to encourage contemplation and understanding of the ponderous changes wrought on the heavens by the long cycle of precession. We remembered that this cycle proceeds at the rate of one degree every 72 years. And we remembered that Angkor lies 72 degrees east of Giza where the Pyramids and the Sphinx stand.

Although there is absolutely no archaeological evidence of any construction at Angkor in 10,500 BC – or indeed of any kind of human habitation – we felt compelled to have a look at the sky in that remote epoch. And since sunrise on the spring equinox in 10,500 BC was the moment at Giza when the entire sky–ground diagram 'locked' into place with Orion at the meridian, we instructed the computer to simulate the sky over Angkor at sunrise on the spring equinox in 10,500 BC.

As we had known it would, Orion lay *due south* at the meridian pretty much as it did at Giza (the only difference being a change in perspective caused by

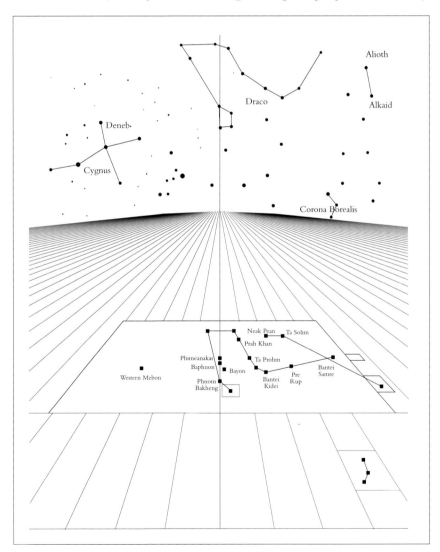

'Right correlation, right time'. The view north from Angkor on the spring equinox 10,500 BC. The constellation of Draco is at the lower culmination, in the right pattern, and is clearly visible above the horizon.

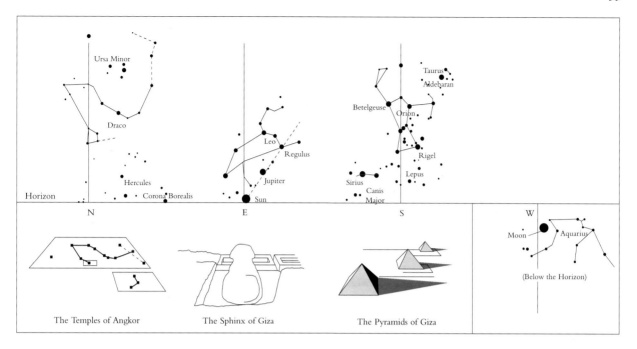

The position of the constellations at the cardinal points and their terrestrial counterparts at the moment of sunrise on the spring equinox 10,500 BC.

Angkor's lower latitude – 13 degrees 26 minutes north as against Giza's 30 degrees 3 minutes north).

As we had known it would, Leo lay *due east* above the rising sun, pretty much as it did at Giza. Again the only difference was a slight tilt of the constellation resulting from the latitude change.

Now we instructed the computer to look north but, because we had never paid a great deal of attention to the northern sector of the sky in the Egyptian correlations, we did not really expect to find anything. We were therefore amazed to discover that at the exact moment of sunrise on the spring equinox in 10,500 BC the constellation of Draco lay *due north* in the middle of the sky, straddling the meridian well above the horizon in precisely the pattern that is replicated on the ground by the principal temples of Angkor.

As at Giza, therefore, a real correlation between sky and ground, representing a real moment in the precessional cycle, does exist. And exactly as at Giza this correlation only 'locks' perfectly at a very remote date. It is remarkable that the same date, indeed the same precise moment, comes up for both sites. It is noteworthy, too, that the temples of Angkor do not model any random constellation, nor do they repeat either of the two constellations used at Giza – namely Orion and Leo, which marked the sky's south and east cardinal directions at dawn on the spring equinox in 10,500 BC; instead, they model sinuous, snakelike Draco, which at that same moment marked the north cardinal direction.

At Giza we have 'Orion temples' in the form of the great Pyramids looking like Orion in 10,500 BC, and 'Leo temples' in the form of the lion-bodied Sphinx and its adjacent structures looking like Leo in 10,500 BC. If there is some sort of covert link between Giza and Angkor then would it not be entirely appropriate for the latter to continue the hidden 'diagram' with an extravagant representation on the ground, across several hundred square kilometres, of the constellation of Draco, 'the Old Serpent',[38] as it looked in 10,500 BC?

CHAPTER 8

CHURNING THE SEA OF MILK

ANGKOR WAT'S WESTERN ENTRANCE is a place of shadows before dawn when the sun still lies invisible in the east beneath the vast mass of the temple. Even in low light it is impossible to ignore the dominant presence of the Naga serpents which, with their stone bodies, and rearing, hooded heads, form sinuous balustrades lining the causeway. The same cobra motif in numerous different forms is frequently, almost incessantly, repeated – leading one authority to conclude that Angkor Wat 'was wholly dedicated to serpent worship. Every angle of every roof is adorned with a seven-headed serpent.'[1] As we shall see, this 'obsession with the Naga'[2] is also expressed in dramatic fashion on the balustrades of Angkor Thom and at many of the outlying temples that coil away to the north and east.

In the Buddhist scriptures we read how Takasaka, one of the fabled kings of the Naga serpents, could, like a dragon, 'cause destruction by the fiery breath of his nostrils'.[3] This is by no means the only ancient text in which the characteristics of serpents and dragons appear to overlap – a state of confusion that extends to the heavens and specifically to the constellation of Draco. Its Greek name means 'Dragon' but it has also been understood, throughout history and in almost all cultures, to be some form of cosmic snake – 'Python' or 'Serpens' in certain ancient astronomical tables, 'the Old Serpent' of more common parlance, the snake snatched by Minerva from the giants and whirled to the sky, the 'man-eating serpent' of the Persians, etc.[4]

Another matter, not necessarily negligible, is that the constellation of Draco does *look* very much like a rearing cobra with its hood extended.

All in all, therefore, it is easy to see how the rearing Naga with its hood extended might have been adopted at the temples of Angkor as a suitable figure for Draco. It also feels somehow 'right', somehow appropriate, that these temples, which are prodigiously decorated with the images and forms of Naga serpents, should also be laid out on the ground in the form of a great coiled Naga.

THE ANCIENT SERPENT

In Indian mythology the Nagas are supernatural beings, cobra-kings who rule on

OPPOSITE: *Statue of the Buddha resting upon the coils of a Naga serpent, upper gallery, Angkor Wat. According to the tradition represented here, the Buddha was sheltered from a week-long rain-storm by the Naga-King Mucalinda who 'enveloped the body of The Blessed One seven times with his folds and spread his great hood above his head … When seven days had elapsed and Mucalinda the serpent-king knew that the storm had broken up, and that the cold had gone, he unwound his coils from the body of The Blessed One. And changing his natural appearance into that of a young man, he stood before The Blessed One…'*

This relief from the outer enclosure wall of the temple of Ta Prohm at Angkor shows Garuda, the mythical birdman of the Hindus, trampling two Naga serpents. According to myth Garuda was the mortal enemy of all snakes. With the limbs of a man and the head, wings, talons and beak of an eagle, his symbolism bears comparison with that of Horus, the falcon-man of Egypt. Garuda is also to be identified with the phoenix and with the constellation of Aquarius (see pages 320–21).

earth but who are ranked amongst the gods.[5] Usually depicted as beautiful five-headed or seven-headed serpents (although the number of heads may be less or more), they are shape-changers who can appear, at will, as fully human or as strange and wonderful hybrids – human above the waist and serpent below.[6]

The earliest references to Nagas are to be found in the Rig Veda, India's oldest-surviving body of sacred texts.[7] They make repeated appearances in the classic writings such as the *Ramayana*, the *Mahabharata* and the Puranas and also feature extensively in Buddhist literature.[8]

Since the antiquity of many of these texts is uncertain and subject to serious dispute, we need not necessarily accept the consensus opinion among Western academics that the Rig Veda dates no earlier than 1500 BC.[9] Learned Indian scholars, notably Lockamanya Bal Gangadhar Tilak, the historian of religion Georg Feuerstein, the Sanscrit scholar Subhash Kak and the Vedic teacher David Frawley, are powerful dissenters from the orthodox view, suggesting that the date of composition may be vastly more ancient:[10]

> The Vedas ... appear to be our best record of the ancient spiritual teachings of humanity. They contain a timeless wisdom, a mantric code in which the wisdom of the race was passed down from World Age to World Age, for how many millennia we can only speculate. The Vedic people were probably in India by 6000 BC and before ... The Rig Veda itself looks back upon earlier ages and reflects the knowledge of long cosmic time cycles.[11]

Likewise, although orthodox Western scholars believe that the *Ramayana* was composed at around 300 BC,[12] Indian traditions (which call the *Ramayana* an 'Adikvaya' or 'Primeval Poem') assert that it describes events that took place 870,000 years ago, that it was originally composed 'soon after that date', and that later versions are merely copies.[13]

It is out of a rather uncertain background, therefore, that the magical Naga serpents emerge. Capable of great benevolence and likewise of great malice, their characters are as ambiguous and as unpredictable as the forces of nature. They are trans-dimensional, crossing the realms of sky and ground, time and space, this world and the next and although they intermingle – and sometimes intermarry – in the material realm of earth and men there is never any doubt that their true identity is as celestial and cosmic forces.

SESHA

The first and greatest of the Nagas is the seven-headed Sesha ('Duration' or 'Remainder') who is also called Ananta ('Endlessness').[14] On his coils, drowned deep in the ocean of unbounded nothingness that preceded the formation of our present universe, India's ancient religious texts tell us that Vishnu, the all-god ('the divine Self-existent'[15]) lay asleep dreaming his forthcoming creation. This 'slumber', following a period of dormancy, is to be understood as a state:

> in which the god's vitality slowly ripens to unfold again in another universe. These alterations of rest and activity, although each of them lasts for thousands of millions of centuries, are as regular and as certain as an organic rhythm – India thinks of them as the god's in-breathing and out-breathing.[16]

Like Atum bursting triumphantly forth from the waters of the Nun – conjuring out of nothingness the Primeval Mound, the Benben Stone and the Phoenix – the all-god of Indian mythology establishes the universe through the strength of his own will, 'appearing with irresistible creative power, dispelling the darkness'.[17] Like Atum, too, the all-god ejaculates semen into the cosmic waters. In the Egyptian tradition that semen solidifies as the Benben (a glowing, reflective pyramidion of stone or iron – perhaps even an iron meteorite[18] – thought of as having fallen from heaven). Similarly, in the Indian scriptures we read how the all-god:

> desiring to produce beings of many kinds from his own body, first with a thought created the waters and placed his seed in them. That seed became a golden egg, in brilliancy equal to the sun; in that egg he himself was born as the progenitor of the whole world.[19]

Perhaps it is a coincidence that the ancient Egyptian name for meteoritic iron – *bja*, meaning 'divine metal', literally 'metal from heaven'[20] – is almost identical to the Sanscrit word for semen: *bija*, 'seed'.[21] *Bja* iron was used in ancient Egypt in rituals aimed at gaining for initiates the 'life of millions of years' through the 'escape of the soul to the stars'.[22] This, too – the escape of the soul 'from the always terrible and constantly changing cycle of births and deaths to which created beings are subject'[23] – was the ultimate objective of all religious ritual, meditation and scripture in ancient India.

India's all-god, the 'Self-existent', is Vishnu and yet more than Vishnu. His nature is complex. According to the *Padama Purana*:

> In the beginning of creation the great Vishnu, desirous of creating the whole world, became threefold: Creator, Preserver, Destroyer. In order to create this world, the Supreme Spirit produced from the right side of his body himself as Brahma; then, in order to preserve the world, he produced from his left side Vishnu; and in order to destroy the world he produced from the middle of his body the eternal Shiva. Some worship Brahma, others Vishnu, others Shiva; but Vishnu, one yet threefold, creates, preserves and destroys: therefore let the pious make no difference between the three.[24]

Shiva, the third partner in the Hindu trinity has as his special emblem the *lingam*. Carved from stone, it is a phallic symbol that is sometimes obviously in the form of a penis, sometimes conical, and sometimes an upright pillar.[25] Once the 'Shivalingam' is linked to Vishnu's shining 'golden egg', formed from divine seed – and additionally said to have 'fallen from heaven'[26] – it is difficult to resist a comparison with Innu, the megalithic 'pillar' of Heliopolis on top of which rested the mysterious pyramid-shaped Benben with its own connotations of 'seed', 'begetting' and renewal.[27]

The event that took place at Heliopolis, according to the Pyramid Texts, was the beginning of a new cycle, a new epoch, a new episode of creation. And this, too, was what followed from the emergence out of the 'darkness' of Vishnu/Brahma/Shiva – the awakening of the all-god that set our current universe in motion.

We are told in the Indian scriptures: 'This universe existed in the shape of Darkness, unperceived, destitute of distinctive marks, unattainable by reasoning, unknowable, wholly immersed . . .'[28] In that darkness suspended in the waters of space–time, 'the Supreme God slept on the lap of the serpent.'[29]

Why is that serpent called 'Sesha', meaning 'Remainder'? According to the French Orientalist Alain Danielou:

THE SEA OF MILK   139

When creation is withdrawn it cannot entirely cease to be; there must remain, in a subtle form, the germ of all that has been and will be so that the world may rise again. It is this remainder of the destroyed universes which is embodied in the serpent floating on the limitless ocean and forming the couch on which the sleeping Vishnu rests.[30]

According to Danielou, Sesha represents the cycles of time.[31] Indeed, in the *Vishnu Purana* we read that when Sesha yawns, 'he causes the earth to tremble with its oceans and forests ... At the end of each cosmic period he vomits the blazing fire of destruction, which devours all creation.'[32]

So the Naga serpents, which we see everywhere around us at Angkor, are connected to the birth and death of world ages and to the eternal regeneration of time.

Detail of Naga balustrade, Angkor Thom.

Buddhist monks walk slowly through the southern gallery on the western side of Angkor Wat, studying the bas reliefs of the 'Churning of the Milky Ocean'. Angkor's reliefs, sculptures, architecture and alignments all seem to have been designed to stimulate reflection and inquiry. The temple is a good teacher and finds many different ways to convey the esoteric knowledge which its builders believed could lead to spiritual transformation.

VASUKI

Angkor Wat's main gallery extends all the way around the temple at the first level. It is 2.5 metres wide with a corbelled roof supported on fine pillars that block little of the external light. Its north and south walls are each 202.14 metres in length.[33] Its east and west walls are each 187 metres in length.[34] These walls are decorated with bas reliefs occupying eight panels 2 metres high, four of them 49 metres long and four of them nearly 100 metres long – amounting to some 1200 square metres of sculpture.[35]

The southern wing of the eastern side of the gallery contains a 49-metre panel showing, in wonderful detail, a famous scene from Hindu mythology in which a

Naga serpent plays a crucial role. This scene is known as the 'Churning of the
Milky Ocean' and the serpent is the five-headed Naga King Vasuki.

We entered the gallery from the southern side, noticing its cloistered hush.
Looking along it to the north it seemed to stretch away for ever in alternating
patches of cool, dark shadow and warm, bright sunlight. Specks of dust hung
suspended in the air.

The scene portrayed in the bas reliefs is too big to take in at one glance.
Instead, it must be walked through frame by frame.

The story begins at the southern end of the gallery with the five rearing heads
of a great hooded cobra – the Naga King Vasuki – who is gripped firmly in the
powerful hands and arms of a high-ranking *asura* or demon. The *asura*, who is of
titanic stature, has his feet planted firmly on the ground and is putting his whole
body weight and energy into leaning back, pulling with might and main on
Vasuki's head. In this task he is assisted by two other gigantic *asuras* and 89 more
of lower rank and stature, who are lined up in front of him along Vasuki's body
like a tug-o'-war team grasping a tow-rope.

At the centre of the panel the body of Vasuki is looped around a thick protu-
berance with sheer sides, rounded at the tip, which represents Mount Mandera –
one of the key landmarks in Indian sacred geography. According to this system of
ideas, which is to be understood symbolically and not literally, the universe consists of:

> seven island-continents surrounded by seven seas. Jambu-dwipa (the
> world) is the innermost of these; in the centre of this continent rises the
> golden mountain Meru, rising 84,000 leagues above the earth ... Meru is
> buttressed by four other mountains, each 10,000 leagues in height. Of these
> one is Mandera.[36]

Suspended in the air beside Mount Mandera is the figure of Vishnu, grasping
Vasuki's body with two of his four hands and seeming to control or direct its

Vishnu suspended beside Mount Mandera, grasping Vasuki's body. The mountain rests on the shell of the turtle Kurma, itself an avatar of Vishnu. Compare the figure of Vishnu and his handling of a multi-headed serpent with similar ancient Egyptian images from the Book of What is in the Duat *(see pages 147 and 84). Note also the Mayan connection of the turtle/tortoise to the Orion sky region and to rebirth symbolism (see pages 35–7).*

The monkey god Hanuman, leader of the divas, *grasping the tail of Vasuki.*

movement. Beneath him the base of the mountain rests on the shell of a gigantic tortoise named Kurma, itself an *avatar*, or manifestation, of Vishnu. The mountain's peak is stabilized by a *deva* or demigod, again thought to be a manifestation of Vishnu.[37]

On the north side of Mandera is the opposing team in the cosmic tug-o'-war, led by three gods of heroic size disposed symmetrically along Vasuki's body and supported by 85 *devas* of lesser stature. At the extreme north of the scene, grasped by the third god – who is shown as monkey-headed – we see Vasuki's tail, which curls into the heavens like a whip.

The counterpoised pulling of Vasuki is all depicted in the central horizontal register of the panel, which is sandwiched between two other registers. The one above represents the heavenly realms and is filled with dancing *apsaras* – sensual celestial nymphs whose gift is flight or 'ascent' and whose forms, endlessly varied, appear in all the temples of Angkor. The one below, which seems to rest on the body of a Naga serpent similar to Vasuki, is a watery region, a great ocean filled with fish and crocodiles and other creatures of the deep. With intense realism, these are shown caught up in an inexorable current that is drawing them towards the base of Mount Mandera, which is being rotated, first one way then the other, by the opposing forces of the *devas* and *asuras*. Close to the mountain the sea creatures are tossed about and cut to pieces by the powerful whirlpool of this cosmic churn which pivots upon the back of the tortoise incarnation of Vishnu.

THE DRAUGHT OF IMMORTALITY

Any pilgrim seeking wisdom who came to the great initiation centre of Angkor in the eleventh or twelfth centuries AD would have been completely familiar with the mythical background to these scenes. The story is found in the *Ramayana*, the *Mahabharata* and several of the Puranas.[38] It tells how: 'at the end of a world age, the deities and demons united to churn the cosmic ocean in order to win the draught of immortality (*amrita*) hidden in its depths'.[39] They did this by uprooting Mount Mandera (with the help of Sesha, we are told in some texts[40]):

> The gods carried the mountain Mandera to the ocean, and placed it on the back of Kurma, the king of tortoises. Round the mountain they twisted the serpent . . . the *asuras* holding its hood and the gods its tail. As a result of the friction caused by the churning, masses of vapour issued from the serpent's mouth which, becoming clouds charged with lightning, poured down refreshing rains on the weary workers. Fire darted forth and enwrapped the mountain . . . [41]

The texts say that Vasuki suffered much from his painful labour, eventually spewing out torrents of venom that 'poured down on earth in a vast river which threatened to destroy gods, demons, men and animals. In their distress they called upon Shiva, and Vishnu joined in their entreaties. Shiva heard them and drank the poison to save the world from destruction.'[42]

Still the churning continued relentlessly, unabated, until, in time, the foaming mass of what is frequently referred to as the 'Sea of Milk' (or the 'Milky Ocean') produced 'butter flavoured by gums and juices ... At length the moon appeared from the ocean; then arose the *apsaras*, who became nymphs in heaven.'[43] These were followed by the goddess Lakshmi, the wife or consort of Vishnu, by the god's white horse, and by the gleaming gem which Vishnu wears on his breast.[44]

> Then came Dhanwantari, the physician of the gods, who carried the golden cup brimming with *amrita*[45] ... Instantly the demons seized the vessel, but before they could drink its priceless fluid and achieve immortality, Vishnu intervened. Assuming the beautiful form of Mohini, enchantress of illusion, he approached the host of demons. They were so bewitched by Mohini's appearance that they willingly handed her the vessel. She immediately gave the nectar to the deities, who drank it and thus secured eternal life for themselves.[46]

Apsaras, celestial dancers, said to have been amongst the benefits conferred upon the world by the churning of the Milky Ocean.

COSMOLOGICAL CLUES

What did the builders of Angkor mean to say when they decided to express this extraordinary myth in the form of sculpture?

For most of the last century, scholars have paid little attention to such questions, dismissing the bas relief – as the leading French archaeologist George Coedes does here – as a work of mere superstition: 'It is easy to see that the churning of the sea by the pivoting mountain represented a magic operation which assured the nation of victory and prosperity.'[47] Perhaps so. Nevertheless, Coedes is the first to admit that the only possible interpretation that can be given to the 'mountains' Mandera and Meru is cosmological in nature and, furthermore, that:

> Angkor Wat with its wall and moats, its central sanctuary, its entrances, its pyramidial temples and its bridges with Naga balustrades, as well as other complicated monuments such as Neak Pean or the Bayon, are actually

representations in stone of the great myths of Hindu cosmology. The purpose of this system was to reproduce on earth a terrestrial model of all or part of the heavenly world, thus ensuring that intimate harmony between the two worlds without which humanity could not prosper.[48]

Coedes shows how the god-kings of the Khmers saw the pyramids that they built and the towers of their temples quite specifically and explicitly as terrestrial copies of Mount Meru: 'The architectural form adopted for this sacrosanct monument [the central temple] was ... a mountain in the form of a pyramid ... Sometimes the pyramid was topped with a quincunx of towers in imitation of Mount Meru which was supposed to have had five peaks.'[49]

Coedes is even aware that one eleventh-century Khmer monarch, Udayadityavarman II, left an inscribed stela beside his principal pyramid – the so-called Baphuon – on which it is related that he had created the monument 'because he thought that the centre of the universe was marked by Meru and he thought it fitting to have a Meru in the centre of his capital'.[50]

What is surprising, having recognized the cosmological character of the 'mountains', is that neither Coedes nor any other archaeologist or orientalist working at Angkor has ever taken the next logical step and considered whether the churning of Mount Mandera might not also be best understood as a cosmological operation of some kind.

MACHINERY OF THE HEAVENS

There *is* a cosmological process that fits the bill: precession – the slow, cyclical wobble of the axis of the earth that inexorably changes the positions of all the stars in the sky and shifts the 'ruling' constellation that lies behind the sun at dawn on the spring equinox. It is this process, according to Giorgio de Santillana and Hertha von Dechend in their landmark study *Hamlet's Mill*, that is the subject of a whole family of myths coming down to us from remotest antiquity. The Churning of the Milky Ocean, they say, is one of these myths.[51]

The great contribution to scholarship made by *Hamlet's Mill* is the evidence it presents – compelling and overwhelming – that, long *before* the supposed beginnings of civilized human history in Sumer, Egypt, China, India and the Americas, precession was understood and spoken of in a precise technical language by people who could only have been highly civilized.[52] The prime image used by these as yet unidentified archaic astronomers 'transforms the luminous dome of the celestial sphere into a vast and intricate piece of machinery. And like a millwheel, like a churn, like a whirlpool, like a quern, this machine turns and turns endlessly.'[53]

The mill, or the pivot, or the churn, is not to be envisaged as a 'straight, upright post'.[54] Instead, the symbol represents it as an axis completely encircled by two hoops, or 'colures', intersecting at the celestial north pole. One hoop connects the equinoctial points on the earth's orbit (i.e the background of stars against which the sun stands on 21 March and 21 September). The other connects the solstitial points – the sun's address against the background of stars on 21 June and 21 December. Since precession causes the polar axis to shift, it follows that it must

The position of the North Celestial Pole denotes the placement of the four 'markers' of the year – the constellations against which the sun rises at the cardinal points of the year. In our present age, with the Pole near Ursa Minor, the markers are Pisces (spring equinox), Sagittarius (winter solstice), Virgo (autumn equinox) and Gemini (summer solstice).

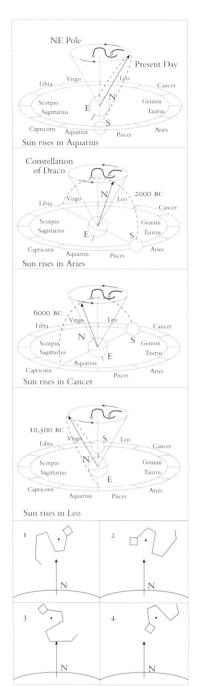

also shift the hoops attached to it – the equinoctial and solstitial colures – 'the great circles that shift along with it in heaven'. This is so because 'the framework is thought of as all one with the axis'.[55] The image therefore explains the process of the precession of the equinoxes as part of an eternal cycle spinning slowly backwards along the ecliptic – the path of the sun – and equally slowly revolving the four constellations that house the sun at the four key moments of the year. The axis and the hoops (or 'props', or 'sky-bearers', etc., as they are sometimes called) thus stand for:

> a system of co-ordinates in the celestial sphere and represent the frame of a world age. Actually the frame defines a world age. Because the polar axis and the colures form an invisible whole, the entire frame is thrown out of kilter if one part is moved. When that happens a new Pole star with appropriate colures of its own must replace the obsolete apparatus.[56]

It is this shifting of the gears of heaven, Santillana and von Dechend argue, that is symbolized at Angkor Wat – the moment of transition between one astrological world age and the next. That is what is happening.[57] That is what the *asuras* and the *devas* are co-operating to bring about as they pull the coils of Vasuki around the pivot of Mount Mandera and churn the Sea of Milk. The implication is that the bas reliefs, like the temples themselves, need not necessarily be understood as mere 'magical operations' designed to ensure national prosperity. They could be another form of mandala or yantra, 'thought tools' designed to focus the minds of initiates on cosmic mysteries – in this case the mystery of precession.

But why should such a focus have been regarded as desirable?

Could it be because 'the churning of the Sea of Milk', which, if Santillana and von Dechend are right, is an allegory for precession, was believed to be linked somehow to the quest for eternal life? After all, the finest fruit of the churning was said to have been the nectar of immortality.

We are reminded of the ancient Egyptian notion that by 'seizing' the sky or 'ascending' to the sky (i.e. by gaining a complete insight into its secrets), and through *knowledge* of precession (the knowledge of 'how to go down to any sky'), the initiate might hope to equip his soul for immortality. Perhaps, among the countless thousands of pilgrims who have made their way to Angkor through the centuries there have been some who have sought that same goal and who have followed the same 'path' of mastery and knowledge?

The change in position of the North Celestial Pole about the North Ecliptic Pole due to precession shifts the constellation against which the sun is seen to rise on the spring equinox. The diagram at the bottom shows the position of the Pole in relation to Draco in each of these epochs.

EGYPTIAN REFLECTIONS

The *asuras* and the *devas* of Angkor find their counterparts in Egypt in the 'contending' deities Horus and Set who, after the murder of Osiris, struggle for 80 years until a new age of the world is established. Though they are commonly portrayed as bitter opponents, like the *asuras* and the *devas*, there are reliefs in ancient Egypt that show Horus and Set co-operating, pulling on either end of a long rope wrapped around a vast drill and thus rotating it.[58]

To all extents and purposes, the message of the scene is identical to that of the churning of the Sea of Milk which is also brought about by an uncharacteristic

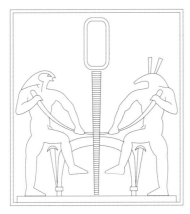

Horus and Set turning the drill.

Set and Horus united as one individual with two heads.

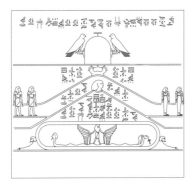

Kingdom of Sokar.

unification of opposing forces. The message could be that the universe *requires* polarity, that everything must have its equal and opposite adversary and that it is only through the creative interplay of these adversaries that new realities can be born.

Other reliefs show Horus and Set united as one individual with two heads,[59] but of greater importance is an intriguing scene in the Fifth Division of the *Book of What is in the Duat*, which contains many of the same symbolic elements as the Churning of the Milky Ocean.[60]

The Duat consists of a total of twelve distinct 'Divisions' each placing a new test or ordeal in the way of the pilgrim soul on its afterlife journey. Many of these tests are symbolized by gigantic hooded cobras rearing up in menacing postures and by fire-breathing serpents, often with multiple heads like the Nagas, and sometimes conspicuously equipped with feathered wings and plumes.

In the Fourth and Fifth Divisions the pilgrim is described as 'the hidden traveller upon the way of the holy country whose secret things are hidden',[61] and his goal is to discover those secret and hidden things: 'the hidden thing which is on this secret way',[62] 'the secret passages which lead to the Aheth chamber',[63] 'the image which is hidden, and is neither seen nor perceived'.[64]

Like the bas relief of the Churning of the Milky Ocean, the entire landscape of Divisions Four and Five (referred to collectively as the 'Kingdom of Sokar') is separated into three parallel registers, with occasional connecting points between the registers. There is all-pervasive serpent imagery in the Egyptian scene, as there is in Angkor, and the lower register of the Fifth Division is shown as being flooded with water, creating a sort of 'contained ocean' very similar to the Milky Ocean portrayed in the Angkor reliefs. Most strikingly, at the heart of the scene, we see the following ingredients:

1 A pyramid-shaped mound, or mountain, comparable to Mount Mandera, out of which emerges the head of a deity.

2 A rope that is passed around the mound, like the body of Vasuki around Mount Mandera.

3 Ranks of pullers, comparable to the *devas* and *asuras* in the Angkor scene, hauling on this rope on both sides of the mountain.

4 Above the mound an enclosed cylindrical object with a rounded tip, surmounted by the hieroglyph meaning 'darkness',[65] or perhaps 'sky' or 'dark sky',[66] on which perch two black birds. These are of particular interest because, in the ancient Egyptian language, a sign showing two birds facing one another was 'the standard glyph for the laying out of parallels and meridians'.[67] In addition it has been persuasively argued by the historian of science Livio Catullo Stecchini that the cylindrical object itself is an *omphalos*, 'a marker for geodetic focal points'[68] of a type that was well known in the ancient world, for example at the oracular temple of Delphi in Greece where a stone *omphalos* can still be seen on site[69] (photograph, page 252). The Delphi *omphalos* is conical and round-tipped, much as the Egyptian *omphalos*, and the sculptor has encased it in what appears to be a decorative mesh or network.

5 Beneath the mound: a) an oval enclosure comparable to the tortoise incarna-

Relief from the temple of Karnak, Egypt. The stylized drilling device between the two lions is continuously mislabelled by Egyptologists as 'the uniting of the two lands' whether Horus and Set turn it, or the two Nile gods (as is more often the case), or the two lions of yesterday and today.

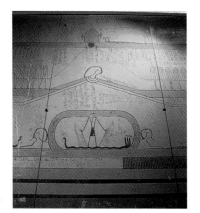

The 'Egg of Sokar', Fifth Division of the Book of What is in the Duat, *tomb of Thutmosis III.*

tion of Vishnu; the mound stands on this enclosure, which is guarded by two man-headed lion sphinxes; b) the channel of waters comparable to the Milky Ocean.

6 Inside the oval enclosure a gigantic three-headed plumed serpent called the 'great god',[70] comparable to the Naga Kings who were also gods.

7 Astride the serpent, hawk-headed Sokar, the 'god of orientation',[71] whose 'work is to protect his own form';[72] in each hand, like Vishnu guiding and directing Vasuki, he holds one of the serpent's feathered wings.

The most striking characteristic of the Fifth Division of the Duat is that the texts describing it contain numerous clues and hints linking it to the three great Pyramids and the Great Sphinx of Giza. Indeed, the respected Egyptologist Selim Hassan goes so far as to state that the Fifth Division 'had its geographical counterpart in the Giza necropolis'.[73] Likewise, Mark Lehner, now Associate Professor of Egyptology at the University of Chicago's Oriental Institute, admits that it is 'tempting' to see the lion sphinxes which guard the mysterious oval of Sokar as:

a representation of the Sphinx at Giza ... The presence of the *omphalos* in the Fifth [Division] adds weight to the suggestion that the symbology therein is focussed on Giza, for the Great Pyramid is situated with exceptional accuracy on the 30th parallel and the prime meridian of Egypt. It is also accurately oriented to true north.[74]

The temples of Angkor display an 'obsessive cardinality' – i.e. high-precision orientations to true north, south, east and west – which could only have been the work of architects and surveyors equipped with advanced astronomical and geodetic skills. We know that architects and surveyors with precisely such skills built the Pyramids and the Sphinx of Giza, and we know that Giza and Angkor are separated by the geodetically significant distance of 72 degrees of longitude. It therefore seems reasonable to wonder whether the strange similarities between the Fifth Division of the Duat and the Churning of the Milky Ocean could be the surviving hints and fragments of an as yet unidentified connection between Egypt and Angkor – a hidden link, attenuated across vast gulfs of time and space, adapted to different cultural contexts, but always essentially returning to the same roots.

TRIUMPH OVER DEATH

In the *Book of What is in the Duat* the pilgrim soul must master the riddles of the Fifth Division, and grasp its hidden meanings, if he is to move on to the next stage of his journey, the Sixth Division, 'the place of the body of Osiris ... these houses which are hidden and which contain the image of Osiris'.[75]

Yama, the Hindu death god, from the judgement scene at Angkor Wat.

In many versions of the books of the netherworld, as we saw in Part II, the Sixth Division of the Duat is the location of the judgement scene in which the eternal destiny of the soul is assessed in the presence of Osiris, god of the dead, lord of resurrection and rebirth.[76] In the bas reliefs at Angkor Wat, in the gallery immediately adjoining the Churning of the Milky Ocean, the counterpart of Osiris is the Indian god of death, Yama, who is also shown presiding in judgement over the souls of the dead.

The Egyptian and Khmer gods, and the scenes in which they are depicted, are point-by-point so closely comparable that it seems far-fetched to ascribe their similarities to coincidence. Like Osiris, Yama is 'a mortal who died [Osiris was murdered] and, having discovered the way to the other world, is the guide to those who depart this life'.[77] Like Osiris, he is often depicted as being green in colour.[78] Like Osiris, it is in his presence that the conscience, deeds and knowledge of the deceased are weighed.[79] And like Osiris, Yama is assisted in his judgements by several other beings, notably Dharma, the god of justice, duty and universal law, who sits to his right and whose duty it is to pronounce the judgement,[80] and Chitragupta, the 'recorder of souls', who 'knows the deeds committed by each person' and who must inscribe the judgement.[81] If the judgement is negative 'the damned are seized by Yama's servants and thrown head-first into the realms of hell'.[82]

Anyone familiar with the Egyptian texts will immediately recognize the affinities between Dharma and the goddess Maat, between Chitragupta and the god Thoth and, indeed, between Yama's servants and the monstrous hybrid Ammit, the 'eater of the dead', into whose hellish jaws would be cast all those souls who failed at the 'weighing of words'.[83]

Somehow the quest to emerge triumphant as a spirit fully vindicated at the weighing of words – which obsessed the Pharaohs of Egypt and the god-kings of the Khmers alike – was seen as mirroring the greater cosmic cycles of the

The lowest of the three registers shows the hell-world of Yama, to which are consigned souls whose conscience, deeds and knowledge have been found wanting in the judgement. Elements of this relief bear close comparison to the judgement scene in the ancient Egyptian Book of the Dead.

'eternal return' and may even have required of the initiate mathematical and astronomical knowledge of these cycles. Again and again the message keeps coming through that it is only by following this straight and narrow path of knowledge – indeed by 'grasping the sky' – that the initiate may be protected from the terrible judgement of the dead.

There is every reason to believe that it was to this end – triumph over death – that the temples of Angkor, like the great Pyramids of Giza, were designed. Indeed, in one of his many inscribed stele, Jayavarman VII, the builder of Angkor Thom and the Bayon, tells us exactly what his objective was in creating these gigantic 'good works'. His purpose, he claimed, was to bestow upon men 'the ambrosia of remedies to win them immortality' and thus 'rescue all those struggling in the ocean of existence'.[84] On another stela Jayavarman invoked the gods to reward his great building works by allowing him to pass freely 'from existence to existence'.[85]

He sounds like a monarch who meant what he said and who saw the temples that he had built as instruments of initiation into an active science of immortality.

MASTER GAME

LEAVING THE GALLERIES AND the 'enclosed, vertical restrictive darkness'[1] of the towers of Angkor Wat, we eventually found ourselves back on the causeway. This led us out through the spacious green gardens and past the large reflecting pools to the western entrance gate, where our driver was waiting.

A detailed survey of Angkor Wat published in *Science* magazine in July 1976 revealed that even the causeway incorporates cosmic symbolism and numbers encoding the cycles of time. After establishing the basic unit of measure used in Angkor as the Khmer *hat* (equivalent to 0.43545 metres) the authors of the survey go on to demonstrate that axial lengths along the causeway appear to have been adjusted to symbolize or represent the great 'world ages' of Hindu cosmology:

> These periods begin with the Krita Yuga or 'golden age' of man and proceed through the Treta Yuga, Davpara Yuga and Kali Yuga, the last being the most decadent age of man. Their respective durations are 1,728,000 years; 1,296,000 years; 864,000 years; and 432,000 years.[2]

It therefore cannot be an accident that key sections of the causeway (see diagram) have axial lengths that approximate extremely closely to 1,728 *hat*, 1,296 *hat*, 864 *hat*, and 432 *hat* – the *yuga* lengths scaled down by 1000. 'We propose', conclude the authors, 'that the passage of time is numerically expressed by the lengths corresponding to *yugas* along the west–east axis.'[3]

CYCLES OF THE AGES

Hindus believe that those of us who inhabit the earth today are living through the unfortunate and tumultuous Kali Yuga, supposedly the last and most decadent of the 'ages of man' in the present Kalpa, or cycle of creation. According to Indian astronomical and calendrical calculations, the Kali Yuga 'began' in 3100 BC,[4] a date that coincides almost to the year with the ancient Mayan computation of the beginning of the Fifth Sun. Exactly like the Kali Yuga, as the reader will recall, the Fifth Sun is our present epoch. The Mayan calendar calculates not only

OPPOSITE: *Looking west, the axis of the arrow-straight causeway of Angkor Wat can be seen in this aerial view to extend beyond the temple's moat and to reach out towards the distant horizon – showing that its builders thought expansively, in very large-scale terms. Within its moat, all the dimensions of the temple are precisely calibrated to express a grand cosmological and numerological scheme related to the precession of the equinoxes. Compare Teotihuacan, Mexico, page 23.*

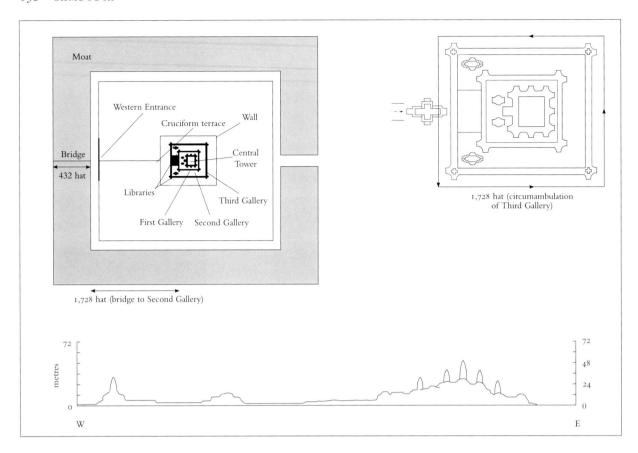

Precessional 'hat' measurements in Angkor Wat.

its beginning but also the date at which it will end in the tumult of a global cataclysm – in AD 2012, on 23 December.

The numbers in the Indian scheme of world ages – 432,000 (or 432), 864,000 (or 864), 1,296,000 (or 1296) and 1,728,000 (or 1728) – all have one thing in common. They belong to the sequence based on 72, which is an important unit of time in the Mayan calendar[5] and which is linked to the precession of the equinoxes (it takes 72 years for the sun's position on the spring equinox to precess one degree against the background of the 'fixed' stars). If you divide the 432 *hat* dimension on Angkor's causeway by 72 you get 6. If you divide the next dimension along the causeway, 864 *hat*, by 72 you get 12. If you divide 1296 *hat* by 72 you get 18 and if you divide the longest dimension of 1728 *hat* by 72 you get 24.

So stepped down by a factor of 1000 in the architectural scale of Angkor, and then denominated by the 'ruling' number 72, the Krita, Treta, Davpara and Kali Yugas can be reduced to a simple mathematical regression: 24 → 18 → 12 → 6.

How likely is it that such consistent 'good order' would have come about by chance? Isn't it also worth considering the possibility that the entire scheme of the Yugas could have been designed with the rhythm of precession in mind? Much suggests this may be so.

One of the governing time frames within the Yuga system is the 'Manvantara' (period of Manu), of which it is said 'about 71 systems of four Yugas elapse during each Manvantara'.[6] This surprisingly vague figure – 'about 71' – is an exception to the general rule in Hindu cosmology. Yet it may be the exception that proves the rule. Modern astronomers have calculated that one degree of preces-sional motion is completed in exactly 71.6 years, a fraction less than the value of 72 favoured in ancient myth. Since myths are stories, the demands of the narra-tive normally rule out the use of anything other than whole numbers (for exam-ple, one could hardly have 71.6 conspirators plotting with Set against Osiris). Quite possibly it was for this reason that Set was given 72 co-conspirators; simi-larly, in India, the value of 71.6 could have been rounded down to 'about 71'.[7]

Also suggestive of precession is the cyclical nature of the Yugas, reminiscent of the world-ages of the zodiac and the eternal return of the 'Great Year'. This is a system in which all things, with the passage of time, come back to their start-ing point and then begin again. The 'First Time' of the ancient Egyptians was such a 'new beginning'. So was the birth of the Fifth Sun in Mexico. So too is the moment of creation described in Hindu traditions in which the all-god Vishnu/Brahma/Shiva awakes from his millennial slumber upon the coils of the celestial serpent Sesha and summons into being the present order of creation. In all cases it is recognized that there have been previous creations – all of which have been destroyed – that this creation too will be destroyed, and that there will be future creations which will also be destroyed.

According to the *Mahabharata*, this universe and all things in it exist in a con-stant state of flux, undergoing first creation – considered to be the work of Brahma – then an epoch of sustainment (it is Vishnu who is the 'sustainer'), and finally destruction at the hands of Shiva. 'After the Universe is dissolved,' how-ever, 'all Creation is renewed and the cycle of the four Ages begins again with a Krita Yuga . . .'[8]

Angkor Wat: statue of the god Vishnu, Lord of the Universe.

Vishnu in his avatar as Rama, riding on the shoulders of the monkey-god Hanuman.

A PALE RIDER ON A PALE HORSE

The Krita Yuga that began our present Kalpa (epoch of creation) is described as one:

> in which righteousness is eternal. In the time of that most excellent of Yugas everything had been done and nothing remained to be done ... [There was] no disease or decline of the organs of sense through the influence of age ... no malice ... no hatred, cruelty, fear, affliction, jealousy, or envy ...[9]

Reading these accounts we are reminded forcefully of the concept of Zep Tepi (the 'First Time') in ancient Egyptian cosmology, the time 'which was born', the Pyramid Texts say: 'before anger came into being; which was born before noise came into being; which was born before strife came into being; which was born before tumult came into being'.[10]

Dharma, god of justice and duty – the direct counterpart of the ancient Egyptian goddess Maat – is described in Indian traditions as having 'walked on four legs' during the Krita Yuga. During the subsequent Treta Yuga,[11] 'a less happy age, in which virtue fell short',[12] Dharma's symbolic legs are said to have been reduced to three.[13] In the Davpara Yuga, Dharma was two-legged and 'lying and quarrelling abounded ... Mind lessened, Truth declined, and there came desire, disease and calamities ... It was a decadent age by reason of the prevalence of sin ... Nevertheless many trod the right path.'[14] In the Kali Yuga – our own – when Dharma must stand on one leg and is helpless, few follow the right path:

> Only one quarter of virtue remaineth. The world is afflicted; all creatures degenerate; men turn to wickedness ... They are unlucky because they deserve no luck. They value what is degraded, eat voraciously and indiscriminately, and live in cities filled with thieves ... They are oppressed by their kings and by the ravages of nature, famines and wars.[15]

It is therefore said that 'in the Kali Age shall decay flourish, until the human race approaches annihilation'.[16]

The agency of that apocalypse has been foretold. He is Kalki, 'the Fulfiller'. According to the *Bhagavata Purana*: 'In the twilight of this age, when all kings will be thieves, the Lord of the Universe will be born as Kalki.'[17] He will come 'riding on a white horse and holding a sword blazing like a comet'. He will punish the evil-doers and comfort the virtuous: 'Then he will destroy the world. Later, from the ruins of the earth, a new mankind will arise.'[18]

AVATARS

Riding his pale horse, Kalki is an avatar – a manifestation in the physical realm – of the high god Vishnu, the 'Sustainer' and 'Pervader', who keeps the created universe in order.[19] As such he is portrayed in Indian tradition as the last in a long line of saviours and guides of mankind who have come to the rescue of goodness and truth in times of darkness.

Several of Vishnu's incarnations have followed *pralayas* – or cataclysms – notably world-destroying floods. The ancient texts tell us that his objective on each of these occasions has been the same: to salvage some fragment of the wisdom accumulated by antediluvian civilizations and pass it on as a legacy for future mankind.

For example, in order to save Manu Satyavrata, the founder of present-day humanity, Vishnu is said to have incarnated in the form of a gigantic fish. Before the flood came he warned Manu to build a great survival ship and ordered him 'to load it with two of every living species and the seeds of every plant, and then to go on board himself'. When the waters rose the god towed Manu's ark for many days and nights until it finally came to rest on the slopes of a high mountain.[20]

Once, when the whole earth was again 'submerged',[21] Vishnu incarnated as a boar, 'dived into the waters and killed the demon who had thrown the earth to the bottom of the sea. He then rescued the earth and re-established it floating over the ocean.'[22] On another occasion, as we have seen, he incarnated as a tortoise at the end of a world age and played his part in the Churning of the Milky Ocean.[23]

Vishnu is also said to have incarnated as a 'man-lion', who tore out the entrails of a genie who was cruelly oppressing the earth, as a dwarf, 'who strode over the universe and in three places planted his step' , as the hero Rama, who, in ancient times, ushered in a golden age of justice and happiness, and as Krishna, who was 'born to teach the religion of love at the beginning of the Age of Strife' (i.e. the so-called 'Kali Yuga', our present epoch).[24]

'At all the crucial moments of the world's history,' sums up the Orientalist Alain Danielou, Vishnu 'appears as a particular individuality who guides the evolution and destiny of the different orders of creation, of species and forms of life':[25]

> Whenever those forms of knowledge that are essential for man's fulfilment of his spiritual destiny happen to be beyond reach, and thus human life fails in its purpose, Vishnu is bound to make this knowledge available again . . . There is, therefore, a new incarnation for each cycle, to adapt the revelations to the new conditions of the world.[26]

In India the reincarnated Messiah comes as Rama and later as Krishna and, in the last of days, will appear again as Kalki, 'the Fulfiller'. The same figure is also found in Mexico as Quetzalcoatl, the once and future king, and in Britain as King Arthur. He appears again in Egypt, manifesting from the beginning of dynastic history in the form of the man-god Osiris – 'the Far-Strider', the Universal Lord – who dies but who will be eternally reborn.

'Osiris', comments R. T. Rundle Clark, 'saves you when you are ill-treated. He is what the Egyptians called a *neb tem* – a "universal master" – human yet mysterious, suffering and commanding.'[27] In myths and scriptures, his is the 'mysterious voice' that every now and then 'calls out with authority, commanding things to be put right when the order of the world is threatened . . .'[28]

What Rundle Clark calls this 'intervention of the commanding deity in the present world'[29] does not differ in any significant degree from the interventions of Vishnu's avatars in the world. Likewise, in both the Egyptian and the Hindu traditions, it is understood that the objective of such interventions is a positive one:'With the purpose of protecting the earth, priests, gods, saints, the Scripture, righteousness and prosperity, the Lord takes a body.'[30]

Quetzalcoatl plays an identical, life-giving role in Mexico, presiding over a golden age and offering his initiates the flower of immortality.

RISHIS

Does human life have any purpose? Or is it purposeless? Does it have any meaning? Or is it meaningless? Is it sublime? Or is it ridiculous?

According to the Rishis – the 'wise men' or 'sages' of ancient India – our lives do have meaning, and a very specific purpose. They called this purpose 'realization' or 'enlightenment' – the ability of the soul, materialized only temporarily in a human body, to understand the true nature of its own existence.

The Rishis are the wise men or sages of ancient India who devote their lives to exploring the mystery of reality. They do not declare themselves but pass quietly through existence, shunning material finery, gathering the knowledge that leads to enlightenment. We encountered this pilgrim outside the sacred city of Dwarka in western India, and were struck by the wisdom and insight in his level gaze. Dwarka is dedicated to Krishna, an avatar of Vishnu.

Dwarkadish, the principal Krishna temple at Dwarka stands five storeys high and is built on 72 pillars. There are local traditions that the original city of Dwarka was swallowed by a great flood at the beginning of the Kali Yuga, our present epoch of the earth. Note astronomical decoration of flag.

What we accept without question as 'reality' the Rishis described as 'the world of form'. They claimed to have discovered that this world is *not* in fact real at all but rather a sinister sort of virtual reality game in which we are all players, a complex and cunning illusion capable of confusing even the most thorough empirical tests – a mass hallucination of extraordinary depth and power designed to distract souls from the straight and narrow path of awakening which leads to immortal life. With a synchronicity that seems strange to anyone who has studied the mysteries of Central America, they named the hallucination 'Maya' and taught techniques to overcome and dispel it. Such techniques, amounting effectively to a 'science of realization', included the single-minded pursuit of spiritual knowledge, meditation, contemplation, concentration of mind through the study of mandalas and yantras, and the correct fulfilment of ritual.

The reader will remember that in Mexico, too, life was understood not to be real but only a dream from which the soul awakes on death.[31] Likewise, in the supposedly unrelated Hermetic Texts, compiled in Alexandria in Egypt at around the second century AD, we read that 'all things on earth are unreal ... Illusion is a thing wrought by the working of Reality.'[32]

The Hermetica teach that the initiate must strive diligently to overcome the material illusion that his consciousness will not survive his physical death and must assiduously 'train his soul in this life in order that, when it has entered the other world, where it is permitted to see God, it may not miss the way which leads to him . . .'[33] Because such training was considered to be a matter of vital importance for the soul, the Hermetic Texts also lament the fact that the cycles of time will bring decay and ruin to the land of Egypt – 'this land which once was holy, a land which loved the gods, and wherein alone, in reward for her devotion, the gods deigned to sojourn upon earth, a land which was the teacher of mankind':[34]

> O Egypt, Egypt, of thy religion nothing will remain but an empty tale, which thine own children in time to come will not believe; nothing will be left but graven words, and only the stones will tell of thy piety. And in that day men will be weary of life, and they will cease to think the universe worthy of reverent wonder and worship . . . Darkness will be preferred to light, death will be thought more profitable than life [and] no one will raise his eyes to heaven . . . As to the soul, and the belief that it is immortal by nature, or may hope to attain to immortality . . . all this they will mock at and will even persuade themselves that it is false . . .'[35]

And 'all this', the Hermetica also make clear, is inextricably linked to the cycles of time, spiralling down to what the Indian sages called the *pralaya* at the end of the world, when evil reigns and the props of heaven must be broken by the churning of the cosmic mill:

> And so the gods will depart from mankind – a grievous thing! – and only evil angels will remain, who will mingle with men, and drive the poor wretches by main force into all manner of reckless crime, into wars and robberies, and frauds and all things hostile to the nature of the soul. Then will the earth no longer stand unshaken . . . heaven will not support the stars in their orbits, nor will the stars pursue their constant course in heaven . . . After this manner will old age come upon the world. Religion will be no more; all things will be disordered and awry; all good will disappear.
>
> But when all this has befallen . . . then . . . God . . . will look on that which has come to pass and will stay the disorder by the counterworking of his will . . . He will cleanse the world from evil, now washing it with waterfloods, now burning it with the fiercest fire, or again expelling it by war and pestilence. And thus he will bring back his world to its former aspect, so that the Kosmos will once more be deemed worthy of worship and wondering reverence . . .
>
> Such is the new birth of the Kosmos, it is a making again of all things good, a holy and awe-striking restoration of all nature; and it is wrought in the process of time by the eternal will of God . . . [36]

Like the Hermetic teachers, the Rishis of India also believed that there had been and would again be prolonged epochs of destruction and darkness – such as our present Kali Yuga – when human beings would be entirely overcome by the sorcery of Maya and would wallow in stupidity, strife and greed for thousands of years. At such times, Vishnu would not only manifest as an avatar offering a revelation for the world, but would exert his influence through a number of 'partial incarnations', said to 'maintain, complete and interpret the revelation. These are mainly seers and sages . . .'[37]

And there would be another pathway to salvation, always available to all who genuinely seek it: the archaic scriptures called the Vedas, India's oldest collection of religious writings.

The Vedas – from the Sanscrit word *veda*, meaning 'knowledge' or 'wisdom'[38] – are believed in India to transmit teachings of primeval antiquity, vastly more ancient than the date at which the surviving recensions were compiled. Indeed it is understood that the Vedas were taught as an oral tradition 'passed down faithfully by special families within the Brahman communities of India' for thousands of years before their codification into the 'books' that have come down to us today.[39] Furthermore, even these oral Vedas were not seen as the original scriptures but rather as the *repromulgation*, after the most recent *pralaya*, of even earlier teachings. This work is said to have been undertaken by seven Rishis who survived the cataclysm and whose desire it was 'at the beginning of the new age ... to safeguard 'the knowledge inherited by them as a sacred trust from their forefathers in the preceeding age'.[40]

Similar traditions exist in Egypt, recorded in the Edfu Building Texts, which also speak of the knowledge of seven wise men – the 'Seven Sages'[41] – and how it was brought to the Nile Valley in the First Time in a spiritual endeavour intended to recreate the former world of the gods:

> An ancient world, after having been constituted, was destroyed, and as a dead world it came to be the basis of a new period of creation which at first was the re-creation and resurrection of what once had existed in the past.[42]

According to the Edfu Texts the method adopted by the Seven Sages in this endeavour was the construction of sacred 'mounds' – by means of which they specified the plans and designs that were to be used for all the future temples in the land of Egypt.[43] These temples, to be built with halls 'resembling the sky', were seen as living beings that could die and be reborn and die and be reborn over and over again, all of them tracing their descent back to a common ancestor – 'a temple that once really existed in the dim past of predynastic Egypt'.[44] We have seen how that ancestral temple was itself believed to have been in some way a 'copy' of a region of the sky.[45] A spirit armed with knowledge of it might hope to gain the life of millions of years – being 'well-equipped both in heaven and in earth, unfailingly, and regularly and eternally'.[46]

A HIDDEN CONNECTION?

Excavations have proved that the Temple of Horus at Edfu, and all of ancient Egypt's other most notable temples and pyramids, were built on sites that were believed to be in some way 'divinely consecrated'. It is agreed, too, that many of these sites were in constant use, and were continuously redeveloped, over thousands of years.

All the major temples of Angkor also show similar traces of having been built directly on top of earlier structures which may in turn have been built on the sites of earlier structures still. If it is not a coincidence, therefore, then the possibility cannot be entirely ruled out that the extraordinary correlation between the Naga

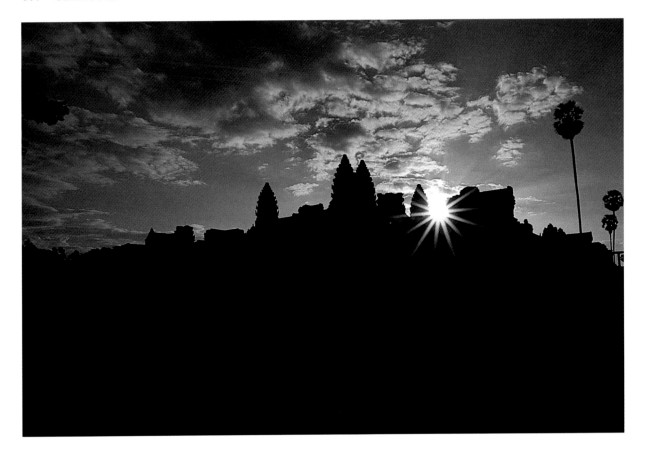

The towers of Angkor Wat playing with the sun at dawn.

temples of Angkor and the stars of the serpent constellation of Draco as they looked at dawn on the spring equinox in 10,500 BC could result from an actual ground plan and original 'mounds' established in Cambodia on that remote date.

This hypothesis does not conflict with the indisputable historical evidence that the temples in the form that we see them today were put up between the ninth and the thirteenth centuries AD by known Khmer monarchs. All that it requires us to do is to accept the already well-established fact that the *sites* of these temples appear to have been 'sacred' long before the temples themselves were built. Perhaps they were held sacred in this way, their locations known, plotted according to 'ancient delineations', not just for hundreds but for many thousands of years?

A slight modification of the hypothesis dispenses with the need for the site-plan to have been physically established in 10,500 BC – something that is anyway hard to prove or disprove. The modification recognizes that the pattern of the temples could have been established without 'primeval mounds' to guide the builders. A detailed sky-map of the constellation of Draco at the meridian at dawn on the spring equinox in 10,500 BC would have been sufficient. If such a sky-map could have been preserved and passed down, either encoded in ritual, or in myth, or in the form of a physical chart or diagram, then in theory it could have been copied on to the ground by the builders of Angkor almost twelve thousand years later.

A further modification refines the hypothesis, calling for the existence neither of physical 'mounds' from 10,500 BC, nor even of plans or charts drawn up in that epoch. All that it requires us to accept is that the builders of Angkor must have been the masters of a precise astronomical science, and this is something that we already know to be true. It really does not take a great leap of the imagination to understand that such brilliant site-surveyors – who aligned their vast temples as accurately to true north as the Great Pyramid of Giza, and who incorporated scientific astronomical observations into almost every key dimension and sightline – are very likely to have had a complete grasp of the phenomenon of precession and of its astronomical effects.

If they had possessed that knowledge, if they had calculated precession, if they had been so 'equipped', then there is little doubt that they could have summoned up mental images of the positions of the stars in the skies in former epochs – the process that the ancient Egyptians referred to as 'going down to any sky'. Such 'mental mandalas' could then have been copied on to the ground in the form of great and enduring works of architecture that would survive, long after their builders were dead, and that would possess the latent capacity to awaken the minds of all who would encounter them in the centuries and millennia to come.

So the real questions at Angkor are not so much a matter of the absolute dates of the construction of the various temples, or even of the many substructures that are known to lie beneath them, but rather:

1   *Why* does the overall site-plan focus so insistently and specifically on the pattern of stars in the sky region surrounding the constellation of Draco as it looked at dawn on the spring equinox in 10,500 BC?
2   How can we explain the fact that this same precise date is signalled by the three great Pyramids and the Great Sphinx of Giza – monuments that are not thought to be linked in any way to the temples of Angkor?
3   Is it not amazing that all three groups of monuments use the same astro-architectural technique to draw attention to that date, i.e. by modelling a prominent constellation that was present at one of the cardinal points of the sky on the spring equinox in 10,500 BC (Draco, to the north, in the case of Angkor; Leo, to the east, in the case of the Great Sphinx; Orion, to the south, in the case of the Pyramids)?
4   Could there be some sort of hidden connection?

THE GAME MASTERS

Inside the Great Pyramid of Egypt there are four narrow, steeply angled shafts, two of which run due north and two due south. We have seen that all these shafts were targeted on significant stars in the epoch of 2500 BC – Sirius and Al Nitak (the lowest of the three stars of Orion's belt) in the case of the southern shafts, Kochab and Thuban in the case of the northern shafts.

Thuban is the tail star in the constellation of Draco and has its counterpart – or 'double' or 'copy' – on the ground in the Draco temples of Angkor. This is the temple of Bentei Samre. Kochab is part of the neighbouring constellation Ursa Minor. Nevertheless, it too is copied on the ground at Angkor by the nearby temple of Ta Sohm.

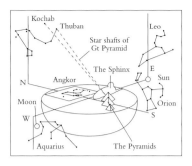

A three-dimensional representation of the position of the constellations at the cardinal points and their terrestrial counterparts (not to scale) at sunrise on the spring equinox 10,500 BC.

Both of the stars targeted by the northern shafts of the Great Pyramid of Egypt therefore turn out to be represented by prominent temples on the ground at Angkor. Perhaps this is just another accident, just another 'coincidence'. There have been times, however, as this investigation has continued, when we have had the eerie sense of stumbling across the fragments of a strange and shadowy archaic master game – a game on a planetary scale that ran for thousands of years and that appears to have been played out in four principal dimensions:

First Dimension: 'Above' – stars in the sky;
Second Dimension: 'Below' – monuments on the ground, scattered around the world like the pieces of an immense jigsaw puzzle, linked to one another through occult astronomical clues;
Third Dimension: 'Time' – measured by the slow cycle of precession, the principal means by which the astronomical signposts pointing from one monument to the next were hidden from the uninitiated;
Fourth Dimension: 'Spirit' – the point of it all, the quest for immortality.

The game – if game it was – has the feel of a beautifully *self-referential* system, one with interlocking and mutually interconnected features that bear all the hallmarks of an intelligent and highly organized design. Thus, the three pyramids of Giza are not only laid out on the ground in the pattern of the three stars of Orion's belt but one of them, the Great Pyramid, also has a narrow shaft oriented due south – which targets its counterpart star in Orion's belt – and two shafts oriented due north, which target the stars Kochab and Thuban. These stars in turn are represented on the ground by two of the temples of Angkor as part of an overall pattern of monuments that depicts the constellation of Draco and some prominent neighbouring stars.

When the dimension of time is added a new level of the game can be accessed and the player discovers that the sky–ground diagrams of Draco and Angkor on the one hand, and Orion and the Pyramids on the other, both lock perfectly, facing each other across the meridian, at sunrise on the spring equinox in 10,500 BC. The player will also realize that *at precisely the same moment* the lion-bodied constellation of Leo has risen due east in line with the equinoctial gaze of the lion-bodied Sphinx. Depth of effect is added to the game by having fully evolved Orion/Osiris myths and wonderful traditions of lion gods and goddesses embedded in the cultural milieu of the Egyptian monuments – linked in all cases to the cycles of time – and wonderful traditions of Naga serpents embedded in the cultural milieu of the Naga temples of Angkor, again linked to great cycles of time and to the creation and destruction of world ages.

Most mysteriously of all, there is a clear sense in which the myths hint at a further series of linkages, pitched at a still higher level of the game – the level of Spirit – between the great 'world-age' theme, which can only be properly grasped by those equipped with precessional knowledge, and the dark and fundamental question of the immortality of the human soul.[47] This, as the Indian texts solemnly put it, is 'the great mystery of what cometh after death [which] even the gods of old knew not ... a matter hard to be learnt'.[48]

What better candidates could there be for the masters of a game with immortality as its goal than the Followers of Horus, the Shemsu Hor, the wielders of magic, the counters of the stars – who are said in the ancient texts to have come to Egypt in the First Time? In the Egyptian language 'Ankh-Hor' really does mean 'the god Horus lives'. In any quest for a link between Angkor's Draco temples and the monuments of Giza it is therefore surely relevant that the myth of Horus, the defining tradition of the Shemsu Hor, makes what Richard Hinkley Allen's authoritative *Catalogue of Star Names* calls 'undoubted reference' to stars in Draco.[49]

In the master game, links between stars and temples are veiled behind precessional changes that can be decoded by those equipped to 'go down to any sky'. Such adepts would have known that the pulse of the precessional cycle beats at the rate of one degree every 72 years and would undoubtedly have been drawn towards temples in the form of Draco located at the geodetic distance of 72 degrees of longitude east of the monumental figures of Orion and Leo at Giza.

But the dimension of time still veils much: 10,500 BC is the astronomical dating of the ground plan of the Pyramids and the Sphinx; 2500 BC is the astronomical dating of the alignments of the Great Pyramid's shafts (supported by undisputed archaeological evidence of intense activity at Giza at around 2500 BC); 10,500 BC is the astronomical dating of the ground plan of the Naga temples of Angkor; and finally AD 1150 is the date of the completion of Angkor Wat, with undisputed archaeological evidence that the entire complex of monuments at Angkor was built over slightly more than four centuries between AD 802 and AD 1220.

What powerful common source of high knowledge and what shared spiritual idea, descending through what underground stream, could have been sufficiently global in manifestation, sufficiently ancient, and sufficiently sustained, to have made such a deep impact on the culture of Egypt at around 2500 BC and, *3500 years later*, on the culture of the Khmers in Cambodia between the ninth and the thirteenth centuries AD?

And why in both cases was the focus of the monuments on precisely the same far-off moment in astronomical time – corresponding with the date of 10,500 BC in the modern calendar?

CHAPTER 10

ELIMINATING THE IMPOSSIBLE

The pyramid-mountain of Phnom Bakheng, its east face flanked by twin lions.

OPPOSITE: *Ground and sky at Angkor. The architecture of the temples seems designed to draw the eyes upwards and encourage contemplation of the heavens.*

WE DROVE DUE NORTH from the western entrance of Angkor Wat for a distance of about one and a half kilometres before coming, on our left, to the pyramid-mountain known as Phnom Bakheng, 67 metres tall, built over a natural rocky knoll.[1] In much the same way, though on a grander scale, the Great Pyramid of Egypt is also built on top of a huge mound of natural bedrock.

We climbed Phnom Bakheng from the east, where the contours of the original pyramid and sacred mound had melted and merged together through centuries of erosion by rain and wind and human traffic. The way up was steep and difficult with the red laterite scree damp and slippery underfoot. At the top was a stone temple in the form of a ziggurat or step-pyramid, 13 metres high with a square base measuring 76 metres on each side, rising in five tiers to a central sanctuary.[2]

We climbed the stairway on the eastern side of the base. Like its counterparts to north, south and west it has a sheer angle estimated to be a little over 70 degrees.[3] Whether or not the precise figure turns out to be 72 degrees, as we suspect it might if accurately measured, it is a fact that the central sanctuary of Phnom Bakheng is surrounded by exactly 108 towers.[4] The number 108, one of the most sacred in Hindu and Buddhist cosmologies, is the sum of 72 plus 36 (i.e. 72 plus half of 72). As such it is a key component in the sequence of numbers linked to the earth's axial precession which causes apparent alterations in the positions of stars, over a great cycle of 25,920 years, at the rate of one degree every 72 years. A specific connection to such a cycle may even be what is meant by an obscure inscription found at the site and hitherto largely overlooked by archaeologists. The inscription was written by Phnom Bakheng's builder, King Yasovarman I (AD 889–900) and tells us that the purpose of the temple is to symbolize 'in its stones the celestial evolutions of the stars'.[5]

From the top of Phnom Bakheng a commanding panorama stretched all around, showing us in the distance, to the north-east, the Kulen Hills with the temple of Phnom Kulen, nearer by but on the same alignment the mound of Phnom Bok with its surmounting temple, and finally the temple and mound of Phnom Krom, also on the same diagonal alignment but south of us and a little to our west.

Angkor Wat: reflections in the sacred lake.

To our north, invisible in the jungle, was the entire vast complex of Angkor Thom. To our east, soaring above the treeline, was the fairytale palace of Angkor Wat. We could make out all four of the towers surrounding its central pyramid and sanctuary, replicating the four celestial mountains (one of which is Mandera) surrounding the central peak of Mount Meru in Hindu cosmology.

Angkor Wat's dominant feature is its long and massive east–west axis which locks it uncompromisingly to sunrise and sunset on the equinoxes. In addition, the temple is cleverly anchored to ground and sky by markers for other key astronomical moments of the year. For example, reports *Science*:

> It is interesting to note that there are two solstitial alignments from the western entrance gate of Angkor Wat. These two alignments [added to the equinoctial alignment already established] mean that the entire solar year was divided into its four major sections by alignments from just inside the entrance of Angkor Wat. From this western vantage point the sun rises over Phnom Bok [17.4 kilometres to the north-east] on the day of the summer solstice .... The western entrance gate of the temple also has a winter solstice alignment with the temple of Prasat Kuk Bangro, 5.5 kilometres to the south-east.[6]

A similar interlocking of sky and ground is achieved at Giza by the Great Sphinx and the three great Pyramids. While the Sphinx is targeted due east to the equinox sunrise, the causeways of the first and second pyramids (targeted respectively 14 degrees north of due east and 14 degrees south of due east) align with the sun's position on the horizon exactly one month before and one month after the summer solstice – in the case of the northern causeway – and exactly one month before and one month after the winter solstice in the case of the southern causeway.[7]

MYSTERIOUS ORIGINS

The origins of Giza are shrouded in mystery but there is no doubt that major developments took place there at around 2500 BC and that ancient Egyptian god-kings – notably the Fourth Dynasty Pharaohs Khufu, Khafre and Menkaure – were intimately connected with these developments.

In like fashion, the origins of Angkor are shrouded in mystery but there is no doubt that major developments took place there between the ninth and the thirteenth centuries AD and that Khmer Devarajas (the word means 'god-kings') such as Jayavarman I, Yasovarman I, Suryavarman II, and Jayavarman VII, were intimately connected with these developments.

Indeed, historians can be precise about the start of the great episode of temple-building at Angkor. On the basis of masses of good archaeological and inscription evidence, they are able to tell us with a fair degree of certainty that it began in AD 802 at the instigation of Jayavarman II, who first underwent an enigmatic, unprecedented and so far unexplained initiation ceremony and then declared himself a 'universal lord'. They also admit that very little is known about the centuries preceding Jayavarman's reign, that very few stone temples at all – and none of high precision and quality – were built prior to the ninth century, and that there is therefore no real evidence of any kind of evolutionary process in architecture leading up to the Angkor monuments.

Jayavarman's own origins are also extremely obscure and historians wrangle over whether or not he was related by blood to the previous ruling dynasty ('a great grand-nephew through the female line',[8] etc.). A matter about which there is no dispute, however, is that later inscriptions style him as having been descended from a 'perfectly pure race of kings'[9] – an epithet that was frequently applied in ancient Egypt to the Followers of Horus, who were thought of as 'superior beings who produced the race of Pharaohs'.[10] We are also told that Jayavarman had become king in order to 'save' his people.[11] This phrase is part of the standard language of the Osirian rebirth cult.

Of course there can be no suggestion of any kind of *direct* influence. The worship of Osiris had been dead for centuries before Angkor rose to prominence and the very little of ancient Egypt still left intact at the end of the Roman occupation in AD 395 had long since been washed away after Egypt's conversion to Islam at around AD 650 – still more than 150 years before Jayavarman II took the throne in far-off Cambodia.[12]

Yet the impossibility of a direct influence does not rid us of the suspicion that some form of indirect 'underground' connection could exist between the stellar temples and pyramids of Angkor and the stellar temples and pyramids of Giza. And obviously we wonder whether this same 'connection' might have been at work in ancient Mexico too. We are not alone in such speculation. As long ago as 1955 the great Mayanist Michael D. Coe commented on the 'many puzzling similarities' between the Khmer empire and the Classic Maya.[13]

What should we conclude about this mounting roster of similarities? One explanation could still be coincidence – although coincidence in this case, with so many interlinked and consistent complexities, is statistically every bit as unlikely as any kind of direct influence. Another explanation, as we hinted at the end of the last chapter, could be an undetected 'third-party' influence, very discreet, very secretive and of very great antiquity. Such an influence – perhaps the long-lived and highly motivated society that referred to its initiates in ancient Egypt as the 'Followers of Horus' – does seem immensely improbable. Nevertheless, as Sherlock Holmes famously reminded Watson in *The Sign of Four*: 'when you have eliminated the impossible, whatever remains, however improbable, must be the truth'.[14]

SHIFTING CAPITALS

How did Jayavarman II become the ruler of Cambodia – and why a 'god-king'?

All that the inscriptions tell us is that this inscrutable figure came from across the sea by boat, having previously spent several years in a far-off land at the court of a monarch named the 'King of the Mountain'.[15] Nothing is known about what he was doing there and although there is speculation that the land was Java,[16] the matter is by no means resolved.[17] He reached the shores of Cambodia in AD 800, finding it in a lawless and dangerous state, collapsed into dark and violent anarchy.[18] According to George Coedes: 'the young prince had to reconquer the kingdom before he could exert his rights or lay claim to the Cambodian throne'.[19]

Jayavarman set up his first capital in a city, then already very ancient, that is referred to in the inscriptions as Indrapura. Its exact location is not known[20] yet it seems to have been a place of great religious learning and authority. Here the king voluntarily put himself under the tutelage of a spiritual teacher named Sivakaivalaya, described in the inscriptions as 'a great Brahman scholar', a man who was 'to follow him in all his moves', according to Coedes, 'and who was to become the first officiating priest of a new cult, the cult of the god-king'.[21]

From the beginning, Jayavarman behaved like a man with a mission, determined to fulfil specific objectives within a specific time-frame. And from the beginning these objectives involved the sacred domain of Angkor, 72 degrees of longitude east of the Pyramids of Giza.

The temple of Bakong at Roluos, which, together with the temples of Prah Ko and Prei Monli, forms the pattern of three stars in the Corona Borealis as they appeared at dawn on the spring equinox in 10,500 BC. Bakong is built on the foundations of an 'artificial mountain' of much earlier origin than the temple itself.

The temples of Angkor, reflecting the stars of Draco and other nearby constellations at sunrise on the spring equinox 10,500 BC.

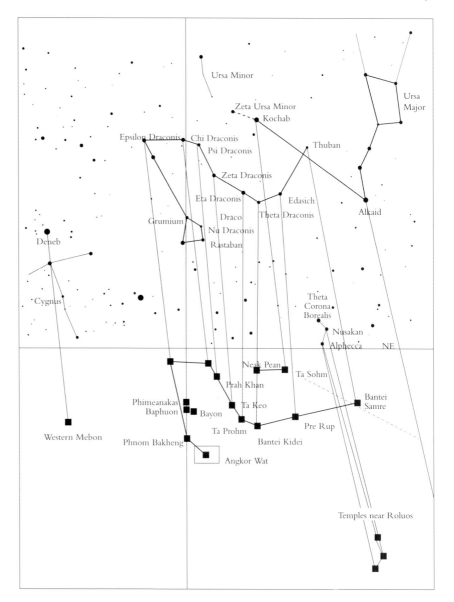

Leaving Indrapura after a stay that could not have been longer than a few months, Jayavarman and Sivakaivalaya first marched north with their armies to the plains on which the temples of Angkor now stand. There they established a city with the melodious name of Hariharalaya at the site that is shown on modern maps as Roluos.

Is it an accident that three temples at Roluos, two of them prominent and one obscure, are laid out on the ground in the pattern of three stars in the Corona Borealis (see diagram)? The Corona, which neighbours Draco, would not have been visible from Angkor in the tenth and eleventh centuries AD when the Roluos temples were built but would, through the effects of precession, have been visible just above the horizon at dawn on the spring equinox in 10,500 BC as Draco reached the meridian.[22] One might say that it was an accident if it were not for the fact that between the ninth and thirteenth centuries AD all the other temples were gradually put into place, mimicking the pattern of the constellation

Temple of Ta Keo, Angkor, built by four Khmer monarchs over the period AD 968 to 1050 and providing the terrestrial counterpart to the star Zeta Draconis in the constellation of Draco. Each of the Khmer rulers seems to have contributed what he could to the completion of the Draco/Angkor sky–ground scheme, some adding merely a temple or part of a temple, others – notably Suryavarman II and Jayavarman VII – completing whole sequences of massive monuments within relatively short periods of time.

of Draco at the meridian at that same precise moment in 10,500 BC – and not only Draco but also parts of other constellations in the same general sky-region such as the stars Zeta Ursa Minor and Kochab in the Little Bear, Alkaid in the Great Bear, and Deneb in the constellation of Cygnus, the swan. There is a sense about the whole enterprise of an ordered plan, of a grand design being methodically completed, making use of natural features of the landscape where possible, incorporating precise astronomical alignments, and reinforcing itself with pervasive Draco symbolism in the form of serpents and myths about serpents.

THE INITIATE AND THE MOUNTAIN

Leaving Hariharalaya, Jayavarman built a new capital at a place called Amarendrapura, thought by archaeologists to have been located to the west of the site that would later become Angkor Thom.[23] Then he moved again, this time 40 kilometres to the east, to the highlands of Kulen, where he established the city of Mahendraparavata on Phnom Kulen, the wooded sandstone hill that overlooks the plain of Angkor.[24]

What happpened next was extremely strange. According to the royal inscriptions: 'His majesty went to rule at Mahendraparavata.' The sage Sivakaivalaya went with him. Later they invited a Brahman scholar more learned even than Sivakaivalaya, a scholar 'well versed in magic' who 'came to establish a ritual . . . that there might be only one king ruling the country. This Brahman recited the texts from beginning to end, to teach them to [Sivakaivalaya], and he instructed him how to institute the ritual of the god-king.'[25]

The Brahmans of Cambodia were wise men, described by the archaeologist Bernard Groslier as 'initiates who were descended from Indian settlers or had studied in India'.[26] These sages:

> kept possession of sacred books which they alone were competent to inter-
> pret, composed the inscriptions and guaranteed the accuracy of astronomi-

cal calculations. Indeed, it sometimes happened that political power passed from the hands of a too weak or youthful monarch to *this veritable priestly oligarchy* ... Until its extinction the dynasty of Angkor surrounded itself with these Brahmans.[27]

It is difficult to ignore the obvious comparison with the oligarchy of astronomer-priests at Heliopolis in Egypt and their role behind the scenes as king-makers who claimed to carry down from the past a knowledge of the divine origins of the Pharaonic line, and of the true purpose for which the civilization of ancient Egypt was conceived. They, too, we know, were involved in the anointing and initiation of monarchs. They too could wield secular power from time to time, they too composed inscriptions and observed the stars ...

The ruling symbol of Heliopolis was the phallic combination of the original Innu pillar and the pyramidial tip of the Benben stone. Precisely the same symbolism dominated the ritual for the enthronement of the god-king in Khmer culture as well. As George Coedes explains in his authoritative study of Angkor:

The 'essence of royalty', or, as it is called in certain texts, 'the essence of self' of the king was supposed to reside in a *linga*, the [phallic] symbol of the creative power of Shiva, ensconced in a pyramid in the centre of the royal city which, in turn, was situated at the axis of the world. This miraculous *linga*, a sort of palladium of the kingdom, was generally considered to have been obtained from Shiva with the help of a Brahman, who presented it to the original king of the dynasty. The communion between the king and the god took place on a sacred mountain, whether natural or artificial ... [Accordingly] Jayavarman II had to receive on a mountain top, from a Brahman, the miraculous *linga* in which the imperial power of the Khmer kings would reside from that time on. This is why he moved to Phnom Kulen ...[28]

A PLAN LONG ESTABLISHED

But Jayavarman did not remain long on the heights of Kulen. After his initiation, the inscriptions say, 'the King went back to rule at Hariharalaya', taking with him the sacred *linga*, a 'miraculous' phallic obelisk, which was then installed in the holy of holies of a new pyramid-temple.[29]

This Benben-like object at the summit of the great central pyramid was called by the name 'Kamrateng Jagat', 'Lord of the Universe', a title often applied to the god-kings themselves.[30] It was transported from temple to temple during the long construction programme at Angkor, even spending some time in the central sanctuary of Phnom Bakheng surrounded by the 108 towers among which we now stood looking down on the plains of Angkor.[31]

It is instructive to trace out the route of Jayavarman's migrations during the half-century between his arrival in Cambodia around AD 800 and his death around AD 850. There is a pattern to it. After first arriving at Indrapura, as George Coedes observes: 'Like a bird of prey soaring over the land, he moved from Hariharalaya, to Amarendrapura, to Phnom Kulen, pivoting in a circle around the

future Angkor.'[32] Finally he turned back to Hariharalaya to end his days, thus completing a circuit.

The orthodox view of these perambulations is that the first great god-king of the Khmer dynasty:

> seemed to be searching out the future location of his capital, near enough to the supply of fish in the Grand Lac, but beyond the reach of the annual floods, convenient to the sandstone quarry of Phnom Kulen, and sufficiently close to the passes giving access to the plateau of Korat and the valley of Menam.[33]

We do not think that Jayavarman was necessarily motivated by such prosaic nuts-and-bolts reasoning, or overly concerned about fish. We suggest it is at least equally probable that this noble god-king, advised by his Brahmans – a shadowy group of astronomer-priests – was pursuing a geodetic and astronomical quest when he circled around Angkor. Perhaps he was trying to feel out the contours of sacred mounds, hoary with age, that may, in remote and far-off times, have been raised up over the plain.[34] Perhaps, as he slowly wheeled around the site, he was establishing and recording its bearings, confirming its longitude in relation to other key locations (such as Giza?), measuring its latitude, perhaps even renewing alignments and orientations that he had retrieved from crumbling documents written in 'ancient delineations' passed down from former ages.

Indeed, not just Jayavarman II, but many of the great Khmer god-kings who followed after him – monarchs of the stature and style of some of the most powerful Egyptian Pharaohs – appear to have been *following a plan* when they created their temples at Angkor, implementing specifications that had long ago been established.

But established by whom, how long ago exactly, and for what purpose?

We left the ruins on the summit of Phnom Bakheng and clambered down the side of the pyramid-mountain to the level of the plains. Ahead of us, due north, a long sweep of road ran between tall trees leading to the southern entrance of Angkor Thom, the huge enclosure and complex of funerary temples and pyramids which seems to stand at the beating heart of this potent sacred domain, pregnant with hidden meaning – every bit as vast and every bit as mysterious as Egypt's Giza necropolis.

COMPLEX SIMILARITIES

The suspicion that there must be some kind of hidden connection between Angkor and ancient Egypt is not new. On the contrary, virtually every traveller who has visited Cambodia during the last century or so has felt compelled to point out that Angkor has something uncannily 'Egyptian' about it, that certain giant works of sculpture resemble the face of the Sphinx or the Colossi of Abu Simbel, that there are of course pyramids everywhere and that the scale of the enterprise is reminiscent of the scale of the construction of the great Pyramids of Giza.

Comparisons of this kind, though plentiful, are generally offered to the reader

Though the surroundings are very different, it has been claimed that certain giant works of sculpture at Angkor resemble the face of the Sphinx or the Colossi of Abu Simbel.

as mere 'impressions' of no scientific value. They have never been seriously pursued because scholars are sure that there is no way the culture of the Khmers could have been 'connected' to the culture of ancient Egypt. All similarities are judged to be coincidences and as such, although quaint, are of no real interest.

This is a reasonable position, given the huge physical distance between Egypt and Cambodia and the historical fact of ancient Egypt's utter extinction long before the emergence of Angkor. In our view, however, the range and sheer extent of the similarities is such that 'coincidence' can no longer be regarded as a safe explanation.

The step-pyramid at Saqqara, Egypt, of the Third Dynasty Pharaoh Zoser, a structure supposedly designed by Imhotep, the divine architect. In the foreground a row of Uraeus serpents – cobras with their hoods extended.

For example, there is an old tradition in Cambodia that the temples and pyramids of Angkor were built by Visvakarma, the architect of the gods, who is said to have taught the art of architecture to men.[35] Imhotep, supposedly the first architect of the pyramid form in ancient Egypt, was said to have 'invented the art of building in dressed stone' – and later became a god.[36]

Likewise, we have seen already how both Egypt and Angkor venerated the serpent. In both cases it was the hooded cobra that was selected as the archetype, in both cases it could be depicted in art as a half-human, half-serpent figure[37] or as fully serpentine. In both cases it was usually shown rearing with its hood extended (taking this form as the *uraeus* worn by the Pharaoh on his crown for example).

In both Egypt and Cambodia the serpent could be a denizen of either sky or ground, frequently terrestrial (or even subterranean) but also equally frequently shown navigating the celestial regions. This ambiguity is expressed in the *Book of What is in the Duat* and, in ancient India, the *Yajurveda* talks of 'snakes whichsoever move along the earth, which are in sky and in heaven'.[38]

In both Egypt and ancient Cambodia, too, the serpent was frequently employed as an image of eternal life and of the cycles of the universe. According to the fifth-century Alexandrian writer Horapollo: 'When the Egyptians would represent the universe they delineate a serpent bespeckled with variegated scales, devouring its own tail, the scales imitating the stars of the universe.'[39] Similarly, on the small golden shrine of Tutankhamun an encircling 'ouroboros' serpent is shown, intended to represent the powers of resurrection and renewal:[40]

> It was thought that the regeneration of the sun god was re-enacted every night within its body. While the ouroboros conveyed a sense of endless spatial length encompassing the universe, another snake called 'Metwi' ('Double Cord') served as a manifestation of the infinity of time.[41]

We are reminded of the great Naga serpent Sesha ('Remainder'), who is also called Ananta ('Endlessness'), winding around the universe. And we are reminded

Serpent amongst the stars, causeway of Unas, Fifth Dynasty, Saqqara, Egypt. The ancient Egyptians envisaged a number of serpents with cosmic functions similar to those of the Nagas.

of how, to this day, Shivalinga, the Indian and Khmer equivalent of the ancient Egyptian Benben, are frequently carved with an encircling serpent.[42]

SPIRITUAL BODY

In ancient Egypt it was believed, and devoutly wished, that the souls of deceased Pharaohs would ascend to the skies: 'O King, you are this great star, the companion of Orion, who traverses the sky with Orion, who navigates the Duat with Osiris.'[43]

In exactly the same way it was believed in Cambodia that when a god-king of Angkor died his soul 'went to the skies'.[44] Traditions likewise asserted that after Visvakarma had built Angkor Wat, 'the gods departed to the land of eternal happiness where they look down from the skies'.[45] In addition, just like the temple of Horus at Edfu in Upper Egypt, Angkor Wat was said to be a 'copy' of an earlier temple which itself was a copy of a cosmic original, 'the plans [of which] had been drawn up by the gods of the Tushita Heaven themselves'.[46]

According to the doctrines of the ancient Egyptians, a successful navigation of the perils of the Duat can only be achieved by 'equipped' spirits who have used the opportunity of physical life to master hidden knowledge of heavenly cycles and to cultivate self-discipline and spiritual insight. They are the ones who are said in the Pyramid Texts to 'seize the sky and take possession of the horizon'.[47] They are the ones, after many rebirths, who may hope to attain practical immortality – the 'life of millions of years'.

The god-kings of the Khmers also sought immortality through knowledge of cosmic cycles. This was why they encoded the minutes and hours and days and months of the 'Great Year' of precession in so many of the significant dimensions of the Angkor monuments, transforming these immense structures into schools of initiation with secrets that could be recovered through diligent quest.

An habitual mistake that historians make when they assess ancient Egypt and ancient Cambodia is to assume that, because the works of architecture are vast and ambitious, the kings who built them must necessarily have been megalomaniacs.[48] In a sense it is interesting that exactly the same charge – of megalomania – is repeatedly levelled against the Pharaohs and the Khmer monarchs, since this does represent scholarly unanimity on at least one apparent similarity between the two cultures. But what is surprising is that the authorities have considered no other explanations for the phenomenal architectural programmes undertaken in Egypt and Cambodia so far apart in time. In particular, they have failed to consider the possibility that the great builders of the Pyramids and the Sphinx, Angkor Wat and Angkor Thom might *not* have been driven entirely by egoism and the desire for self-aggrandizement but rather by a kind of altruism – perhaps even by a desire to initiate the entire human race into the gnostic system of enlightenment that they themselves practised.

Yet this possibility merits serious consideration. After all, the great pyramids of Egypt bear no inscriptions linking them to the Pharaohs who are supposed to have built them – hardly a sign of megalomania! Angkor, by contrast, is filled with inscriptions but these do not support the 'self-aggrandizing' theory either. As we

Statue of King Zoser in a chapel by the north face of his step-pyramid at Saqqara. The ancient Egyptians regarded statues as vessels for the life-force of the deceased.

The name of Khufu, the supposed builder of the Great Pyramid, is not inscribed anywhere inside that monument, and appears only in graffiti, so-called quarry marks, sealed away out of sight.

saw earlier, Jayavarman VII, who ruled from AD 1181 to 1219, stated quite explicitly on a stela that his extensive building programme was undertaken, 'full of deep sympathy for the good of the world', so as to 'bestow on men the ambrosia of remedies to win them immortality ... By virtue of these good works would that I might rescue all those who are struggling in the ocean of existence.'[49]

Another monarch, King Rajendravarman, who was perhaps more modest, said that he had built his temples out of 'a passion for *dharma*' (the 'law', 'balance', 'justice', 'right order', etc.,) a concept very similar to the ancient Egyptian idea of *maat*, cosmic justice. 'This supplication', he added rather enigmatically – referring to his building programme – 'is for the immortality one should *try* to achieve.'[50]

The delicate and beautiful temple of Neak Pean, with its islands and pools, was built by Jayavarman VII, who hoped that it would serve as a boat in which those souls who might come into contact with it could 'cross the ocean of existences'.[51]

On a stela from another pyramid-temple at Angkor the King finally sends the lifeboat out to rescue himself:

> Through the merit of this good work may I pass from existence to existence. May those who protect this my work, be they kindred, friends or strangers, be taken to the abode of the gods; at each rebirth may they be granted a smiling countenance ...[52]

In the Fifth and Sixth Dynasty pyramids at Saqqara in Egypt, just 10 kilometres south of Giza, there are remarkably similar texts dating back at least to the twenty-third century BC and associated with the great wisdom school at Heliopolis. Here the all-god Atum, the ancient Egyptian counterpart of Vishnu/Brahma/Shiva, is invoked to set his arms 'about the King, about this construction, and about this pyramid ... that the King's essence may be in it enduring for ever ... Protect this construction of his from all the gods and from all the dead and prevent anything happening evilly against it for ever.'[53]

A few lines further on in the same passage we discover that the deceased Pharaoh is in some mysterious manner identified directly with his Pyramid, and with the god Osiris, as though man and stone have become fused into a single spiritual body – a body of glory in which 'this King is Osiris, this pyramid of the King is Osiris, this construction of his is Osiris'.[54]

Such ideas, which are extremely curious, appear already mature and fully formed at the *very beginning* of ancient Egyptian civilization almost 5000 years ago. But even more curious is the fact that exactly the same ideas surface magically out of the ether some 4000 years later in Cambodia. According to Paul Mus and George Coedes, the funerary pyramid-temple was seen in Angkor 'not so much as a shelter for the dead as *a kind of new architectural body*, substituted for the mortal remains of a deceased "cosmic man" where his magic soul will live on'.[55]

LIVING IMAGES

The unusual notion of the pyramid as a new body for the deceased was not only elaborated by the god-kings of ancient Egypt and Cambodia but was also

ABOVE AND OPPOSITE: *Statues in Angkor were believed to be capable of containing the living essence of deceased individuals. The cult of statues, called 'living images', is still active in Angkor to this day.*

The name 'Sphinx' is derived from the ancient Egyptian words sheshep ankh, *meaning 'living image'.*

extended by both, in exactly the same way, to include the functions and the cult of statues.

In the funerary rituals of Angkor a statue of the deceased king was referred to as his 'body of glory'[56] and was believed to be animated with his spiritual essence[57] – 'a magical projection into the future of the king's destiny'.[58] Similarly, in ancient Egypt, statues were accorded enormous importance as vessels for the life-force of the deceased – with which they were of course believed to be animated.[59] In both cultures statues were referred to as 'living images' (*sheshep ankh*).[60]

A greater level of complexity often underlies such superificial similarities. For example, in both Egypt and Angkor it was believed that a ritual must be conducted in order to make a statue 'alive'.

In Angkor this ritual was called the 'opening of the eyes', a ceremony involving purification, censing and anointing in which a number of instruments were used. The ritual culminated with a symbolic 'opening' of the statue's eyes by pricking them with an iron pin.[61] Only then was the statue considered to be 'imbued with the vital principle, the divine essence of the departed king ... acting as a bridge between the immortal realm and this'.[62]

In ancient Egypt a similar ritual was used to bring statues to life. There it was called the 'opening of the mouth and the eyes'.[63] It involved purification, censing and anointing of the statue and the use of a number of instruments, some of meteoritic iron, some of stone.[64] The ritual culminated with the mouth of the statue being touched – 'opened' – with an instrument known as the *peseshkaf*.[65]

Frequently the two eyes were also 'touched with special instruments'.[66] Thereafter the image became animated and could function as a 'framework' for the soul's immortality, as secure 'as the body had been for the short span of mortal existence'.[67]

How could Egyptian and Cambodian rituals for bringing statues to life end up being so similar – indeed with almost the same names – unless some connection exists between the two? We agree that any form of direct connection is impossible and must be eliminated from our inquiries. But if coincidence, too, is 'impossible' then what is left?

THE LINEAMENTS OF THE SOUL

Ancient Egyptian conceptions of the soul, which, like so much else, were already fully developed at the start of the historical period, reveal an amazingly sophisticated system of ideas which divided the immortal essence of the individual into at least four principal manifestations or entities. These were, respectively:

1 The *ka*, the 'double' or 'twin' – the guardian-angel and spirit-guide of the deceased – which was 'independent of the man and could go and dwell in any statue of him'.[68] According to James Henry Breasted, the *ka* 'was a kind of superior genius intended especially to guide the fortunes of the individual in the hereafter where every Egyptian who died found his *ka* waiting for him'.[69]

2 The *ba*, or 'heart-soul', was 'in some way connected with the *ka*' but existed as a *person*, possessing powers that might enable it 'to subsist and survive in the life hereafter'.[70] The characteristic attribute of the *ba* soul was its gift of un fettered movement. The *ba* is frequently depicted in ancient Egyptian art as a swallow in flight, or as a human-headed swallow, a 'metaphor for freedom that cannot be bettered', as the Egyptologist Stephen Quirke has noticed.[71]

3 The *ab*, or heart, was closely associated with the soul. According to Sir E. A. Wallis Budge: 'The preservation of the heart of a man was held to be of the greatest importance, and in the Judgement it is the one member of the body which is singled out for special examination; here, however, the heart is regarded as having been the centre of the spiritual and thinking life . . .'[72]

4 'Justified' in the Judgement, the highest stage of the soul's evolution was the *sahu*, or spiritual body, within which dwelt the *akh*, or transfigured spirit, 'an ethereal being which under no circumstances could die' and would therefore possess the coveted 'life of millions of years'.[73] In the ancient Egyptian language, the word *akh* (found also in *akhet*, 'horizon') always conveys a concept of 'light', 'brightness', 'brilliance', or 'shining'.[74]

We think it very likely that the god-kings of Angkor had something similar to the *sahu* and the *akh* in mind when they declared that they wished, after death, to be 'clothed in a divine body' which would 'illuminate the spiritual glory' belonging to them.[75]

The goal of the initiate in the wisdom schools of ancient Egypt was to equip himself for eternity as a shining and transfigured *akh*. Before his final spiritualization could take place he knew that he must pass beyond death, endure the terrors of the Duat and emerge pure and justified from the weighing of the heart – the 'weighing of words' – in the Judgement Hall of Osiris. In order to do that, as we saw in Part II, more was required of him than merely moral and decent behaviour – necessary but not sufficient – which might gain for him a moral and decent rebirth but would not ensure the transfiguration of his spirit. What was required appears to have been knowledge, pure knowledge, cosmic knowledge, because – for reasons that are nowhere fully explained – it was understood that only this could lead him on the path to enlightenment.

Could this be the same 'high path to supreme enlightenment' that Jayavarman VII is said to have pursued with wholehearted dedication – 'the unique doctrine without obstacle to attain a comprehension of reality ... the law which the immortal honour in the three worlds'?[76]

Again and again the sense of a gnostic quest for immortality that is so pervasive in the ancient Egyptian texts surfaces in the Cambodian inscriptions as well. We read, for example, that Jayavarman's wife Jayarajadevi 'followed in the serene path of the sage',[77] while her older sister 'surpassed the wisdom of the philosophers in her knowledge' and was particularly commended for having extended 'to women whose great desire was for science ... the favours of the king, like a delicious nectar, in the form of knowledge'.[78]

As we have seen, the knowledge that was so highly prized at Angkor, and that the monarchs sought to encode in the dimensions and symbolism of their great temples, was believed to have the capacity to rescue souls from the 'ocean of existence'. The power of such knowledge – which could only be acquired by arduous searching – was that it could dispel Maya, the terrible illusion of the 'reality' of the material world. The ancient Egyptians and the Khmer kings alike therefore believed it to be the sacred responsibility of all sentient creatures that they should seek out the meaning and attempt to penetrate to the depths of the mystery of their existence. In the process some would discover the fundamental truth that it is only: 'When all desires that linger in the heart are driven forth, that mortal is made immortal ... When every knot of the heart is loose then doth he win immortal Being ...'[79]

So exactly as in ancient Egypt, the spiritualization of the soul came to be seen by the Khmer monarchs as a process of enlightenment – of the gradual stripping-away of layer after layer of illusion until the clockwork of the universe itself was exposed and the adept was equipped with perfect knowledge.

RAINBOW BRIDGE

We had reached the bridge over the wide, lichen-covered moat that led to the southern entrance-gate of Angkor Thom. Ahead of us, 108 titanic figures – forming balustrades in two parallel rows of 54 figures each – engaged in a tug-o'-war using a Naga serpent as their rope.

In ancient times this bridge was likened to the rainbows that were thought to arch between the worlds of gods and men.[80] It was said that those who crossed it and passed under the gateway had entered the realms of the celestials and it was believed that a great secret lay concealed here.

CHAPTER II

STILL POINT IN HEAVEN

LIKE THE ARGONAUTS' golden fleece, which was guarded by a dragon, a mystery of heaven lies wrapped in the coils of Draco. Fixed eternally in its place in the depths of the sky, its location is almost impossible to detect within one or many lifetimes. Modern astronomers call it the 'ecliptic north pole'.[1] It is entirely separate and just 23.5 degrees distant from the 'celestial north pole' – an extension into the heavens of the earth's axis which, as every schoolchild knows, is tilted away from the vertical by 23.5 degrees.

Imagine an extremely long pencil passed through the centre of the tilted earth, entering at the south pole, exiting at the north pole and then continuing onwards. The 'celestial north pole' is the 'mark' that such a pencil would make on the vault of the northern half of the sky. In our epoch it lies close to a convenient star – Polaris – which we call the 'north star' or 'pole star'. But because of the precessional wobble of the earth's axis the pencil point will not always coincide with Polaris. Instead, over the cycle of 25,920 years, it will gradually trace out a vast circle in the heavens, passing close to some stars and far away from others.

The 'ecliptic north pole' is the still, fixed point at the centre of this circle, the pole around which even the celestial pole revolves. Poised in space at an infinite distance above the exact centre of the earth's orbital plane, it is, in a sense, the pole of the gods. It may well have been what the ancient Egyptians had in mind when they spoke in their rebirth texts of a 'great mooring post' in heaven.[2] And its location, for all eternity, is in the heart of Draco, behind the hood of the Naga, between the stars Grumium and Chi Draconis.

It is reasonable to ask whether this important point in the sky has a counterpart on the ground amongst the 'Draco' temples of Angkor.

MODEL OF PRECESSION

We walked slowly across the bridge over the lichen-covered moat leading to Angkor Thom's southern entrance gate. On either side of us, sculpted in titanic style, stretched parallel rows of 54 *devas* and 54 *asuras*, leaning back, muscles bulging, hauling on the body of the Naga serpent Vasuki and thus symbolically 'churning the Milky Ocean'.

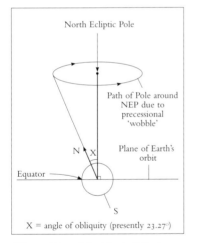

The position of the North Ecliptic Pole at the centre of the circle formed by the path of the North Celestial Pole.

OPPOSITE: *Giant faces of the Bodhisattva Lokesvara, gazing towards the cardinal directions, southern entrance gate, Angkor Thom. It is thought that the features are modelled on those of the Khmer god-king Jayavarman VII, the builder of Angkor Thom.*

Bridge to Angkor Thom with Naga balustrades and parallel rows of devas *and* asuras *symbolically 'churning the Milky Ocean'. Photograph is taken from the south looking north towards the southern entrance gate.*

The scene conveys exactly the same ideas and interprets the same myths as the bas reliefs of Angkor Wat's south-eastern gallery. Nevertheless, the effect in three dimensions is especially powerful, and thought-provoking. Pilgrims well-versed in the Milky Ocean story would immediately have felt curious about the whereabouts, in this architectural scheme, of a counterpart for Mount Mandera – the celestial peak supposedly used by the gods and demons as their 'churning stick'. For higher initiates equipped to 'go down to any sky' the numbers of statues might likewise have called to mind the occult cycle of precession which moves at the rate of one degree every 72 years (72 plus 36 – i.e. 72 plus half of 72 – equals 108, a number that divides equally into two 'teams' of 54).[3]

A pilgrim prepared to make the 16-kilometre circuit around the outer perimeter wall of Angkor Thom would discover that the southern gateway is only one of five and that each is built to the same design. This consists of a tall narrow corbelled vault surmounted by four serene and gigantic sculpted faces (intuitively described in 1861 by the French explorer Henri Mouhot as 'immense heads in the Egyptian style'[4]). The faces are oriented with high precision towards the four cardinal directions. Angkor Thom's east wall has two such gates, while its north, west and south walls have one gate each. Each gate is approached by a bridge and each, like the bridge that we were now crossing, is lined by parallel rows of 54 *devas* and 54 *asuras* – '108 per avenue . . . altogether 540 statues'[5] – a number that is again significant within the precessional sequence. Accordingly, Giorgio de Santillana and Hertha von Dechend conclude in *Hamlet's Mill* that 'the whole of Angkor' is 'a colossal model set up' of the precessional cycle.[6]

ENLIGHTENED ONES

Unlike Angkor Wat, which was dedicated to the Hindu god Vishnu, Angkor Thom is a Buddhist monument. The giant faces that surmount the gateways are not those of any of the gods in the Hindu pantheon but of a 'Bodhisattva' known in Cambodia as Lokesvara, and to Buddhists elsewhere as Avalokitesvara, 'the Lord who looks in every direction'.[7] The serene and meditative features, which show both masculine power and a curious, rounded femininity, are thought to be modelled on those of the god-king Jayavarman VII (ruled AD 1181–1219), the builder of Angkor Thom. Scholars agree that Jayavarman did identify himself with Lokesvara: 'He was an ardent partisan of Buddhism, which he professed in its active, Mahayana, form, and the Bodhisattva Avalokitesvara, the Merciful, the "Master of the World", became under his reign the most widely spread subject of worship.'[8]

The concept of the Bodhisattva is a central one in the Mahayana ('Greater Vehicle') school of Buddhism. At the heart of it is the notion of the continuous evolution of the soul, through countless thousands of rebirths, until freedom from the miserable round of material existence is at last achieved:

> 'Bhante Nagasena,' said the king, 'are there any who die without being born into another existence?'
>
> 'Some are born into another existence,' said the elder, 'and some are not born into another existence.'
>
> 'Who is born into another existence, and who is not born into another existence?'
>
> 'Your majesty, he that still has corruptions is born into another existence; he that no longer has the corruptions is not born into another existence.'
>
> 'But will you, bhante, be born into another existence?'
>
> 'Your majesty, if there be in me any attachment, I shall be born into another existence; if there shall be in me no attachment, I shall not be born into another existence.'[9]

The attachment that must be broken is to 'the illusory city of samsara'.[10] This is the world of material forms in which the soul is obliged to incarnate repeatedly until, through conscious self-improvement and detachment, it reaches *samadhi*, 'total self-collectedness ... the highest state of mental concentration that a person can achieve while still bound to the body ... a precondition of attaining release from the cycle of rebirths'.[11]

Having arrived at this stage of perfection, however, not all souls choose to pass on to Nirvana. According to Mahayana Buddhism there are some, the Bodhisattvas, who, out of selflessness and love for fellow human beings, postpone their transfiguration and reincarnate again and again in the material world where they act as teachers and guides, showing others how to escape from the 'ocean of existence'.

A Bodhisattva is a Buddha-to-be. Siddhartha, the most recent of the Buddhas, who is thought to have lived around the sixth century BC, was therefore a Bodhisattva prior to his enlightenment ('Buddha' is a title, not a proper name, and means 'Awakened One' or 'Enlightened One'). Bodhisattvas can also incarnate, without becoming Buddhas, specifically in order to help mankind at times of particular need:

It was assumed that many thousands of years had to pass between the appearance of each individual terrestrial Buddha. So as not to leave man during this long period completely deprived of assistance and succour in his endeavour to preserve the pure doctrine, celestial Bodhisattvas were imagined . . . [12]

Although there are differences there are also striking similarities between these 'celestial Bodhisattvas' – such as the 'compassionate and merciful' Avalokitesvara – and the Hindu concept of the avatars of Vishnu. In both cases a fully self-realized being, immortal and 'equipped', choses to incarnate amongst men to assist them through some great spiritual and physical crisis. It is perhaps a sign of how small the differences really are that the Hindus regard the Buddha himself as an avatar of Vishnu.[13] Moreover, both Hinduism and Buddhism envisage one further incarnation – Kalki, in the case of the Hindus, and Maitreya in the case of the Buddhists – who will wipe the world clean of evil and repromulgate the pure teachings of the ancients.

STELLAR COPULATION

The Phimeanakas, 'Palace of Heaven'. Was it the scene for astronomical rituals involving the constellation of Draco memorialized in Khmer myths?

Angkor Thom is a great, sacred compound enclosing many structures, of which three are paramount: the Phimeanakas, the Baphuon and the Bayon.

The Phimeanakas – 'the Palace of Heaven' – was built by King Suryavarman I

(AD 1002–1050). It is a tall, rectangular stepped pyramid 35 metres × 28 metres at the base[14] bearing a resemblance, as a glance at page 27 will show, to Mayan stepped pyramids.

We climbed the monument up the steep but still largely intact western stairway to the high sanctuary at the summit. Open to the wind and skies, legend has it that this sanctuary was once the scene of an extraordinary symbolic intercourse between the god-king and a Naga serpent.

The tradition was recorded by Chou Ta-Kuan, an emissary of the Chinese Emperor, in the late thirteenth century. Chou tells us that within the Phimeanakas were 'many marvellous sights, but these are so strictly guarded that I had no chance to see them'.[15] He then describes the sanctuary as:

> a golden tower, to the top of which the ruler ascends nightly. It is common belief that in the tower dwells a genie, formed like a serpent with nine heads, which is Lord of the entire kingdom. Every night this genie appears in the shape of a woman, with whom the sovereign couples. Not even the wives of the King may enter here. At the second watch the King comes forth and is then free to sleep with his wives and concubines.[16]

This strange story is often cited as an example of primitive superstition and 'mumbo-jumbo'. But when we remember that the short axis of the Phimeanakas is oriented perfectly due north–south – like any modern observatory – another possibility presents itself. Perhaps the king's 'coupling' with the 'serpent' is a metaphor for astronomical observations of the Naga constellation of Draco?

After all, we know from Chou Ta-Kuan's travelogue that ancient Angkor was well provided with 'men who understand astronomy and can calculate the eclipses of the sun and the moon'.[17] From another source we learn that astronomy was called 'the sacred science' at Angkor.[18] And from another we may gather that the destruction of astronomical manuscripts and records was regarded as a crime that would be 'punished with eternal damnation'.[19] It is therefore not difficult to see how Draco could have come to be called the 'Lord of the entire kingdom' – particularly so since the pattern of its stars appears to have dictated the pattern of the temples. Moreover, a lofty tower such as the Phimeanakas oriented to the 'meridian', the north–south line that divides the sky directly above the observer's head, would have been the natural place from which an astronomer-king might have sought to view a northern constellation like Draco (even today the majority of astronomical observations are made at the meridian).

Traditions also speak of the original founders of the civilization of Cambodia, the demi-gods Kaundinya and Kambu, who lived in the time of myth before history began. They came separately, we are told, by boat from across the sea. When Kaundinya first sought to land he was attacked by a beautiful Naga princess whom he overwhelmed and later married.[20] Likewise, the story of Kambu also ends with his marriage to the daughter of a Naga king: 'Kambu married her and founded a kingdom in the river valley. The people were called "Kambujas", or the children of Kambu. In time the name changed to Cambodge and then to Cambodia.'[21]

So there is a sense, from the very beginning, in which the 'ground' of Angkor, represented by the sacred temples and the person of the god-king, could be said to have been allegorically 'coupled' with the sky-region of the Naga constellation of Draco.

Is it an accident that a system of ideas which focussed on different constellations but was otherwise almost identical flourished in ancient Egypt at least as far back as the Old Kingdom?

A glimpse into this system is provided by Utterance 366 of the Pyramid Texts in which the Pharaoh is urged to identify himself with the constellation of Orion, the celestial counterpart of the high god Osiris, and then to 'couple' with Sothis (i.e Sirius) the star-form of his immortal sister – the goddess Isis: 'Your sister Isis comes to you rejoicing for love of you. You have placed her on your phallus and your seed issues into her, she being ready as Sothis.'[22]

In *The Orion Mystery* Robert Bauval presents persuasive evidence that the Old Kingdom Pharaohs may have participated in symbolic re-enactments of this scene inside the Great Pyramid of Egypt. Indeed, he shows that the southern shaft of the so-called Queen's Chamber, which targeted the meridian-transit of Sirius in 2500 BC, could have been specifically designed to play the obvious role in such a 'stellar copulation'.[23]

We suspect that what happened in the meridional chamber at the top of the Phimeanakas pyramid in Angkor Thom may not have been so very different – a symbolic 'coupling' of the god-king with an astronomical figure, this time not with Sirius, the star of Isis, but with Draco the constellation of the celestial Naga.

What was meant by such symbolism was not, necessarily, some form of bizarre and improbable physical intercourse between men and stars naïvely believed in by credulous primitives. It is equally possible – at any rate for initiates if not for the masses – that the symbols were intended to imply a communion of ideas between sky and ground. Through this sustained congress the initiate might have hoped to elevate his intellect to such a high degree that he would at last be able to declare: 'I take possession of the sky, its pillars and its stars . . . I am a snake, multitudinous of coils; I am the scribe of the god's book, who says what is and brings about what is not.'[24]

AS ABOVE, SO BELOW

Leaving the Phimeanakas, we walked south to the Baphuon, the pyramid-mountain of King Udayadityavarman II (ruled AD 1050–1066). This structure stands on a rectangular base measuring 120 metres × 90 metres and reaches a height of over 50 metres.[25] Its appearance particularly impressed Chou Ta-Kuan, who described it as 'the Tower of Bronze . . . a truly astonishing spectacle, with more than ten chambers at its base.'[26]

The central core of the pyramid is a lofty artificial hill of compacted earth on top of which a massive temple once stood as the sanctuary for the Shivalinga. Again there are obvious and unmistakable parallels with the 'High Sand' of Heliopolis – the mound of Atum surmounted by the Temple of the Benben containing the sacred obelisk of the Benben.

The Baphuon collapsed in antiquity because:

insufficient time was allowed for the earth to settle and find its foundations. The massive, heavy blocks of stone were hoisted into position and the building went up with excessive haste. For a few years all went well . . . Then ominous cracks began to appear. The building could not be shored up. Its terrific weight hastened its end.[27]

Despite this disaster, the central core of the monument retains the distinctive pyramid shape that it was given when it was originally built as a 'small replica of Mount Meru'.[28] And like Mount Meru, explains the French Orientalist George Coedes, the Baphuon was believed to have had 'a subterranean part that extended as far into the earth as the visible part rose into the sky'.[29]

Now compare this notion to the words of the Roman traveller Aelius Aristides, who visited Egypt in the second century AD:

> With admiration we behold the tops of the Pyramids, *but we do not know that which is equal to it and opposite to it underground*. I speak of what I have received from the priests.[30]

So at Angkor and at Giza it seems that exactly the same peculiar notion was in circulation about pyramids, i.e. that the foundations of these monuments extend as far below the ground as their summits tower above it – or even that each visible pyramid stands upon an 'equal and opposite' inverted pyramid, invisible beneath the ground.

Is it reasonable to suppose that such original and distinctive ideas could have arisen 'independently, by spontaneous generation',[31] in both these widely separated centres of ancient culture?

In their study of myths, Giorgio de Santillana and Hertha von Dechend conclude that any 'complex of uncommon images' cropping up in supposedly unconnected traditions should – at the very least – arouse our curiosity. 'Though the reservoir of myth and fable is great,' they point out, 'there are morphological markers for what is not mere storytelling of the kind that comes naturally.'[32]

SEVENTY-TWO

Many travellers and scholars who have studied Angkor are aware that some sort of number symbolism is expressed in the monuments and that certain numbers recur repetitively. But before Santillana and von Dechend cracked the 'precessional code', nobody had the faintest idea what these numbers were for. If not disregarded entirely they were taken merely as:

> evidence of an obsession with the magic of numbers and of the dignifying, under artistic forms, of primeval superstitions. One feels that the Khmer must have reasoned that if it was a good thing to erect one statue of Vishnu then it was fifty times better to have fifty of them . . .'[33]

Of course it would not have been 50 statues but 54 (or 72, or 108, or 216, etc.) – apparently a minor detail unless one is equipped with the numerical code by which the cosmic cycle of precession was evoked in ancient cultures.

Unfortunately, however, even today the startling findings of *Hamlet's Mill* are poorly understood by academics. In consequence the many obvious and perhaps even overriding precessional characteristics of the monuments of Angkor have rarely been considered except in obscure scientific papers.[34]

We therefore cannot blame Wilbur E. Garret, editor of the meticulously accurate *National Geographic* magazine, for failing to note the possible significance of a statistic that appears several times in a special report on Angkor that he published in May 1982.[35] Yet, as he himself informs us in the introduction to that report, there are '72 major stone and brick temples and monuments of Angkor'.[36]

The fact that there are 72 structures on a site that repeatedly makes use of other numbers in the precessional sequence such as 54 and 108 (and that, as a bonus, is located 72 degrees of longitude east of the Pyramids of Giza) is, in our view, highly suggestive of an overall plan. Moreover, if such a plan did exist, then it must have been in place *from the beginning until the very end* of the historically isolated phase of temple construction at Angkor – which started abruptly with the reign of Jayavarman II in AD 802 and ended equally abruptly with the death of Jayavarman VII in 1219.

MASTERPLAN

In an otherwise good book published in 1963, the Polish scholar Miroslav Krasa maintains that 'one hundred years after its discovery, the mystery of Angkor no longer exists'.[37] This is a statement with which most of the academic authorities would still be happy to agree, and it is true that a great deal is now known about the temples and their builders. Nevertheless, several extremely important and, we would have thought, glaringly obvious parts of the puzzle do still remain completely unsolved. These include:

1 an explanation for the amazing suddenness with which the sacred domain of Angkor was brought to life at the beginning of the ninth century AD;

2 an explanation for why it was developed so methodically and so industriously, at such vast expense, for approximately 420 years;

3 an explanation for why this staggering and unprecedented burst of temple-building, greater in magnitude and quality than anything in India, took place in a remote backwater of rural Cambodia; and

4 an explanation for why all new temple-building at Angkor suddenly ceased in the thirteenth century after the death of Jayavarman VII and never resumed – even though the site continued to be occupied until at least the sixteenth century.

The notion that the rulers of Angkor were working to an imported master-plan that they were for some reason obliged to fulfil within a specific time-frame provides a complete explanation for all of these mysteries. The existence of a similar plan at Giza in 2500 BC would also explain the mystery of the sudden appearance there of the Great Pyramids of Egypt and of the associated smaller structures at Saqqara containing the Pyramid Texts. These massive cultural

ABOVE, *entrance, and* BELOW, *interior of Ta Prohm, the terrestrial counterpart of the star Eta Draconis in the Draco/Angkor sky–ground plan. Ta Prohm was built by the god-king Jayavarman VII.*

achievements of the Fourth, Fifth and Sixth Dynasties were without precedent and without sequel. And just like the pyramids, temples, bas reliefs and inscriptions of Angkor, they were completed within a span of approximately 420 years (from 2575 to 2152 BC).

Jayavarman II might have brought the plan with him when he came to Cambodia 'from across the seas' in AD 800. Or he could perhaps have received it from the learned Brahman, 'skilled in magic', who initiated him into the cult of the god-king in 802. We can only speculate. His forty-year 'circling' of the site, as well as the behaviour of subsequent monarchs, is consistent with the working out of a plan. Indeed, each of the Khmer rulers seems to have contributed what he could, according to his own resources, some adding merely a temple here or there, others – notably Suryavarman II and Jayavarman VII – completing whole sequences of massive monuments within relatively short periods of time. In addition, few of these monarchs enjoyed peace: most were obliged to defend their frontiers against the inimical forces of invading barbarians while stubbornly and methodically continuing to carpet the flood-plains of the Mekong with the pre-ordained total of 72 major structures that would truly qualify the land of Kambu for the enigmatic title 'similar to the sky' given to it in the inscriptions.

King Jayavarman VII, the monarch of Angkor from AD 1181 to 1219, shown on horseback in this bas relief from the Bayon.

FOR THE BENEFIT OF MANKIND

The life and works of Jayavarman VII are worthy of particular study. He behaved like a driven man throughout his reign of 38 years, furiously completing not only

the immense perimeter wall of Angkor Thom, but also the temples of Ta Prohm, Bantei Kidei, Neak Pean, Ta Sohm, Srah Srang, the so-called Terrace of the Elephants and Terrace of the Leper King (both inside Angkor Thom), Krol Kro, Prah Palilay, Prasat Suor Prat, Prah Khan and, last but not least, the Bayon.[38]

Understandably it is the consensus of scholars that Jayavarman was a megalomaniac and that what drove him to sustain this gargantuan construction project was only the siren-song of his own over-inflated ego: 'inexorable will-power at the service of mania',[39] as one critic puts it. Yet when we look at his monuments we find that a larger number of them than we would expect by chance are amongst those that match the principal stars of the constellation of Draco. In addition, several seem to mirror important stars in the neighbouring constellation of Ursa Minor — and all in the positions that they occupied at dawn on the spring equinox in 10,500 BC.

BELOW, LEFT AND RIGHT: *Ta Prohm.* BOTTOM: *Fishermen at work in the moat of the Bayon, an 'infinitely mysterious temple'. Its location marks the Ecliptic North Pole in the Draco/Angkor sky–ground scheme.*

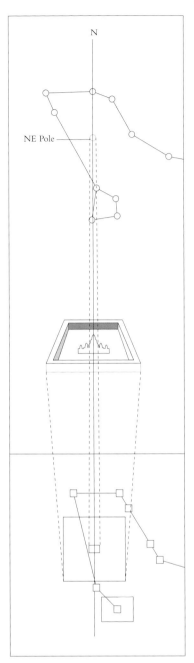

The Bayon in Angkor Thom as the terrestrial counterpart of the North Ecliptic Pole.

To be specific, the structures built by Jayavarman VII that contribute directly to what appears to be an ambitious and extensive sky–ground plan are Angkor Thom, Ta Prohm (AD 1186), Bantei Kidei (thought to be the earliest of his temples), Neak Pean, Ta Sohm, Srah Srang, Prah Khan (AD 1191), and last of all the spectacular and bizarre Bayon which he completed just before his death in the year 1219.[40]

It might all be an accident and modern historians might be 100 per cent correct when they allege that this high-speed building programme was just megalomania, resulting in the random construction of many temples here and there: 'an orgy of building, a brief yet sustained period of hectic, almost crazy architectural creation'.[41] Yet in inscriptions from which we have already quoted in earlier chapters, Jayavarman sounds far from crazy or egocentric. On the contrary, he tells us explicitly that his temples were part of a grand scheme to win the 'ambrosia of immortality' for 'all those who are struggling in the ocean of existence'.[42] We know, too, that he saw the Angkor monuments as effective instruments in this quest because of their special qualities as 'mandalas of the mind'.

The possibility therefore cannot be ignored that this remarkable monarch may, *for the benefit of mankind*, have taken it upon himself to complete in one reign the sacred 'mandala' of the Draco sky-region that he had inherited from the god-kings before him. They in turn had received the plans for the mandala from an unknown source at the time of Jayavarman II.

There are some hints about the nature of this source. The reader will recall that inscriptions concerning the reign of Jayavarman II speak of the existence of a group of wise men – 'Brahmans', 'sages', 'astronomer priests', 'mystery teachers of heaven' – who knew how to conduct initiations and who could function as king-makers. If the purpose of the entire Angkor building programme had been to fulfil the preordained plan of such a powerful behind-the-scenes brotherhood, and if Jayavarman VII, 'full of deep sympathy for the good of the world',[43] did indeed put the last temples envisaged by that plan into place, then this would explain the 'mystery'[44] of why the building programme stopped so abruptly after his death. Quite simply, with the sky–ground diagram complete, there would have been no more temples left to build.

'THIS INFINITELY MYSTERIOUS TEMPLE ...'

When the star-map of 10,500 BC is transposed to the ground, the perimeter wall of Angkor Thom delimits a sacred enclosure drawn around the breast or heart of the Naga constellation of Draco. At the exact geometrical centre of this enclosure, where the diagonals cross over the 'heart' itself, looms the breathtaking edifice known as the Bayon – which is regarded as Jayavarman VII's finest architectural achievement.

Is it an accident, within the overall celestial plan of the temples, that the 'heart' of Draco thus marked by the Bayon correlates extremely closely with the location of the ecliptic north pole? The reader will recall that this is the point in the sky around which the celestial north pole circulates, as a result of precession, at the rate of half a degree every 36 years, three-quarters of a degree every 54 years,

One of the 54 stone towers of the Bayon, each with four faces of Lokesvara.

1 degree every 72 years and 30 degrees every 2160 years. The most notable and distinctive architectural characteristic of the Bayon – a squat step-pyramid that stands on top of a much older and as yet unexcavated structure[45] – is that it is surmounted by 54 massive stone towers, each of which, in the same manner as the entrance gateways of Angkor Thom, is carved with four gigantic faces of Lokesvara ('in the Egyptian style') oriented due north, south, east and west, making a total of 216 faces in all. According to Jean Boissellier, former curator of the National Museum of Phnom Penh, the faces have been sculpted with 'the typical expression of the Buddhist in the "active state of mind" which the scriptures call *brahmavira*, the "things pleasing to Brahma", the "sublime state" leading the mind to charity, compassion, joy and tranquillity'.[46]

The French traveller Pierre Loti, who visited Angkor during the rainy season in 1901, said of the Bayon:

> Through an inextricable tangle of dripping brambles and creepers, we have to beat a path with sticks in order to reach this temple. The forest entwines it tightly on every side, chokes it, crushes it; and to complete the destruction, immense 'fig-trees of the ruins' have taken root there everywhere, up to the very summit of its towers, which act as their pedestal . . .

> My Cambodian guide insists that we should leave. We have no lanterns, he tells me, on our carts, and we must return before the hour of the tiger. So be it. Let us go. But we shall return just for this infinitely mysterious temple.

Before I leave, however, I raise my eyes to look at the towers which over-hang me, drowned in verdure, and I shudder suddenly with an indefinable fear as I perceive, falling upon me from above, a huge, fixed smile; and then another smile again, beyond, on another stretch of wall; then three, then five, then ten. They appear everywhere and I realize that I have been over-looked from all sides by the four-faced towers. I had forgotten them though I had been warned about them. These masks carved in the air are so far exceeding human proportions that it requires a moment or two fully to comprehend them. They smile under their great flat noses, and half close their eyelids, with an indescribable air ...[47]

In 1912 the French diplomat and writer Paul Claudel visited Angkor and described the Bayon in his journal as 'one of the most accursed, the most evil, places that I know ... I came back ill ...'[48] But Claudel, who died in 1955, was a man with unnecessarily rigid views about spiritual truth who believed passion-ately that the dogma of the Catholic Church and unconditional faith in Christ are the only roads to salvation. Perhaps, therefore, what really sickened him about this dark and eerie temple, its air of 'evil' as he saw it, was his own intuitive recogni-tion that he was confronted here by the workings of a great spiritual power that he could not understand.

The Bayon was always intended to transform – a matter over which there can be little doubt when we remember that its name is derived from 'Pa yantra', the 'father' or 'master' of *yantra*.[49] This is a Sanscrit word, meaning literally 'instrument',[50] defined as a form of mandala: 'a diagram used as a support for meditation ... The component parts of the *yantra* take the believer along the different steps leading to Enlightenment ...'[51]

We suspect that those who fully understood the Angkor monuments were not 'believers' but 'adepts', high initiates in a lost system of cosmic wisdom, who would have come to the Bayon in search of the final mysteries. As such, through diligent inquiry, they would of course have already been equipped to 'go down to any sky' – i.e. to make the precessional calculations that would allow them to visualize the positions of important stars in former epochs.

In a general way, they would have long since realized that the layout of the Angkor monuments was intended to draw their attention to the sky-region around the celestial north pole – notably, as we have seen in previous chapters, to stars in the constellations of Cygnus, Ursa Minor, Ursa Major, the Corona Borealis and Draco ... especially Draco. In order to have discovered so much they would have had to work their way back, again just as we have done, to the spring equinox in 10,500 BC (although, of course, they would have used a different dating system). And they would have realized that an observer looking north at the moment of sunrise would have seen a perfect meridian-to-meridian match between the patterns of the stars in the sky and the temples on the ground.

In the process of mentally 'winding the stars back' until the correlation was achieved such adepts would inevitably have discovered what we can so easily con-firm on our computer screens today: the slow, cyclical rotation of the celestial north pole around the 'heart' of the constellation of Draco, i.e. the ecliptic north pole. It is this 'heart', this abstract point in space that finds its terrestrial counterpart at Angkor in the great pyramid of the Bayon – a monument that rises in three terraces from a square base measuring 80 metres along each side and that culminates in an unusual circular central sanctuary at a height of 45 metres above ground-level.[52]

Coedes rightly calls the Bayon 'the mystic core of the Khmer Empire'[53] while Bernard Groslier describes it as the '*omphalos*' ('navel') of Angkor's 'stone cosmos'[54] and John Audric notes that to this day rumours persist of some great treasure long ago concealed within it.[55]

Such a treasure need not be gold or jewels.

It might equally be knowledge, *gnosis*, the elixir that all true initiates must seek – in all lands, in all epochs – if they are to attain the life of millions of years . . .

TIME-KEEPERS OF THE SKY

The ancient Egyptians did not depict Draco as a snake or dragon but as another reptilian monster, the crocodile, which they oddly conflated with various parts of a hippopotamus and of a lion. The result was a composite astronomical deity named Taweret[56] who is referred to in the Pyramid Texts, who appears frequently in the *Book of the Dead* and who takes centre stage in the remarkable 'circular zodiac' of the temple of Dendera in Upper Egypt.

A matter of particular interest about this deceptively simple star-map, other aspects of which we considered in Chapter 3, is that it not only correctly locates Draco in relation to other northern constellations such as Ursa Minor (which the ancient Egyptians knew as 'the Jackal') and Ursa Major ('the Thigh') but also, according to the French mathematician R. A. Schwaller de Lubicz, that it 'shows the pole of the ecliptic, located in the breast of the hippopotamus, or constellation of Draco'.[57] Schwaller points out that the mythological figures at Dendera which represent the zodiacal constellations are not arranged in a single circle, as we might expect, but instead 'are entwined in two circles – one

A Buddhist monk offers prayers in an inner shrine of the Bayon, overlooked by one of the 216 giant stone faces of Lokesvara. The temples of Angkor have never ceased in their function as instruments of spiritual initation.

Circular central tower of the Bayon, 'the omphalos *[navel] of Angkor's stone cosmos'.*

around the celestial north pole and one around the ecliptic north pole'[58] – in a noticeably off-centre spiral flow. In this way, argues Schwaller, the zodiac expresses a clear knowledge of what happens in the sky as the celestial north pole gradually precesses around the pole of the ecliptic.[59]

A number of scholars have observed that Taweret's crocodile-hippo-lion characteristics are identical to those of Ammit,[60] the terrible 'Eater of the Dead' who attends the weighing of the heart in the Judgement Hall of Osiris.[61] This hybrid, confirms Dr Stephen Quirke, Curator of the Department of Egyptian Antiquities at the British Museum, incorporates: 'the three animals of voraciousness who could be depicted in Egyptian formal art, the crocodile for the head, lion for the forepart and hippopotamus for the rear'.[62]

To all extents and purposes, therefore, the monster of the Judgement Scene *is* Draco, standing for the annihilation of the soul just as Osiris stands for rebirth and resurrection. But there is a strange ambiguity, very similar to the ambiguity that we find in ancient Indian texts concerning Naga serpents – which were sometimes dangerous and sometimes benevolent. Thus, although Draco in the form of Ammit was viewed by the ancient Egyptians as a voracious and unpitying destroyer, Draco in the form of Taweret was seen as a benign guide and protector of souls[63] and as the patron of childbirth.[64] Indeed, so strong was this more positive perception of Draco that amulets of Taweret were frequently placed in ancient Egyptian tombs so as to 'protect the rebirth of the deceased into the [Osirian] kingdom of the dead'.[65]

The sense of a subtle link between the functions of Orion-Osiris and Draco-Taweret-Ammit is enhanced by Egyptian traditions in which a crocodile is said to have swum to Osiris in the Nile (after his drowning by Set) and to have carried his corpse on 'its back safely to land'.[66] In some accounts Osiris himself is mysteriously described as a 'great Dragon' lying on sand,[67] while in others more closely related to the symbology of Angkor we read that the god transformed into

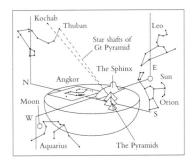

The constellations at the cardinal points and their terrestrial counterparts at sunrise on the spring equinox 10,500 BC.

a serpent when he passed into the Netherworld.[68] In the *Book of the Dead* we are even told that Osiris, as 'Lord of the Duat', resides in a palace whose walls are 'living cobras'.[69]

Such ideas transpose very well to the sky of the spring equinox in 10,500 BC – a sky that has about it the sense of a strange, heraldic device:

- At the moment of sunrise, if we look due west, our computer shows us that Aquarius has set and that 'the Fishes' – Pisces – are following.
- Due east Leo the lion is on the rise, seemingly drawing up the sun beneath him.
- Straddling the meridian due south stands the giant figure of Orion-Osiris, known in ancient India as Kal-Purush, 'Time Man',[70] who states in the ancient Egyptian *Book of the Dead*: 'I am time and Osiris. I have made my transformations into the likeness of divers serpents.'[71]
- Straddling the meridian due north, facing Orion, is Draco the celestial dragon – or serpent, or crocodile or hippopotamus – the secret guardian of the ecliptic north pole.

It is easy to see how the behaviour of Orion and Draco, and thus their cosmic functions, could have come to be seen as linked by the ancients. Indeed, as scientific observations have confirmed, they *are* linked, by the cycle of precession, in a great cosmic see-saw which swings up and down like the pendulum of time itself. Computer simulations covering thousands of years show us that as Orion's altitude at the south meridian steadily rises Draco's altitude at the north meridian steadily falls. When Draco reaches its lowest point, Orion reaches its highest point. Then the opposite side of the cycle begins with Draco steadily rising and Orion steadily falling. The 'up' motion takes just under 13,000 years. The 'down' motion takes just under 13,000 years. And so it proceeds, up for 13,000 years, down for 13,000 years – to all extents and purposes for ever.

What is particularly intriguing is that the sky-ground plans of Angkor and Giza have succeed in capturing the *highest point* in Draco's trajectory and the *lowest point* in Orion's – the end in other words, of one half-cycle of precession and thus the beginning of the next. This last happened, we know, around the year 10,500 BC, in which epoch the ecliptic north pole lay due north of the celestial north pole at dawn on the spring equinox and the pattern of the stars in the sky was taken as the template for the pattern on the ground of the monuments of Angkor and Giza.

Since that golden age, rotated by the churn of precession, the celestial pole has travelled a full half-circuit around the pole of the ecliptic. The pendulum of Orion and Draco has likewise swung back almost as far as it can go – with Draco now at its lowest point and Orion at its highest.[72]

As in 10,500 BC, in other words, the time-keepers of the sky, who stand at the gates of immortality,[73] are poised to go into reverse again. Any initiate steeped in the Hermetic dictum 'as above so below' would be bound to interpret this configuration as a sign that some great change is imminent – a change that could be for the better, or greatly for the worse, depending on humanity's own choices and behaviour.

PART IV

THE PACIFIC

CHAPTER 12

FRAGMENTS OF A BROKEN MIRROR

CAMBODIA LIES ON THE western rim of the world's largest ocean, the Pacific. Since the Angkor monuments are positioned 72 degrees of longitude to the east of Giza – a significant precessional number – we naturally wondered whether other sites might be identified using the same system of geodesy.

The precessional number that recurs most insistently at Angkor is 54 – notably in the form of the 54 towers of the Bayon and the 54 *devas* and 54 *asuras* on either side of the causeways leading into Angkor Thom. It is therefore a curious fact that a most unusual and mysterious archaeological site with unknown origins and function does lie in the Pacific Ocean 54 degrees of longitude to the east of Angkor. The name of this site is Nan Madol and it consists of approximately 100 artificial islands, constructed out of basalt and coral, which lie in the blue waters of a lagoon off the south-eastern coast of the Micronesian island of Pohnpei.

Although the setting is very different, Nan Madol has a number of features in common with Angkor. Scholars believe that the bulk of the temple-islands were completed between AD 800 and AD 1250, precisely the period of Angkor's flores-cence, but have also detected the traces of an earlier layer of construction[1] – as is again the case at Angkor. The largest structure, Nan Douwas, is oriented to the cardinal directions, with its principal entrance facing west.[2] Adopting the classic 'mandala' form, it consists of two concentric perimeter walls separated by a sea-water moat and enclosing a central pyramidial mound. The walls reach 7.6 metres in height and are made from crystalline basalt megaliths, some of which weigh 50 tonnes and are more than 6 metres in length.[3]

A complex system of canals dredged out of the lagoon is part of the overall design of Nan Madol, calling to mind the miles of canals and moats that criss-cross the landscape of Angkor. Moreover, the temples of Angkor mimic the sky-image of the constellation of Draco whilst the people of Pohnpei remember a legend that the canals separating their temples were originally dredged by a 'dragon'[4] which offered its assistance to Olosopa and Olosipa, the two mythical founders of the city.[5]

Said to have been brothers, Olosopa and Olosipa were 'Ani-Aramach', pri-mordial god-kings,[6] who arrived in boats 'from a land to the west' bringing with them a 'sacred ceremony', which they instituted in their new homeland with the

PREVIOUS PAGE: *The fifteen gigantic Moai of Ahu Tongariki, Easter Island.*

OPPOSITE: *Underwater monument thought to be at least 10,000 years old – Yonaguni island, south-west Japan*

Plan of the islands of Nan Madol.

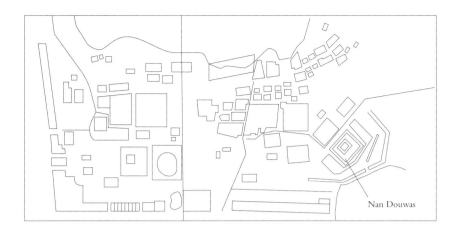

Nan Douwas

help of wise magicians.[7] Likewise, King Jayavarman II came to Cambodia by boat – in his case from the mysterious land of the 'King of the Mountain' – bringing a sacred ceremony which he instituted with the help of wise Brahmans 'skilled in magic'.[8]

A notable feature of Jayavarman's reign was the manner in which he seems to have marked out, or geodetically 'prospected', the site of Angkor. The reader will remember that in the process he established the three temporary capitals of Hariharalaya, Amarendrapura and Phnom Kulen, moving constantly – 'like a bird of prey soaring over the land' as George Coedes puts it, 'pivoting in a great circle around the future Angkor . . .'[9]

Parallel traditions concerning Olosopa and Olosipa report that these 'wise and holy men' built and then abandoned four capitals on Pohnpei: the first in the district of Sokehs in the north-west of the island, the second in Net, the third in U, and the fourth in the unpronounceable Madolenihmw.[10] The existence of these capitals has been confirmed by archaeology and if we trace out their pattern (see diagram) we find that it is not random but moves constantly in the same direction

Aerial view of Nan Madol, with the basalt temple of Nan Douwas, centre, behind massive breakwaters forming an arrow-shaped rampart oriented east. Beyond the breakwaters the ocean floor sinks away steeply and there are legends of a city under water.

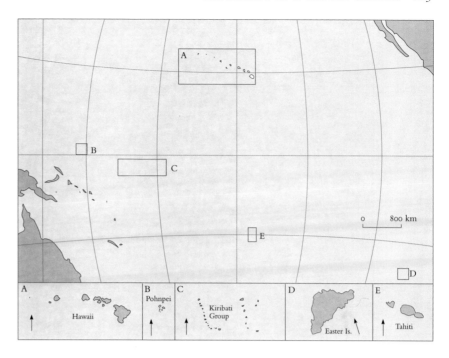

– from north-west, through north, to north-east, to south-east – like the hand of a clock travelling half-way around a dial.

The place where the hand 'comes to rest' is Nan Madol, a site that was traditionally known as 'Sounhleng or "Reef of Heaven" …'[11] Olosopa and Olosipa are said to have selected it after climbing to the top of a high peak in order to survey the island. From this vantage point, looking down to the blue Pacific Ocean far below, 'They saw a city under water. They took this as a sign that they should build their own city there and, further, built Nan Madol as a "mirror image" of its sunken counterpart.'[12]

The name of the submerged 'city of the gods' under the 'Reef of Heaven' is Khanimweiso and it is not a figment of myth. As part of an archaeological conservation project led by Dr Arthur Saxe of Ohio State University a thorough mapping-survey of large parts of Nan Madol has now been undertaken and has confirmed the existence of extensive undersea ruins, some of them lying at very great depths.[13] The majority have so far been discovered to the east, and a little to the south of the massive breakwaters around Nan Douwas and have included what appear to be a series of tall pillars or columns, standing on flat pedestals on the sloping sides of the island and reaching heights of up to 8 metres.[14]

CITY OF BASALT

Roughly circular in shape, with a diameter of no more than 20 kilometres, Pohnpei is a small, unusual, beautiful island. Its interior, which rises steeply towards a series of 800-metre peaks, is heavily forested and its shores are densely overgrown with mangroves. The island is surrounded by an almost continuous coral reef, through which a number of passages give access to the sea.

We reached Nan Madol in a flat-bottomed boat with two powerful outboard

Map of Pohnpei, showing clockwise shifting of the island's capital over time.

motors. Skimming across the turquoise waters inside the reef, we passed rich green hillsides, isolated mesas, and strange pyramid-shaped mounds. One of these, the double hummock of Takiun Peak, is particularly striking and we wondered whether it might have been from this vantage point that Olosopa and Olosipa had caught their first glimpse of the legendary underwater city of Khanimweiso.

We passed the wreck of a Chinese fishing vessel that had been driven up into the shallows in a recent typhoon. Then we came to tiny Nakapw island and saw evenly spaced rows of basalt boulders running along its shore and disappearing under the water towards the west – the direction of Nan Madol. Finally we crossed Nakapw Harbour and came to the curiously angled basalt breakwater, shaped like three sides of a pentagon, that surrounds the great temple of Nan Douwas. The blocks used in the breakwater were enormous and had been brought here – according to local traditions – by the magic powers of Olosopa and Olosipa.

We cut the engines and allowed the boat to glide to a halt just outside the breakwater, then gingerly edged it through the shallow, boulder-strewn channel of the gateway. Beyond lay a harbour, sheltered by a maze of megalithic walls. Overlooking the harbour to the west was the extraordinary edifice of Nan Douwas. Rising out of the green jungle canopy, its alternating courses of rough-hewn basalt blocks and finely shaped basalt crystals seemed predominantly black and dull-red in colour and radiated a massive, almost oppressive solidity. At the corners, however, the walls became light and elegant, with the stonework drawn up into raised peaks like the prows of seagoing boats.

All in all our first impression – which never changed – was that Nan Douwas is an eerie and numinous place. Most Pohnpeians do not like to go there, even in daylight, and will not under any circumstances visit it after dark.

It is believed to be possessed by spirits.

Pyramid-shaped peak near Nan Madol. Could this have been the vantage point from which Olosopa and Olosipa saw the legendary underwater city?

Channel through the megalithic breakwater, Nan Douwas.

AFTERLIFE JOURNEYS

We entered Nan Douwas by its western gate, passing through the series of geometrical courtyards around the temple's heart. Exactly as at Angkor Wat, the courtyards enclose one another and each leads up to a higher level. At the highest level we came to the sanctum, a semi-subterranean rectangular enclosure, cut down into the earth to a depth of about a metre and a half and roofed over with five-tonne basalt logs.

All around Nan Douwas are shallow canals lined with megalithic blocks, running amongst the 100 artificial islands of the sacred city of Nan Madol. We explored these canals but found none of the other temples and structures to be as well preserved as Nan Douwas. Some had almost completely disappeared beneath the waters and the encroaching mangroves.

What was the function of this once extensive sacred site?

Dr Rufinio Mauricio, a US-trained Pohnpeian archaeologist, has been investigating this question for more than 20 years. He told us that the temples of Nan Madol are linked to ancient local beliefs in life after death. According to these beliefs, which are astonishingly similar to those of the ancient Egyptians, the soul must make a perilous afterlife journey during which it will face many trials and tests. In Egypt this journey takes place in the Duat, a region of the sky; in Pohnpei

Entrance causeway, Nan Douwas.

the netherworld is under the waves – perhaps even in Khanimweiso, the underwater city. Like the pyramids of Giza there is much which seems to suggest that the temples of Nan Madol could have been built as physical models, copies – 'mirror images' – of the afterlife realm and might have been designed to serve as places of preparation and initiation for the soul's ordeal.

CITIES BELOW

We scuba-dived several times in the deep bay east of the Nan Douwas breakwater looking for the underwater city of Khanimweiso – which, legend said, could only be entered through a gateway guarded by two monstrous sharks.

Our first dive was off the boat to the south of the bay, in warm murky waters. We sank down, quite rapidly, to a depth of almost 40 metres and then levelled off on the bottom. The visibility was extremely bad, like thick fog, and we groped around for half an hour, finding nothing at all.

We surfaced and took the boat closer to the Nan Douwas breakwater, just outside its gateway. Diving there in depths of 3 metres or less we found a thick carpet of scattered boulders and crystalline basalt logs. These had clearly been laid down at the time of the construction of the breakwater.

A little further out, the sea-bed fell away steeply to more than 30 metres. Here, swimming north, we found two of the columns identified by the archaeologists. Thickly encrusted with multi-coloured corals they soared up from the darkness below to the light above – reaching towards the surface like fingers.

The columns are part of the underwater city of Khanimweiso. But there is a curious complication. As Arthur Saxe points out, the legends of Pohnpei tell not just of one but of 'two such cities'.[15] The name of the second is Khanimweiso Namkhet. It is said to lie 'outside the reef' and to have its entrance in a 'deep and sandy place'.[16] A number of local people, some in living memory, have claimed that they were dragged down to it when spear-fishing for turtles. And in an inter-

Inner courtyard and sanctuary, Nan Douwas.

Megalithic ruins, Nan Douwas.

Detail of basalt wall, Nan Douwas.

view with the American researcher David Hatcher Childress a Pohnpei elder reported the following strange tale:

> Some years ago a fisherman was dying. His spirit took a trip to the city outside the reef. After he had seen the city, which goes all the way to Kosrae [a high volcanic island, 550 kilometres south-east of Pohnpei, which also contains mysterious megalithic ruins], he returned to his body, told the people what he had seen, and then he died.[17]

Although the connection may prove to be spurious, there is an oddly familiar ring to the word 'Namkhet', which is phonetically close to the ancient Egyptian, *akh* ('light'), *akhu* ('transfigured spirit') and in particular *akhet*, 'horizon'. The latter was a frequently used prefix in Egyptian place names – for example in the city of Akhetaten and indeed in 'Akhet Khufu', the 'horizon of Khufu', one of the titles of the Giza necropolis. It is therefore tempting to speculate about Khanimweiso Namkhet. Since *khanimweiso* is the word for 'city' in the Pohnpei language, might not the full title be taken as meaning something like 'City of *akhet*' or 'Horizon City'?

DELUGE

Pohnpei is not the only Pacific island in which place names, traditions and religious beliefs sometimes have a peculiarly Egyptian flavour and nor is it the only one with legends of drowned cities off its shore. It would be a mistake to read too much in to such material. Nevertheless, what we find particularly noteworthy about the site of Nan Madol is that it is specifically said to have been 'modelled' on an earlier 'city of the gods', now under water, and to stand directly above that earlier city. We cannot see how this concept differs from that of the Egyptian temples such as the Temple of Horus at Edfu which were also 'copies' of earlier originals, built on primordial foundations, with the object of bringing about the resurrection of the 'former world of the gods'. We further note that the earlier divine world, which the Edfu Building Texts refer to as 'the Homeland of the Primeval Ones', is said to have been an island that was swallowed up in the waters of a great earth-destroying Flood.[18]

We have shown elsewhere that such a deluge did sweep the earth at the end of the last Ice Age when continental ice caps kilometres thick, which had smothered northern Europe and North America for more than 100,000 years, began to melt and lose their grip.[19] During this period of rapid change, which came to an end about 9000 years ago, it is generally accepted by geologists that the level of the world's oceans rose by more than 100 metres. If this assumption is correct then it is easy to see that at the date of 10,500 BC signalled by the astronomical correlations at Angkor and Giza, the island of Pohnpei might not have looked at all as it does today. Its coral reef is a relatively recent arrival which has developed since the rise in sea levels when the current shoreline was established. The island itself, however, is volcanic and consists of a hard basaltic core – the peak of a towering mountain – which juts above the surface of the Pacific Ocean. Beyond the reef the sides of this great sea-mount slope away steeply to an abyssal depth, connecting it far below to a submerged continent.

Western wall of Nan Douwas. Note the manner in which the stonework is drawn up at the corners into raised peaks like the prow and stern of a sea-going boat. The structure seems to sail on the waters of the canal.

A SKY-MONSTER KILLED THE EARTH

The epoch of 10,500 BC is of great interest to earth scientists. Three thousand years before the end of the last Ice Age, but long after the start of the great melt-down, it was a period marked by a series of cataclysms, *almost certainly of astronomical origin*, which completely changed the face of our planet. Studies of palaeomagnetism have confirmed that approximately 12,400 years ago there was a 180-degree reversal of the earth's magnetic poles.[20] Just 800 years later, in 9,600 BC, the earth was in collision – and not for the first time – with several fragments from a disintegrated comet. According to Professor Alexander Tollman of the University of Vienna: 'The consequence of the impact explosions appears to have included a chain of up to a dozen individual catastrophes, including earthquakes, geological deformation, a vapour plume and tidal waves.'[21]

It is an instructive coincidence that the epoch of the year 2000, in which we ourselves live, happens to be one in which extraordinary astronomical and geological phenomena of the type experienced some 12,000 years ago continue to manifest themselves.

Particularly troubling is the fact that a remorseless decay of the earth's magnetic field has been underway during the last 2000 years and has been rapidly accelerating during the past century. Indeed, scientists now expect the field's energy to drop to zero, initiating a rapid north–south magnetic 'pole reversal' before the year 2300.[22]

This does not necessarily imply a reversal of the *geographical* poles – i.e. the planet physically tumbling in space. Nevertheless, as the geologist S. K. Runcorn

THE AFTERMATH OF COMETS

Victor Clube, an astrophysicist at Oxford University, his colleague Bill Napier of the Royal Edinburgh Observatory, Dr Duncan Steel of Spaceguard Australia, and Benny Peiser of Liverpool John Moore's University are amongst a growing number of scientists who propose that a giant comet entered the solar system some time before 20,000 years ago and disintegrated, 'leaving a debris-strewn orbit into which the earth periodically blunders. Most of the time, the only visible sign of our passage through this debris is a slight increase in the number of "shooting stars" ... But every so often the earth runs into a much denser part of the debris – triggering an apocalyptic storm of impacts and devastation.'

Clube has linked one such storm to the cataclysmic meltdown of the last Ice Age between 14,000 and 9,000 years ago, an event of sufficient magnitude to have wiped out almost all traces of any 'antediluvian' civilization. Sea-levels rose by 100 metres, enough to bury forever an entire maritime culture. Together with his colleagues, Clube has also argued that lesser cataclysms, caused by collisions with smaller fragments of the same original giant comet, may have occurred at around 2350 BC and AD 500. During both periods thriving and successful historical civilizations mysteriously and simultaneously disappeared. According to Benny Peiser: 'There is very strong evidence to suggest that massive meteor storms are the real scientific reason why these ancient civilizations collapsed.'

If civilizations have been destroyed by cosmic impacts in historical times then it is surely logical to investigate the possibility that a civilization might have been destroyed in *prehistoric* times, before 2350 BC – perhaps by the terminal Ice Age cataclysm more than 9,000 years ago? That cataclysm and the two that followed it at 2350 BC and at AD 500 show a steady decline in magnitude, but this cannot be taken as a reliable trend. On the contrary, the belt of cometary debris which Clube and his colleagues believe has been orbiting the solar system for the last 20,000 years may still contain several extremely large fragments of the original object.

has observed, 'there seems no doubt that the earth's magnetic field is tied up in some way to the rotation of the planet'.[23]

The field is a mysterious force, of uncertain origin, but is most probably generated by the Earth's inner core – a solid iron sphere about three-quarters the size of the moon, suspended within a shell of seething liquid iron that is in turn encased within multiple layers of rock, mineral deposits and lubricating slurry thousands of kilometres thick. Recent research by Xiaodong Song and Paul G. Richards has established that the inner core has a rotational movement all of its own – in the same direction as the rest of the planet but about one per cent faster. This means that the surface of the core, at its equator, is moving at a rate of about 20 kilometres a year relative to the outer layers in which it is enclosed. As Richards points out: 'That's 100,000 times faster than the types of motion we normally associate with properties of the solid earth.'[24]

Some researchers, notably Rand and Rose Flem-Ath in Canada and the late Professor Charles Hapgood in the United States, have proposed that such a relatively rapid 'skidding' or 'slippage' of one layer over another could also have occurred – perhaps several times – at the level of earth's crust. Known technically as the 'lithosphere', this rigid, stony outer shell floats on top of a lubricating layer called the 'asthenosphere'. Were the crust to slip in one piece over the asthenosphere, 'much as the skin of an orange, if it were loose, might shift over the inner part of the orange all in one piece',[25] the result would be apocalyptic worldwide devastation – quite literally the end of the world as we know it.

An Earth-Crust Displacement 535 Million Years Ago?

The earth-crust displacement theory advocated by Hapgood, the Flem-Aths and others has not been well-received by orthodox planetary scientists. Geologists in particular have poured scorn on the theory, linked it to the lunatic fringe, and failed to give it any serious peer-review. The impression has been successfully conveyed that it is simply an 'impossible' process, advocated by lunatics and therefore not worthy of consideration by genuine scientists. Yet behind the scenes the evidence has been slowly mounting up that displacements of the earth's crust *do* occur from time to time and that there is no physical or geological reason why such a displacement should not have occurred around 12,000 years ago – precisely as the Flem-Aths allege.

On 25 July 1997, whilst managing to avoid the use of the phrase 'earth-crust displacement', evidence was published in the orthodox scholarly journal *Science* which effectively proves that the earth's crust can and does shift. The evidence was gathered by researchers at the California Institute of Technology and focusses on the period from 550 to 535 million years ago. This period immediately preceded what evolutionary scientists refer to as the 'Cambrian explosion' – the greatest diversification and expansion of life that this planet has ever seen.

The Caltech group reports

The Flem-Aths believe that a number of such events have taken place in the past, most recently between 11,000 and 12,000 years ago:

> The earth's crust ripples over its interior and the world is shaken by incredible quakes and floods. The sky appears to fall as continents groan and shift position. Deep in the ocean, earthquakes generate massive tidal waves which crash against coastlines, flooding them. Some lands shift to warmer climes, while others, propelled into polar zones, suffer the direst of winters. Melting ice caps raise the ocean's level higher and higher. All living things must adapt, migrate, or die . . . [26]

If a planetary event of such magnitude has taken place, as the Flem-Aths argue, then it is clear that enormous force would have been required to set it off. In 1953 Albert Einstein, an early supporter of the earth-crust displacement theory, suggested an answer:

> In the polar region there is continual deposition of ice, which is not symmetrically distributed about the pole. The earth's rotation acts on these unsymmetrically deposited masses, and produces centrifugal momentum that is transmitted to the rigid crust of the earth. The constantly increasing centrifugal momentum produced in this way will, when it has reached a certain point, produce a movement of the earth's crust over the rest of the earth's body. [27]

Einstein did not consider other trigger-factors. It seems likely, however, that he might have sought to calculate the possible effects of a head-on collision with an asteroid, meteoroid or comet if he had had at his disposal the information that is available to scientists today.

For instance, in February 1994 US President Bill Clinton was woken by his staff with news of a possible military attack on America detected by six orbiting satellites. It soon became clear that what the satellites had picked up was not incoming enemy missiles but a massive meteoroid that ultimately exploded high in the atmosphere. [28] In May 1996 another gigantic meteoroid, with a diameter of more than 300 metres, skimmed by the earth just three days after it was first detected by military satellites. [29]

These were small members of a class of heavenly objects that we have recently been encountering with increasing frequency. Since 1990 more than 12 new comets have been discovered every year, [30] the most spectacular of which – the awkwardly named Comet Hale-Bopp – tore brilliantly through the earth's northern skies in March and April of 1997, reaching its closest point on the spring equinox, 21–22 March.

The ancients would have seen such a perigee at such a moment of the year, so soon before the end of a millennium, as ominous to say the least – an interpretation which does not differ greatly from that of modern scientists. Indeed, the latest research confirms that increased cometary activity is part of a great cosmic cycle, as the solar system moves in a slow, undulating fashion within our home galaxy, the Milky Way. Approximately every 30 million years we pass through the galaxy's dense central plane. There we encounter large numbers of meteoroids, that 'this evolutionary burst coincides with another apparently unique event in earth history – a 90-degree change in the direction of Earth's spin axis relative to the continents … Regions that were previously at the north and south poles were relocated to the equator, and two antipodeal points near the equator became the new poles … The geophysical evidence that we've collected from rocks deposited before, during and after this event demonstrates that all the major continents experienced a burst of motion during the same interval of time.'

The Caltech researchers insist that this event is to be distinguished entirely from 'plate tectonics' – the process that very slowly and gradually causes continental land-masses to drift apart or move together at a rate of no more than centimetres per year. What their evidence points to is a titanic rotation of the entire crust of the earth *in one piece* and at a cataclysmically fast rate. According to Dr Joseph Kirshvink, a geologist at Caltech and lead author of the study: 'The rates … were really off the scale. On top of that everything [seems to have been] going the same direction.'

The Caltech researchers point out (as the Flem-Aths did several years before them) that during the period in which this rapid rotation of the crust was underway, the existing lifeforms would have been 'forced to cope with rapidly changing climatic conditions as tropical lands slid up to the cold polar regions, and cold lands became warm.'

asteroids and comets and some of them – each and every time this has happened – have collided catastrophically with the Earth. Since studies of impact craters show intense, sustained and violent bombardments between 96 million and 94 million years ago, between 67 and 65 million years ago, and around 35 million years ago,[31] it would be foolish to ignore the possibility that we could already be five million years overdue for another very bad collision.

Indeed there is much to suggest that we may have been under intense bombardment for at least the last million years and that the frequency of collisions is increasing. It is not disputed that during this period the earth has been afflicted by massive unexplained fluctuations of climate, seismic instability, the onset and relapse of Ice Ages and floods, large oscillations in surface temperature, and regional extinctions of animals.[32] It is possible that all these phenomena could be connected to collisions with cosmic debris and it is also possible that they could be the harbingers of a bigger collision to come – perhaps even as big as the one that caused the worldwide extinction of the dinosaurs 65 million years ago.

We can only hope that the cosmos will spare us such a fate. The 10-kilometre-wide comet or asteroid that crashed into the Gulf of Mexico at that time produced a blast 1000 times as powerful as all the nuclear explosives currently stockpiled on earth. The resulting dust cloud blotted out the sun for more than five years and the entire planet was shaken for decades by aftershocks and volcanic eruptions.[33]

Even with the advanced remote-sensing capabilities of modern satellites, we would have very little time to prepare for impact. It could be just a few days. But it would not, under any circumstances, be longer than two years – even in the case of the largest and most visible comets. Big asteroids, and some reach 30 kilometres in diameter, are almost impossible to detect at very great distances in the abyss of space.[34]

Could a sky-monster like this kill the earth?

Taking the example of an object just 10 kilometres wide – with which we would collide at a speed close to 100,000 kilometres per hour – Professor Emilio Spedicato of the University of Bergamo in Italy has calculated that 'the atmospheric disturbance would be colossal and extended over hemispheric areas'. On the basis of a rather conservative estimate of 10 per cent of the initial energy going into the blast wave, Spedicato calculates 'that at 2000 kilometres from the impact point the wind velocity would be 2400 kilometres per hour and the air temperature increase 480 degrees Centigrade; at 5000 kilometres the velocity would be 400 kilometres per hour and the temperature increase 60 degrees'. Even at a distance of 10,000 kilometres the wind velocity would still exceed 100 kilometres per hour and would blow without any let-up for 14 hours, increasing air temperature by 30 degrees. 'Additional effects in the atmosphere would be chemical reactions leading to the formation of poisonous substances, like cyanogen or nitrous oxide, which would completely remove the protecting layer of stratospheric ozone.' Last but not least, 'landwaves metres high' would tear across the continents, tectonic plates would shift, volcanos would spew forth along the shattered fault lines and there would be 'a global catastrophic *tsunami* [tidal wave] with substantial continental flooding'.[35]

Looking back into the history of the earth, Alexander Tollman and others have

Kihachiro Aratake, discoverer of the Yonaguni underwater monument.

gone so far as to suggest that the ancient global 'myth' of the Deluge may well have its origins in a real event of this nature.[36] Opinions vary as to the precise date of the cataclysm. But all seem to agree that somewhere between 14,000 and 9000 years before the present (12,000 to 7000 BC) something very bad, indeed something terrible, did happen to the earth.

GOOSEBUMPS ON BOTH ARMS

Diving beneath a heavy surface swell, almost strong enough to break the waves into whitetops, we sank down through murky blue waters into a world of submarine silence.

It was late March 1997, soon after the Hale-Bopp comet's closest approach to earth, and we had come to an island that lies exactly one degree north of the Tropic of Cancer and 19.5 degrees east of the monuments of Angkor. This is the remote island of Yonaguni, which prides itself on being 'the most western point of Japan'. Its main claim to fame has traditionally been its possession of a unique species of large moth. In addition, there is a point along the coast where scuba divers can cling to a rocky outcrop at a depth of 30 metres and gaze up at hundreds of fully grown hammerhead sharks schooling above them in a food-rich current close to the surface.

Yonaguni's most experienced diver is Kihachiro Aratake, a bearded salt, crippled by polio in his left leg but massively strong in his chest and upper arms, who makes much of his living from the hammerhead business. Hampered in his movements on land, he turns into a dolphin under water, swimming powerfully into the depths. When pressed he told us that he had once descended in a normal wet-suit and scuba unit to more than 60 metres, almost 200 feet below sea level, in order to free a ship's trapped anchor.

It is Aratake's lifetime project to explore every square inch of the coast of Yonaguni. In 1987, in pursuit of this objective, he was diving off the southeastern shore near a wind-blown headland called Arakawa Bana. He told us that he was hoping he might find another schooling point for hammerheads in these difficult waters. Instead, he made a discovery which some scholars believe could be of immense and disturbing historical significance – a discovery, he remembers, that brought him out in uncharacteristic 'goosebumps on both arms'.

What Aratake had found was an apparently man-made structure, carved out of solid rock in complex shapes and patterns, that lay with its base on the ocean bed at a depth of 27 metres. More than 200 metres long, it rose gracefully before his eyes in a series of pyramid-like steps to a summit platform just 5 metres beneath the surface.

In 1996 we read about the monument in academic papers written by Professor Masaaki Kimura, a leading Japanese geologist from Okinawa University (the University of the Ryukyus). Kimura has studied the monument intensively, making hundreds of dives to it over many years of research. Against the opinion of a number of scientific opponents he adamantly insists that it is a man-made object.

Our interest was piqued and in March 1997 we arranged our first visit to the site . . .

Map of Yonaguni.

MEGALITHS AND PLATFORMS

At first, as we sank beneath the waves, there was confusion, a mass of sensations, streams of rising bubbles. Then, little by little, all became still and we found ourselves drifting weightlessly into blue space, floating over the edge of sheer cliffs and plunging underwater ravines, falling into deepness and darkness.

The sea bed beneath us, sloping downwards to the south, was jumbled with outcrops of bedrock and coral. There seemed to be no special rhyme nor reason to it. Then, quite suddenly at a depth of 20 metres, we came to two huge monolithic blocks – weighing perhaps 200 tonnes each – standing parallel, thrusting upwards almost to the surface. Like the sarsens of Stonehenge in England they appeared to have been deliberately cut and aligned.

A strong current swept around the megaliths and carried us off towards the east, along the side of a perfectly vertical, perfectly straight rock wall. Pushing hard with our fins we swam up this wall, finally pulling ourselves over its edge on to a flat plateau at a depth of 12 metres. Here the waters were calmer and we were able to cling on to crevices and corals, seizing the opportunity to look around and take stock.

Stretching away in all directions we saw the surface of a stone platform which seemed to have been hand-carved into large triangular and rhomboidal patterns with intricate steps and terraces leading down to lower levels and up to higher levels. At the eastern end of the platform we found a straight channel, approximately three-quarters of a metre wide and half a metre deep, running for 8 metres through a raised plinth. At the centre of the monument are four terraces in the rock, each one curiously shaped, and all seeming to lead inwards to a point a little to the west of centre, where four massive stepped courses were set into a corner. We swam down these, pausing on each level, crossed another wide platform further below and then launched ourselves off the edge of the great structure where the last terrace fell sheer away into an open trench that sank vertically to the sea bed at a depth of 27 metres.

We noted that the bearing of the trench was very insistently east–west, perhaps even precisely east–west once the discrepancy between magnetic compass and astronomical directions was taken into account. By contrast, our compasses indicated that the main body of the structure was set along a due north–south axis. Assuming that it was built at a time when sea-level was significantly lower, we

realized that the stepped alcove down which we had just swum would once have looked due south over the waters of the Pacific Ocean.

Between its north and south walls the 4-metre-wide floor at the foot of the trench was littered with the debris of large, apparently quarried blocks that seemed to have fallen from above. Swimming down, we found that one of these blocks lay on top of a shallow spiral stairway rising out of a central basin. Further to the west we came to a series of rectangular niches carved at regular intervals at the base of the north wall.

Filtered through cloudy skies and then through seawater flowing with particles and plankton, the colour of light takes on a cathedral gloom at 27 metres. In these conditions, at the bottom of the trench, it was hard to gain any kind of perspective on the mysterious monument – if it was a monument at all – which loomed above us.

Professor Kimura has staked his considerable professional reputation on the assertion that it is a monument. He suggests that it may be linked in a deliberately contrived geodetic triangle to an ancient shrine on the peak of Yonaguni's highest mountain, lying to the north-west, and to another point along the eastern shore beneath which he believes that further submerged ruins are likely to be discovered. His argument is based on the well-established geological fact that more than 9000 years ago Japan's widely scattered Okinawa island chain was part of a narrow but continuous peninsula linked to the Chinese mainland. Rising sea-levels at the end of the last Ice Age caused the peninsula to disappear almost completely beneath the waves, leaving only scattered highland remnants.

Yonaguni is one of those remnants and if its underwater structure really is a man-made monument as Kimura asserts then history will have to be rewritten. On the other hand several experts have disputed his opinion, arguing hotly that the monument must be natural because no civilization is known to have existed anywhere in the world 9000 years ago with sufficient technology or organization to create such a wonder.

SCHOCH AND WEST

In the hope of settling this issue once and for all we returned to the island in September 1997 with Professor Robert Schoch – the Boston University geologist whose work has stirred up an international debate over the age of the Sphinx (see Part II). Also along on the trip was John Anthony West, the Egyptologist who encouraged Schoch to study the Sphinx.

Before their first dive both men were convinced from video footage and photographs available on the Internet that the Yonaguni monument must be man-made. First-hand acquaintance with it, however, had the effect of making them less and less sure.

After the fourth dive Schoch announced that he was finding it impossible to reach a final opinion. 'If you were to just put me down in isolation on any one part of it,' he said, 'I would instantly conclude that it was a natural rock formation, but when I stand back and look at the whole thing in context I'm more inclined to feel that it has been worked on.'

John West was equally nonplussed. 'I thought this thing was going to be the smoking gun that would prove the existence of a lost civilization,' he lamented, 'but it's turned out ambiguous. Some bits of it really do look like they've been cut into shape. Others seem totally natural.'

'Maybe it's both,' we suggested. 'Maybe we're dealing with a religious cult that saw symbolic significance in natural regularities and faultlines in the local rock and that then set about enhancing those features in an artistic way.'

At the end of our three-day visit to Yonaguni, after several more dives, Schoch's opinion had hardened against the notion that the monument could be artificial: 'I believe that the structure can be explained as the result of natural processes ... The geology of the fine mudstones and sandstones of the Yonaguni area, combined with wave and current actions and the lower sea-levels of the area during earlier millennia, were responsible for the formation of the present Yonaguni Monument about 9000 to 10,000 years ago.'

KIMURA'S EVIDENCE

After leaving Yonaguni we flew to Okinawa for a meeting with Professor Masaaki Kimura, the Japanese geologist at the University of the Ryukyus who argues that the monument is artificial. Kimura's case rests on a number of powerful pieces of evidence which he explained to Schoch over charts and photographs. For example:

1 Blocks carved off during the formation of the monument are not found lying in the places where they should have fallen if only gravity and natural forces were operating; instead they seem to have been artificially cleared away to one side and in some cases are absent from the site entirely.

2 In relatively small local areas of the monument it is common to find several completely contrasting features in close proximity to one another – for example a raised edge, two metre-deep circular holes, a stepped, cleanly angled geometrical depression, and a perfectly straight narrow trench. If only natural erosional forces had been at work, one would expect them to have acted uniformly on the same member of rock in the same locality of the monument. That such stark differences of topography can be observed side by side is therefore strong evidence in favour of artificiality.

A series of steps rises in regular intervals up the south face of the monument.

Deep symmetrical trench on the upper surface of the monument.

Paired monolithic blocks.

3   On the higher surfaces of the structure there are several areas which slope quite steeply down towards the south. Kimura points out that deep symmetrical trenches appear on the northern elevations of these areas which could not have been formed by any known natural process.

4   A series of steps rises at regular intervals up the south face of the monument from its base, 27 metres under water, towards its summit less than 6 metres below the waves. A similar stairway is found on the monument's northern face.

5   A distinct 'wall' encloses the western edge of the monument. It is difficult to explain its presence as the result of natural processes, because it consists of limestone blocks not indigenous to the Yonaguni area.

6   What looks like a ceremonial pathway winds around the western and southern faces of the monument.

'After meeting with Professor Kimura,' Professor Schoch later reported, 'I cannot totally discount the possibility that the Yonaguni monument was at least partially worked and modified by the hands of humans. Professor Kimura pointed out several key features that I did not see on my first brief trip . . . If I should have the opportunity to revisit the Yonaguni monument, these are key areas that I would wish to explore.'

On the face of things it seems unbelievable that leading scientists, who have studied the monument at first hand, are unable to reach instant agreement on whether it is a natural rock-formation or whether it has been carved out of the solid rock by human beings. Yet as Schoch himself pointed out to us, such debates and uncertainties have occurred before. For example, when Stone Age axe-heads, arrowheads and flint knives began to be unearthed in the nineteenth century by Victorian geologists, most initially believed that the objects were entirely natural. It took many decades before they were universally recognized as tools that had been made and used by humans.

Precisely the same sort of 'recognition' problems – this time involving an unfamiliar architectural tradition rather than an unfamiliar tool-making tradition – could be responsible for the continuing academic confusion surrounding Yonaguni's underwater monument. For this reason the scientists studying it should take account of all its known and observable features.

THE FINGERPRINTS OF ASTRONOMERS

In our view it may well be significant that the structure incorporates characteristics that would unhesitatingly be recognized as astronomical if it stood above sea-level. It is oriented due south, towards the meridian, and features a massive east–west trench targeted to sunrise and sunset on the spring and autumn equinoxes. It also stands at a latitude that feels 'non-random' from the astronomical point of view – 24 degrees 27 minutes north, i.e. exactly one degree to the north of the dotted line labelled on modern maps and schoolroom globes as the 'Tropic of Cancer'.

The tropics exist because the earth does not revolve vertically on its axis in relation to the plane of its orbit around the sun. Instead, the axis lies tilted away

RIGHT: *One of 2-metre-deep circular holes on the top of the structure. When Yonaguni stood on the Tropic of Cancer more than 9,000 years ago, the sun's rays at midday would have shone directly into such vertical 'wells'. A similar technique using circular wells was used by ancient Egyptian astronomers to observe the sun's zenith passage at Aswan in upper Egypt.*
BELOW: *Curious rhomboidal protruberance at the apex of the Yonaguni monument. Could it have been a shadow-casting device in the epoch when the structure stood above sea-level? Compare the Intihuatana, Machu Picchu, pages 293–5.*

from the vertical by an angle which presently stands at 23 degrees 27 minutes.[37] A consequence of particular importance to astronomers is that the 'celestial equator' – i.e. the extension of the earth's geographical equator into the celestial sphere – intercepts the 'ecliptic' (i.e the plane of the earth's orbit around the sun) at the same angle of 23 degrees 27 minutes. In technical terminology, this angle is referred to as the 'obliquity of the ecliptic'. From any given latitude, it governs the extreme northward and southward positions along the horizon at which the sun is observed to rise during the course of the year. It also determines the precise locations of the tropics. In our epoch these lie respectively 23 degrees 27 minutes north and 23 degrees 27 minutes south of the .earth's geographical equator – latitude belts that 'correspond to the northernmost and southernmost declinations of the ecliptic to the celestial equator'.[38]

The jargon is confusing but the effects are straightforward:

1   The sun achieves its 'northernmost declination' on 21 June, the northern hemisphere's summer solstice. The earth is then at the point in its annual orbit at which the *northern* end of its polar axis (tilted away from the vertical by 23 degrees 27 minutes) points most directly towards the sun. At *23 degrees 27 minutes north latitude* – the present location of the Tropic of Cancer – the sun on the June solstice is seen to stand vertically overhead at midday (see diagram) with its rays making an angle of 90 degrees to the observer and thus casting no shadow.

2   The sun achieves its 'southernmost declination' on 21 December, the northern hemisphere's winter solstice. The earth is then at the point in its annual orbit at which the *southern* end of its polar axis (tilted away from the vertical by 23 degrees 27 minutes) points most directly towards the sun. At *23 degrees 27 minutes south latitude* – the present location of the Tropic of Capricorn – the sun on the December solstice is seen to stand vertically overhead at midday (see diagram) with its rays making an angle of 90 degrees to the observer and thus casting no shadow.

The reader is already familiar with the manner in which the precessional 'wobble' of the earth's axis slowly changes the background stars against which the sun is seen to rise on any given day in the year. The northern tropic is known as the 'Tropic of Cancer' because, at the time when these effects were supposedly first observed and recorded by astronomers – about 2000 years ago – the sun on the June solstice rose against the background of the constellation of Cancer. The southern tropic is called 'Capricorn' for the same reason: 2000 years ago on the December solstice the sun rose against the background of the constellation of Capricorn. Because of precession, however, Cancer and Capricorn are no longer solstitial today. Indeed, strictly speaking – since the sun presently rises in Gemini on the June solstice and in Sagittarius on the December solstice – the 'Tropic of Cancer' should be renamed the 'Tropic of Gemini' and the 'Tropic of Capricorn' should be renamed the 'Tropic of Sagittarius'. Furthermore, the solstitial constellations will continue to shift because of the precessional cycle: within less than one thousand years from now the sun will be rising in Taurus on the June solstice and in Scorpio on the December solstice.

Precession has no bearing on the sun's extreme northwards and southwards rising positions along the horizon during the course of the year and no bearing on the geographical location of the Tropics. What these co-ordinates are affected by – rather significantly over long periods of time – are minute but cumulative changes which modern instruments have detected in the angle of the earth's obliquity. Painstaking observations published in the journals *Science* and *Astronomy and Astrophysics* have confirmed that this angle is not fixed at the present value of 23 degrees 27 minutes.[39] Instead it varies within a range of slightly under two and a half degrees from a minimum of 22 degrees 6 minutes to a maximum of 24 degrees 30 minutes.[40]

Astronomers warn that the *rate* of change across this range may not remain constant, but are satisfied that for at least the last four thousand years:

> the obliquity of the ecliptic has been steadily decreasing ... by approximately 40 seconds of arc per century ... Thus the obliquity has decreased by nearly half a degree between 2000 BC and the present, enough to result in a sizable shift in the azimuth of sunrise and sunset.[41]

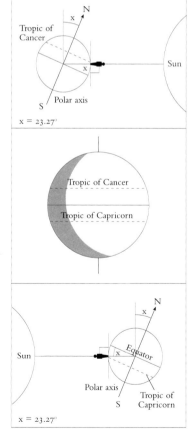

ABOVE: *The formation of the Tropics.*
RIGHT: *How changes in obliquity change the position of the Tropics.*

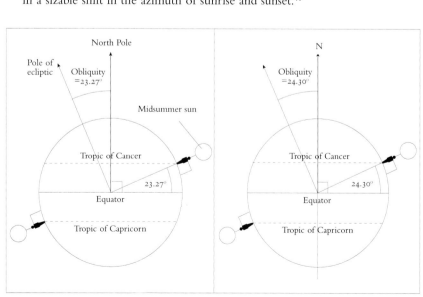

By the same token the latitude of the Tropics has also decreased by nearly half a degree during this same 4000-year period to reach its present value of 23 degrees 27 minutes north of the equator in the case of the Tropic of Cancer and 23 degrees 27 minutes south of the equator in the case of the Tropic of Capricorn. It follows, at some point in the past, that the Tropics must have been located at 24 degrees 30 minutes north and south of the equator (maximum obliquity) and that at some point in the future their latitudes will fall as low as 22 degrees 6 minutes north and south of the equator (minimum obliquity). It also follows that there have been times, accessible to us by calculation, when the Yonaguni 'monument', located at 24 degrees 27 minutes north latitude, would have stood exactly astride the Tropic of Cancer.

A TENTATIVE DATE

Such calculations are invited by some of the monument's apparently astronomical characteristics and, in the absence of a precise geological consensus, could be of great assistance in refining the date of its construction. Moreover, if we accept the modern benchmark figure of 40 arc seconds per century for the rate of change of the earth's obliquity, we find – not necessarily by coincidence – that we are in familiar numerical territory.

Varying between 22 degrees 6 minutes at its minimum and 24 degrees 30 minutes at its maximum, the range of the obliquity cycle is 2 degrees and 24 minutes. Each degree is subdivided into 60 arc minutes. The value of 2 degrees 24 minutes therefore amounts to a total of 144 arc minutes – each of which is in turn subdivided into 60 arc seconds giving a total of 8640 arc seconds (144 × 60) for the complete range.

If we now divide these 8640 seconds by 40 seconds (the estimated change of obliquity in a century) we find that 216 centuries – i.e. 21,600 years – is the period required for the tilt of the earth's axis to shift from its minimum to its maximum value. It also follows that a further period of 21,600 years will pass as the obliquity falls once again from maximum to minimum.

We know that the earth's obliquity today is 23 degrees 27 minutes, that it is declining, and that the maximum figure that can be reached by the cycle is 24 degrees 30 minutes. Since that maximum was last registered, in other words, the obliquity has declined by 1 degree 3 minutes – i.e. by 63 minutes or 3780 seconds. Using the accepted rate of change of 40 arc seconds per century, we can see that the earth's last period of maximum obliquity must have occured 94.5 centuries ago (3780 divided by 40 = 94.5), i.e. 9450 years before the present.

Since Yonaguni's latitude (24 degrees 27 minutes) is 3 minutes less than the figure for maximum obliquity (24 degrees 30 minutes) we can now easily calculate when the monument would have marked the Tropic of Cancer: 3 minutes = 180 seconds, requiring the passage of 450 years at the rate of 40 seconds per century. If the earth's obliquity was indeed at its maximum 9450 years before the present, it would therefore have declined to 24 degrees 27 minutes – and the Tropic of Cancer would have passed through Yonaguni – some 9000 years ago. The Tropic would also have been at the same latitude 900 years earlier, i.e. 9900 years before

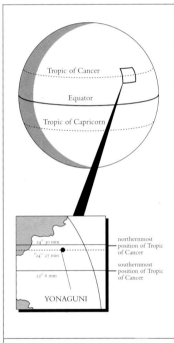

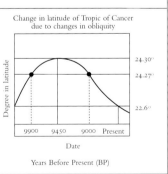

Position of Yonaguni on the Tropic of Cancer circa 9900–9000 BC.

the present, towards the end of the *previous* half-cycle (when the obliquity was rising towards the maximum rather then declining from it).

Astronomical considerations, therefore, suggest that the underwater monument of Yonaguni is likely to have been constructed – and thus to have stood on dry land – somewhere between 9900 and 9000 years before the present. The reader will recall that precisely the same epoch was identified by geologists as the last time that the monument would have stood above sea-level.

A NETWORK?

Suppose that a network of monuments was set up around the world in prehistoric times by unknown navigators and architects – a network marking out the tropics, gridding the earth with set co-ordinates of longitude and latitude, and relating sky to ground through the sequence of 'precessional numbers': 54, 72, 108, 144, 180, 216, etc. Such a network would, of course, have been subjected to disruption by significant earth changes such as rising sea-levels and sinking landmasses, and might, if the changes were sufficiently severe, have required the reconstruction – or even replacement at the nearest suitable location – of particular monuments.

Certainly it is strange that the latest scientific information on the earth's obliquity should suggest a half-cycle of 216 centuries (21,600 years) – a precessional number that would undoubtedly have been of great interest to ancient astronomers, who believed that changes in the heavens must be copied or duplicated on the earth below. Also strange is the geodetic relationship between the monuments of Angkor and Giza, which are separated by 72 degrees of longitude, and the monuments of Angkor and Pohnpei which are separated by 54 degrees of longitude. Stranger still, as we look further east into the Pacific, is the fact that astronomically aligned megalithic structures, of unknown origin, have been found on the islands of Kiribati, 72 degrees of longitude east of Angkor (and thus 144 degrees east of Giza), and Tahiti, 108 degrees of longitude east of Angkor (and thus 180 degrees from Giza).[42] Is it a coincidence that many of these monuments are linked to religious ideas concerning the afterlife journey of the soul – ideas very similar to those expressed in the great temples and pyramids of Egypt and in the hieroglyphic texts of the Books of the Dead?

Precisely engineered megaliths in the Pacific are not confined to Pohnpei, Kiribati and Tahiti, but are also found as far afield as Tonga, Samoa, the Marquesas and Pitcairn Island at longitudes that seem to bear no significant relationship to Angkor and Giza in terms of the precessional scale. The greatest concentration of such unexplained structures, however, occurs on Easter Island – which, at present sea-level, stands as close as it is physically possible to get to 144 degrees of longitude east of Angkor.

Before being 'discovered' on Easter Sunday 1722 by three Dutch ships commanded by Admiral Jacob Roggeveen, Easter Island had been known to its inhabitants by two evocative names – Te-Pito-O-Te-Henua, 'The Navel of the World', and Mata-Ki-Te-Rani, 'Eyes Looking at Heaven'.[43] As we shall see in the following chapters, these ancient names carry clues going to the heart of the mystery of this lost island on the wind-blown edge of nowhere.

CHAPTER 13

ISLAND OF THE SORCERERS

OPPOSITE: *The inscrutable gaze of one of Easter Island's great Moai. The enigma of the island's megalithic sculptures and the origins of the obviously advanced culture that once flourished on this remote spot have not yet been satisfactorily explained by scholars. Could the answer be connected to local traditions that Easter Island was once part of a 'much larger country'?*

KNOWN TO ITS INHABITANTS since ancient times as Te-Pito-O-Te-Henua, 'The Navel of the World', and as Mata-Ki-Te-Rani, 'Eyes Looking at Heaven', Easter Island stands at latitude 27 degrees 7 minutes south of the equator and at longitude 109 degrees 22 minutes west of the Greenwich meridian. These co-ordinates put it just a fraction over 147 degrees of longitude east of the great temple complex of Angkor Wat in Cambodia. Since there is no other habitable land for over 3000 kilometres in any direction in the surrounding wastes of the Pacific, this is as close as it is physically possible to get at present sea-level to the magical precessional figure of 144 degrees of longitude east of the 'Angkor meridian'. The island, moreover, is part of a massive subterranean escarpment called the East Pacific Rise, which reaches almost to the surface at several points. Twelve thousand years ago, when the great ice caps of the last glaciation were still largely unmelted, and sea-level was 100 metres lower than it is today, the Rise would have formed a chain of steep and narrow antediluvian islands, as long as the Andes mountain-range. One link in that precipitous chain would have extended more than 300 kilometres to the west of the peak later named Te-Pito-O-Te-Henua and would have reached out towards a point in the ocean located exactly 144 degrees east of Angkor. Is it possible that some sort of solar observatory or temple of the stars might have been located at this geodetic centre in remote prehistory, later to be drowned by rising sea-levels?

Such speculation is fuelled by the fact that when the American nuclear submarine *Nautilus* made her round-the-world voyage in 1958, scientists on board 'called attention to the presence of an exceedingly lofty and still unidentified underwater peak close to Easter Island'.[1] It is a fact that Professor H. W. Menard of the University of California's Institute for Marine Resources has identified 'an important fracture zone in the neighbourhood of Easter Island, a zone parallel to that of the Marquesas archipelago', together with 'an immense bank or ridge of sediment'.[2] It is also a fact, not easy to explain away as coincidence, that the oldest local traditions describe Easter Island as having once been part of a 'much larger country'.[3] These traditions contain confusing and contradictory elements but all agree that in the most distant mythical past:

a potent supernatural being named Uoke, who came from a place called Hiva ... travelled about the Pacific with a gigantic lever with which he pried up whole islands and tossed them into the sea where they vanished under the waves. After thus destroying many islands he came at length to Te-Pito-O-Te-Henua, then a much larger land than it is today. He began to lever up parts of it and cast them into the sea [but] the rocks of the island were too sturdy for Uoke's lever, and it was broken against them. He was unable to dispose of the last fragment, and this remained as the island we know today.[4]

Other legends of the Easter Islanders tell us more about 'Hiva', the mysterious land from which Uoke is said to have come. We learn that it was once a proud island of enormous size, but that it too suffered in the 'great cataclysm' and was 'submerged in the sea'.[5] Afterwards, a group of 300 survivors set out in two very large ocean-going canoes to sail to Te-Pito-O-Te-Henua, having magically obtained foreknowledge of the existence of the island and of how to steer a course towards it using the stars.[6]

MAGIC AND GEODESY

In both Angkor and Pohnpei the arrival and installation of the god-kings (Jayavarman II in the case of the former and the brothers Olosopa and Olosipa in the case of the latter) was orchestrated by a man of high learning described as a 'magician'. In both Angkor and Pohnpei the god-kings were said to have journeyed by boat from a faraway land. And in both Angkor and Pohnpei the site of the sacred city appears to have been selected – perhaps rediscovered would be a better word – by a process of 'geodetic prospecting' in which the monarch made a physical journey around his future capital in a great clockwise circle.

From the dawn of its strange history, almost up to the time of first contact with Western civilization in the eighteenth century, Easter Island too was ruled by a dynasty of god-kings.

The founder of this dynasty was Hotu Matua, the leader of the fleet of two great canoes full of survivors that set sail from Hiva shortly before it sank beneath the sea.[7] Easter Island traditions tell us that this god-king, whose name means 'prolific father',[8] was accompanied by his Queen Ava-Reipua and was instructed by a certain Hua Maka, a magician, who foresaw the destruction of Hiva and made an out-of-body journey in which he located Te-Pito-O-Te-Henua as a place of refuge:

Hua Maka had a dream in which, in spirit, he travelled over the whole of the island[9] ... Having looked over all the bays ... the spirit stopped at Anakena [on the north-east coast] and cried, 'This is the place, and this is the great bay, where King Hotu Matua will come and live.'[10]

Following Hua Maka's magical vision-quest, the physical exploration of Easter Island is said to have been undertaken by seven sages – 'kings' sons, all initiated men'[11] – who travelled 'from Hiva in a single canoe'.[12] It was their mission to 'open the way' for Hotu Matua and to prepare the island for settlement.

Let us not forget that ancient Egypt, too, had its 'Seven Sages', said in the Edfu Texts to have fled to the Nile Valley from a far-off island – the 'Homeland of the Primeval Ones' – that had been destroyed by a flood (see Part II). Sometimes referred to as the 'Builder Gods', the Edfu inscriptions leave us in no doubt that the principal task of the Sages was to construct 'sacred mounds' at key locations throughout the land of Egypt with the long-term objective of bringing about the 'resurrection' of their destroyed 'former world'. Is it a coincidence, therefore, that one of the first tasks performed by the seven initiated men from Hiva after their arrival on Easter Island, was the construction of 'stone mounds'?[13]

It was only after these mounds had been prepared that Hotu Matua's two canoe-loads of refugees from Hiva came in sight. Before landing at Anakena, however, it was for some reason necessary for the settlers first to circle the island, as though fulfilling the demands of a ritual:

> Then the two canoes separated. Hotu Matua's went round the island eastwards [clockwise]. Queen Ava-Reipua's went round it westwards [anticlockwise]. They met again at the opening of the bay of Anakena and each canoe went towards one of the two rocky headlands that bound it. The King went to the point called Hiro-Moko; the Queen landed at that called Hanga-Ohiro.[14]

Ahu Ature Huki: the squat and bizarre eighth Moai at Anakena.

MYSTERIES

One afternoon around the June solstice – mid-winter in the southern hemisphere – we stood on the white sand beach between the two rocky headlands of Anakena Bay. Behind us was Ahu Nau Nau, a massive stepped pyramid of hulking stone blocks culminating in a long, flat platform. Mounted on this platform, with their backs to the sea, towered seven extraordinary statues, one a torso only, one headless, one intact but bare-headed, and four wearing gigantic red stone crowns.

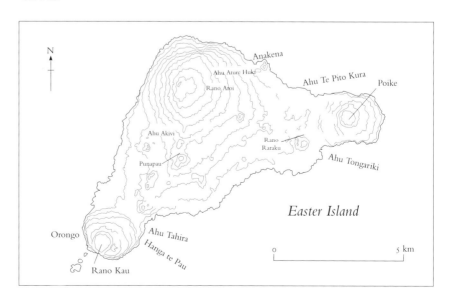

Map of Easter Island.

*Ahu Nau Nau: the seven sages of
Anakena Bay?*

Some scholars have speculated that these seven 'Moai' (literally 'images'[15]) represent Easter Island's original Seven Sages, the forerunners of Hotu Matua. There can be no certainty of this, particularly since an eighth statue, squat and bizarrely formed, stands off to the side of the bay on the nearby Ahu Ature Huki. Indeed, there is no certainty about the purpose or significance of *any* of the more than 600 great statues of Easter Island.[16] They represent a nearly pristine mystery, one that has been repeatedly challenged by generations of investigators in the past three centuries but that has never been satisfactorily solved.

The mystery concerns a vanished primeval homeland – the legendary island of Hiva that was swallowed up by the sea – and the claim that a small band of people survived the cataclysm of 'Uoke's lever' and eventually settled on the rocky peak, still above the waves, that they called Te-Pito-O-Te-Henua. Are we to dismiss such traditions as pure fancy? Or could there be something to them?

The mystery concerns a people who at one time must have been accomplished seafarers – for only consummate navigators and sailors could have succeeded in bringing their vessels intact to so remote a spot as Te-Pito-O-Te-Henua.

And the mystery also concerns a people who must already have possessed a well-developed tradition of architecture and engineering when they arrived at the 'Navel of the World' – for there is little trace of experimentation and trial-and-error in the execution of the great Moai. On the contrary, the consistent and carefully thought out artistic canon expressed in these unique works of sculpture appears to have been fully elaborated *at the very beginning* of Easter Island's remarkable statue-making phase – with the best Moai often being the earliest ones.[17]

The same goes for the massive stone platforms known as 'Ahu' on which many of the Moai stand: once again the earlier examples tend to be superior to those built later.[18]

Archaeologists believe, probably correctly, that they have got Easter Island's chronology pretty well worked out:

* The earliest accepted evidence of human settlement comes in the form of reeds – carbon-dated to AD 318 – from a grave at the important Moai site of Ahu Tepeu.[19]
* The next evidence is charcoal, carbon-dated to AD 380, found in a ditch on the Poike peninsula.[20]
* The next carbon-date, AD 690, comes from another important Moai site, Ahu Tahai – from organic materials apparently incorporated at the time of building into the Ahu platform itself.[21]

Ahu Tahai is therefore regarded by archaeologists as 'the earliest such structure dated so far'.[22] Its Moai, on the other hand, which cannot be directly dated by radio-carbon, are thought to have been added much later. This is because what is described as Easter Island's 'earliest known classical statue'[23] stands alone just to the north of Tahai. 'Contextual' evidence, and radio-carbon tests on associated organic materials, have persuaded archaeologists to assign this 20-tonne, 5-metre-tall Moai to the twelfth century AD.[24] Paradoxically, however, they do admit that 'the classic statue form was already well developed' by that time.[25]

Ahu Akapu.

Thereafter 'classical' Moai continued to be sculpted in large numbers for approximately half a millennium until the last, 4 metres tall, was erected at Hanga Kioe at around AD 1650.[26] Seventy-five years later, after a series of genocidal wars between the two principal ethnic groups on the island (the so called 'Long-Ears' and 'Short-Ears'), the much diminished population had its first, fateful contact with European sailing vessels. Predictably, random murders, kidnappings, systematic slave raids, and epidemics of smallpox and tuberculosis followed with such intensity that by the 1870s the population of Easter Island had been reduced to just 111 individuals. This tiny group of survivors included not a single member of the island's hereditary cast of teachers and initiated men, the Ma'ori-Ko-Hau-Rongorongo, all of whom had been abducted and carried off during a ferocious Peruvian slave raid in 1862.

HIEROGLYPHS

The little that is known about the Ma'ori-Ko-Hau-Rongorongo forms part of the enduring mystery of Easter Island, for the word Ma'ori in this context means 'scholar' or 'master of special knowledge'.[27]

The first of these Ma'ori masters (not to be confused with the Maori people of New Zealand) were said to have come to Easter Island with Hotu Matua himself. They were scribes, literate men. Their function was to recite the sacred words written on 67 wooden tablets that Hotu Matua had brought with him from Hiva[28] and, when the originals became rotten or worn out, to recopy the writings on to replacement tablets.

This is not a myth: 24 of the so-called 'Rongorongo tablets' have survived to this day. Their old and complete name was Ko Hau Motu Mo Rongorongo, meaning literally 'lines of script for recitation'.[29] Generally they take the form of flat wooden boards, somewhat rounded at the edges, shiny with use and age. Inscribed on these boards, in neat rows a centimetre high, are hundreds of different symbols – animals, birds, fish, and abstract shapes. Linguists point out that there are far too many of these symbols 'to suggest any sort of phonetic alphabet or syllabary'[30] and argue that they are a fully developed hieroglyphic script somewhat similar to that of ancient Egypt or of the Indus Valley civilization.[31]

The sequence of the writing on the Rongorongo tablets has attracted particular attention because it is:

> a rare and curious one called 'reversed boustropedon' – that is, each line of script when it reaches the edge of the board turns back upside-down to form the next line. This means that to read the script one must turn the board around at the end of each line. There is no doubt that this writing was inscribed by experts and that it represents a work of art as well as a script.[32]

Oral histories collected on Easter Island make it clear that the knowledge of how to read and write this script was transmitted from generation to generation – and indeed was formally taught at a special circular schoolhouse established at Anakena – until 1862 when the slave-raiders took away the last of the Ma'ori-Ko-

The Rongorongo script of Easter Island.

Hau-Rongorongo.[33] Up until that time, when the golden thread of tradition that attached Easter Island to its past was so brutally severed, Anakena had also been the scene of an important annual festival at which 'the people were assembled to hear all the tablets read'.[34]

Some brief recitations were given to European and American investigators in the nineteenth century but the script was not deciphered. Since then several scholars have claimed to have 'cracked' the code – most recently in 1997 – but none of these claims have come to anything. The truth is that today we can only guess at the contents of these sacred tablets and wonder why, for so long, they were accorded such importance by the Easter Islanders. It is guesswork, too, when we attempt to explain something even more fundamental – how and why the script and the tablets ever came into existence *at all* in such an unlikely spot. Father Sabastian Englert, a Bavarian Capuchin monk and part-time archaeologist who lived on Easter Island for more than 30 years, saw the problem very clearly:

> Written languages, wherever found, are almost always the product of large societies and complicated cultures which have great masses of information that require recording. They result from this need and are indeed unusual as products of small and isolated groups. That a script would be needed or invented by the tiny community of Easter Island is genuinely astonishing. Yet no source away from the island from which this script could have been derived has yet been identified.[35]

'THOSE OLD WORKERS ...'

The mystery of Easter Island so far seems to have at least four distinct ingredients:

- the mystery of Hiva, the legendary homeland of the gods, supposedly destroyed by a flood;
- the mystery of the master mariners who first guided a fleet of refugees from Hiva to the remote shores of Te-Pito-O-Te-Henua;
- the mystery of the master architects who first conceived the great Ahu and Moai;
- the mystery of the master scribes who understood the Rongorongo inscriptions.

Such sophisticated skills are the hallmarks of an advanced civilization. To find them brought together and *focussed* on a remote island in the Pacific, apparently all at once, is extremely hard to explain in terms of the normal 'evolutionary' processes usually ascribed to human societies. An alternative that many scholars have considered, therefore, is the possibility that the Easter Islanders did *not* in fact develop these skills in splendid isolation but rather received them as an influence – as a legacy – from elsewhere.

We do not wish to dwell here on the old and tired debate about whether Easter Island was first peopled (and thus culturally influenced) from the west, i.e. from Polynesia, or from the east, i.e. from South America. Since it is obvious that the first settlers of Easter Island were master navigators and seafarers, it should also be obvious that such a people in their heyday could have travelled extensively, not only to the islands of Polynesia but also much further afield to Latin America and perhaps even beyond. We think it is for precisely these reasons that Easter Island shows clear signs of prehistoric contact with both the South American mainland and with Polynesia – the chicken and the banana, for example, could only have been introduced from Polynesia whilst the sweet potato, the bottle gourd and the totora reed could only have been introduced from South America.

At least during the early phase of Easter Island's settlement, when the people still remembered how to navigate and sail ocean-going vessels, such items are likely to have flowed quite frequently in both directions – together, we assume, with many other valuable commodities, including skills, knowledge and artistic and religious ideas. We are therefore not surprised that monolithic statues that are superficially similar to the Moai (though in far fewer numbers) have been found in the ruins of Tiahuanaco in the Andes mountains of South America more than 4000 metres above sea-level,[36] in the Marquesas islands of Polynesia,[37] and in several other locations.[38] Likewise, it does not surprise us that the Ahu of Easter Island have been compared to the *marae* platforms of Polynesia[39] and, in the case of Ahu Tahira to 'the finest Inca masonry'.[40]

We are quite sure that at least some of these comparisons are valid and that mutual influences will ultimately be proved to have been at work – though not necessarily very frequently – in the prehistoric cultures of Easter Island, South America and Polynesia. Nor is this a controversial proposition, since the majority of orthodox archaeologists would be willing to support it. What is far less certain, however, is the question of Easter Island's role in the larger scheme of things – which may have been much more than just that of a passive 'recipient' of external influences. Its efficent cadre of literate architects and sculptors, whose predecessors had found the 'Navel of the World' through extraordinary feats of astro-navigation, were clearly people of the highest determination and calibre. Until the time when evil entered their community, not long before the first contact with Europeans, they had dedicated themselves single-mindedly for hundreds of years to the creation of transcendant and awe-inspiring works of religious art.

We are not the first to suspect that they must have been driven to do all this by an overwhelming sense of *purpose* which, if it could only be fathomed out, might offer the key to the whole labyrinthine mystery. In the words of Mrs Scoresby Routledge, an intrepid British traveller and researcher who spent a year in Easter Island between 1914 and 1915:

> the shadows of the departed builders still possess the land. Voluntarily or involuntarily the sojourner must hold commune with those old workers; for the whole air vibrates with a vast purpose and energy which has been and is no more. What was it? What was it?[41]

The possibility that we intend to pursue here is that the purpose of Easter Island's high initiates may have been connected to the same underground stream of archaic spiritual gnosis that we have identified at Angkor in the first millennium AD and at Giza in the third millennium BC – and that appears to have originated outside both those areas and before recorded history began. We wonder also whether the very real similarities that have been noted linking Easter Island,

The fine megalithic architecture of Ahu Tahira (detail, above) has frequently been compared with the Inca and pre-Inca masonry of Peru (compare page 272).

Tiahuanaco in South America, and various anomalous megalithic structures in the Pacific, might have as much to do with such an *ancient and indirect, third-party influence* – touching all these cultures – as with the direct contacts that did also undoubtedly occur between them.

FROM HEAVEN TO EARTH

The oldest evidence that we have thus far considered of a shadowy 'influence' at work behind the scenes of history has been in Egypt. There it was associated with traditions concerning a mysterious group of 'semi-divine' beings called the Shemsu Hor – the 'Followers of Horus' – who are said to have settled in the Nile Valley in a remote epoch, thousands of years in the past, referred to as 'the early primeval age'. As we saw in Part II, the influence of these settlers, and their crucial role in shaping and directing the later civilization of the historical Pharaohs, is spelled out in many of the Egyptian funerary and rebirth texts and is a particular focus of the remarkable Building Texts carved on the walls of the temple of Horus at Edfu in upper Egypt.

Easter Island's earliest known classical Moai on Ahu Teriku, just north of Ahu Tahai.

The fragmentary legends concerning the settlement of Easter Island by the seven sages and by the dynasty of the god-king Hotu Matua contain elements that are reminiscent of the Edfu Building Texts. In both cases we have an original island of the gods – called 'Hiva' by the Easter Islanders and the 'Homeland of the

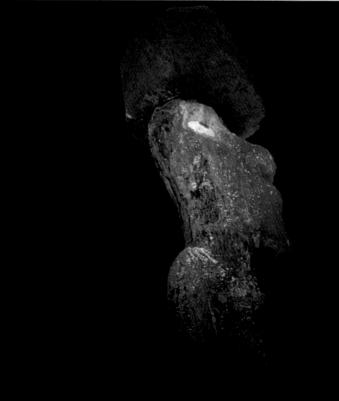

Boat 'grave' and Moai, Ahu Vaiuri

Detail of construction, Ahu Nau Nau. The Ahu is not the work of one epoch but of many.

Primeval Ones' by the ancient Egyptians. In both cases the island is destroyed by a violent storm and flood – attributed to 'Uoke's lever' by the Easter Islanders and depicted as 'a great serpent'[42] in the rich symbolism of Edfu: 'The aggression was so violent that it destroyed the sacred land with the result that the divine inhabitants died.'[43] In both cases the gods of the former homeland were said to have been 'drowned beneath the waters'.[44] In both cases there were survivors. In both cases they fled by boat and came eventually to a land in which they settled. In both cases they were led by god-kings. In both cases their numbers included scribes and architects and astronomers. In both cases, also, these survivors made a particular point of building sacred 'mounds'.

In the case of Egypt we are told that the purpose of the mounds was to serve as the foundations and define the orientation of temples to be built in the future, that they were laid out according to a plan that would in some way 'resemble the sky', and that the object of the whole exercise was to bring about the 'resurrection of the former world of the gods'. The effect of such ideas was that the great temples in the Nile Valley were always built on the foundations of earlier sacred structures. Is it an accident that all of the Ahu of Easter Island also turn out to stand on the foundations of earlier Ahu?[45] Or is it possible that they, too, could be connected to a grand and mysterious scheme, set in motion all around the globe and played out over thousands of years, intended to bring about the 'rebirth' of a 'former world'?

Walking slowly around Ahu Nau Nau at Anakena Bay it was easy to see that the huge platform was not the work of one epoch but of many and that it had been enlarged on several different occasions.[46] Much of the core masonry of the Ahu itself seemed rather rough, with several great boulders, covered with petroglyphs, that had clearly been placed in different orientations in earlier monuments. One of the blocks, on closer examination, even proved to be the decapitated head of an ancient Moai, very weathered and the worse for wear. Under several metres of topsoil to the landward side we also knew that a megalithic stone wall had been excavated in 1987 by Thor Heyerdahl. It was made of large, beautifully cut ashlar blocks and he judged it to belong to a much older layer of construction that had long predated the raising-up of the Ahu.[47]

In addition, the excavators discovered an enormous, stone-lined, boat-shaped enclosure immediately to the landward of Ahu Nau Nau – one of a number of similar structures that have been found near Ahu at various points around the island (there are particularly well-preserved examples at Ahu Vaiuri and Ahu Tepeu). Archaeologists assume that all such structures must have been the foundations of boat-shaped houses in which the early settlers lived. But there are traditions that refer to them as 'boats of bones' and associate them with a builder-god named Nuku Kehu who supposedly came to Easter Island in primeval times with Hotu Matua.[48] There are also seven boat-shaped platforms known as Ahu Poepoe, which were used as tombs. The best example, 21 metres long and 4 metres high, with the bow elevated over a metre above the stern, lies just to the west of Anakena close to the ocean, 'as if it were ready', comments Father Sabastian Englert, 'to carry its deceased passengers to some far away coast'.[49]

The Ahu Poepoe and the so-called 'boat house' foundations are forcefully reminiscent of the 'boat graves' commonly associated with pyramids and tombs in ancient Egypt. In some cases that we have studied – for example at Abydos in Upper Egypt – these are stone or brick replicas of boats. In others they are full-sized sailing vessels – the most famous being the 143-foot-long 'solar boat' found buried in one of the several 'boat-graves' beside the Great Pyramid of Egypt (see page 45).

The symbolic language of the ancient Egyptian funerary and rebirth texts describes the souls of deceased kings passing between earth and heaven in such boats. We were therefore interested to discover an Easter Island legend concerning the god-king Hotu Matua which states: 'He came down from heaven to earth ... He came in the ship ... came to earth from heaven.'[50]

THE GODS WHO STAND UP

In the ancient Egyptian language the word *akh* or *akhu*, sometimes also written *ahu*, means variously 'being of light', 'horizon-dweller', 'shining one' or 'transfigured spirit'.[51] In Easter Island the word *aku* means 'supernatural spirit'.[52] Turning back to Egypt we find that the same word was used regularly as an honorific of the Shemsu Hor, the Followers of Horus – Akhu Shemsu Hor being the full title given to the mysterious cult of divine kings that was believed to have ruled the Nile Valley for thousands of years before the first Pharaoh of the first historical dynasty took the throne.[53] We also came across a curious passage in the ancient Egyptian *Book of What is in the Duat* which tells the initiate that he must 'stand up with the Gods Who Stand Up ('Ahau')'.[54] These were supernatural beings said to have been 9 cubits or approximately 6 metres in height.[55]

That rainy afternoon on Anakena beach we stood up under the tallest of the seven Moai of Ahu Nau Nau. It towered more than 6 metres above us, an 18-tonne monolith carved out of the characteristic red-and-grey volcanic tuff of Easter Island's Rano Raraku quarry – the same material used for all Moai. Surmounting its head, and cut from a different stone (red scoria from the Puna Pua quarries), was its conical crown or top-knot, estimated to weigh an additional 6 tonnes.

Such topknots are found on only a relatively small number of Moai. They are called *pukao*.[56] The largest of them, 1.8 metres high, 2.1 metres in diameter, and weighing approximately 11 tonnes,[57] is to be seen at Ahu Te Pito Kura, 2 kilometres east of Anakena, where it was originally placed on top of the most massive Moai ever successfully erected on a platform. This figure, now fallen, has been calculated to weigh a little under 81 tonnes and was transported roughly 6.5 kilometres from the Rano Raraku quarry.[58] Another Moai, a real monster still inside the quarry, may weigh as much as 90 tonnes and would have stood around 23 metres high had it ever been erected.[59]

The transportation of scores of these gigantic statues to Ahu all around the island, their erection, and then the astonishing 'crowning' of a select few with

The crowning of certain Moai with gigantic topknots was a formidable feat of engineering. Local traditions attribute such feats to sorcery.

unwieldy topknots weighing many tonnes apiece, have rightly been described as 'formidable feats of engineering'.[60] A great deal of scholarly ink has flowed over the vexed issue of precisely how these feats were accomplished on a remote island with a population that has never, even at its peak, exceeded 4000 individuals.[61] Because Easter Island is a subject bedevilled by an intense academic phobia of what are seen as 'cranky' ideas and of the 'lunatic fringe', every archaeologist strives to outdo his or her colleagues in appearing to be completely sane, rational and 'scientific'. No doubt this is the reason why not a single orthodox scholar has ever for a moment taken seriously the numerous old traditions of Easter Island which state, quite matter-of-factly, that the Moai were moved and raised up by the power of *mana*, which means, literally, 'sorcery' – the force of charisma and mental power that the ancient Egyptians called *hekau*.

What the Easter Island traditions preserve is the jumbled memory of an episode in the past when 'great magicians' knew how to move the statues with 'words of their mouths'.[62] The magicians made use of a round stone called Te Pito Kura 'to focus their *mana* power and so command the statues to walk'.[63] Chiefs, too, were said sometimes to possess sufficient *mana* to command the statues to walk or to float through the air: 'The people had to work hard to carve the Moai, but when they were finished the king provided the *mana* to move them.'[64]

Once again a near-parallel situation exists in Egypt, where many of the most spectacular monuments are linked to traditions of the use of magic. In one representative papyrus we read of Hor, an 'Ethiopian magician' who:

> made a huge vault of stone, 200 cubits long and 50 cubits broad, to be above the heads of Pharaoh and his princes, and it threatened to crash down and kill them all; when the king and his people saw this they uttered piercing shrieks. Hor, however, uttered a spell, which caused a great phantom boat to come into being, and he made it carry the stone vault away.[65]

Very similar traditions of miraculous construction techniques were recorded in South America by the first Spanish visitors to the mysterious Andean city of Tiahuanaco (see Part V) with its megalithic statues and gigantic walls and pyramids. The traditions speak of the great blocks coming down from the mountain quarries 'of their own accord, or at the sound of a trumpet' and taking up 'their proper positions at the site'.[66] Much further north, in the Mayan city of Uxmal in Central America, almost identical stories are told about the so-called Pyramid of the Magician. It was said to have been miraculously heaped up 'in just one night' by a dwarf with magical powers who only had to 'whistle and heavy rocks would move into place'.[67] Likewise, there are well-attested traditions that the megalithic city of Nan Madol on the Micronesian island of Pohnpei was built by the sorcery of Olosopa and Olosipa, its god-king founders: 'By their magic spells, one by one, the great masses of stone flew through the air like birds, settling down into their appointed place.'[68]

Perhaps it is a mistake to dismiss all such legends as mumbo-jumbo. Perhaps historians and archaeologists should devote a little less effort to the diligent search for humdrum and prosaic explanations of the mysteries of the human past and pay a little more attention to the extraordinary possibilities that also exist there.

Whether amidst the Pyramids of Egypt, the temples of Angkor, the stone cities of Central and South America, the ghostly basalt walls of Nan Madol, or the Ahu and Moai of Easter Island, the fact is that we know almost nothing about our own prehistory. It could have been a period of long, slow, unremarkable evolution, as most scholars would like to believe. Or it could have been very different, much more subtle and complicated, filled with vitality and imagination, hope and despair. Perhaps one – or several – former high civilizations lie forgotten in the darkest valleys of our collective past, wiped out by nameless cataclysms aeons ago. Perhaps they used advanced technologies, quite different from our own today. Perhaps they had even learned how to transcend technical solutions and to manipulate the physical world by focussed mental power – thus finding it easy to accomplish tasks such as the lifting and moving of huge blocks of stone.

We are already certain, and have said so in previous books, that there has been at least one major forgotten episode in human history – a lost civilization destroyed in the great cataclysms at the end of the last Ice Age. Much connects that civilization to the epoch of 10,500 BC. But the possibilities we are considering here are even more remarkable: the possibility that the system of knowledge once practised by that civilization may have been salvaged from its wreckage by survivors and the possibility that ways may have been devised to distribute the knowledge around the world and transmit it to the future, down through the generations, perhaps even into modern times. This would explain why what appears to be the same well-thought-out system of spiritual initiation using sky–ground dualities in the quest for the immortality of the soul – a system of unknown origins and antiquity – is able to resurface, refreshed on each occasion, in ancient Egypt in the Pyramid Age, in the Hermetic texts of the early Christian era, in Cambodia and in Central America at the end of the first millennium AD, perhaps in Micronesia, as we saw in the last chapter, and perhaps also in Easter Island with its strange aboriginal names: Te-Pito-O-Te-Henua, 'the Navel of the World', and Mata-Ki-Te-Rani, 'Eyes Looking at Heaven'.

CHAPTER 14

SPIDER'S WEB

ON THE LATE AFTERNOON of the June solstice, towards sunset, we reached Ahu Akivi near the centre of the western side of Easter Island. This is an inland site, 3 kilometres from the coast. Like Ahu Nau Nau at Anakena, it has seven Moai, but in this case none of them have topknots and, uniquely, all face west towards the sea – which is clearly visible from the high point on which they stand.

There is a curious tradition concerning these grizzled, otherworldly statues, solemn and powerful, with their blank, aloof eye-sockets gazing out over the limitless ocean. Like most of the other Moai of Easter Island the local belief is that they died, long ago, at the time when *mana* – magic – supposedly fled from the island never to return.[1] However, in common with only a very few of the other Moai, it is believed that these particular statues still have the power, twice a year,[2] to transform themselves into *aringa ora* – literally 'living faces'[3] – a concept that is startlingly similar to the ancient Egyptian notion that statues became 'living images' (*sheshep ankh*) after undergoing the ceremony of the 'opening of the mouth and the eyes'. Statues at Angkor were likewise considered to be lifeless until their eyes had been symbolically 'opened'.

The great stone Moai of Easter Island were at one time equipped with beautiful inlaid eyes of white coral and red scoria.[4] In a number of cases – though not at Ahu Akivi – sufficient fragments have been found to make restoration possible, showing that the figures originally gazed up at an angle towards the sky.[5] It is therefore easy to guess why this island was once called Mata-Ki-Te-Rani, 'Eyes Looking at Heaven'.[6] On a moonlit night its hundreds of 'living' statues scanning the stars with glowing coral eyes would have seemed like mythic astronomers peering into the cosmos. And in the heat of the day those same eyes would have tracked the path of the sun, which the ancient Egyptians called the 'path of Horus' or the 'Path of Ra'. This was also the 'path' pursued by the Akhu Shemsu Hor, the 'Followers of Horus', for whom the exclamation Ankh'Hor – 'the god Horus Lives' – would have been an everyday usage.

The principal astronomical alignments of the great temple of Angkor Wat in Cambodia are towards sunrise on the December solstice and sunrise on the March equinox – respectively midwinter and the beginning of spring in the northern hemisphere. The two moments in the year when Easter Island traditions

OPPOSITE: *Three of the seven grizzled and otherworldly statues of Ahu Akivi.*

The Moai of Ahu Akivi towards sunset on the June solstice. Compare to the second shrine of Tutankhamun's tomb, see page 88.

say that the Moai of Ahu Akivi come alive and are 'particularly meaningful' are the June solstice and the September equinox[7] – respectively midwinter and the beginning of spring in these southern latitudes.

Rigorous archaeoastronomical studies by William Mulloy, William Liller, Edmundo Edwards, Malcolm Clark and others have confirmed that the east façade of Ahu Akivi does have a very definite equinoctial orientation and, indeed, that 'the complex was designed to mark the time of the equinoxes'.[8] Equally firm equinoctial and solstitial alignments have been found at many of the larger coastal Ahu (for example, at Ahu Tepeu, at Ahu Hekii on the north shore, at Ahu Tongariki, and at two particularly striking megalithic Ahu with finely cut trapezoidal blocks at Vinapu).[9] It has also been established that several of Easter Island's other 'well inland Ahu were oriented with the rising winter solstice'.[10] In the case of Ahu Akivi no such solstitial alignment has been detected. Nevertheless as we stood beside its row of ancient Moai and looked west with them on the late afternoon of the June solstice, we thought we could sense a powerful connection between sky and ground. For a moment the rays of the setting sun seemed to penetrate directly into the foreheads of the statues, reminding us vividly of golden figures on the second shrine of the ancient Egyptian Pharaoh Tutankhamun which are also connected through their foreheads to celestial orbs[11] (see page 88). 'These gods are like this,' reads the accompanying inscription, 'the rays of Ra enter their bodies. He calls their souls.'[12]

OPPOSITE: *Eyes looking at Heaven.*

Ra, the name of the Egyptian sun god, appears frequently in connection with Easter Island's sacred architecture, its mythical past and its cosmology. The word *raa* actually means 'sun' in the Easter Island language.[13] The fourth son of the god-king Hotu Matua was called Raa, from whom descended the Raa clan.[14] The syllable *ra* appears in the names of two other clans – Hitti-*ra* (meaning 'sunrise') and U*ra*-o-Hehe ('red setting sun')[15] – and in the names of the island's three principal crater lakes: *Ra*no Kao, *Ra*no A-Roi, and *Ra*no *Ra*raku. There is an also an Ahu at Hanga Papa called Ahu *Ra*'ai. Its orientations have been studied by the archaeologists Edmundo Edwards and Malcolm A. Clark, who agree that at the very least its name is 'suggestive'.[16] According to their calculations Ahu Ra'ai was carefully aligned by its builders to two conveniently located volcanic peaks in order to act as a marker and an observatory for the path of the sun on the day of the December solstice.[17]

BIRDMAN

On the south-western tip of Easter Island, at Orongo, up near the ragged edge of the Rano Kau crater, are four small holes very precisely pecked through the bedrock just beside a large Ahu. Since Orongo is known to have been an important ritual centre, these holes attracted the attention of the Norwegian Archaeological Expedition which visited the island in 1955–56. They were studied by Dr Edwin Ferdon. After making detailed observations at the solstices and the equinoxes he concluded: 'it can definitely be stated that the complex of four holes constituted a sun-observation device'.[18]

Rainbow bridge over Rano Kau crater.

As well as the one Ahu, Orongo also formerly had one Moai, a unique speci-men, carved out of basalt, that was removed to the British Museum in 1868.[19] Perched on a headland with a precipitous drop to the ocean on one side and the gigantic, reed-filled crater of Rano Kau on the other, the main remaining feature of the site is a conglomeration of 54 squat oval houses with massively thick walls of horizontal stone slabs and domed corbel-vaulted ceilings.[20]

The ritual that took place in this setting was the annual 'birdman' contest which was held each September – the month of the spring equinox in the south-ern hemisphere.[21] The origins of this apparently bizarre ceremony are entirely unknown. Its centrepiece was a physical quest for the egg of a sooty tern and specifically for the first tern's egg of the season to be laid on the bird island of Moto-Nui which stands offshore just under a mile to the south-west of the Orongo headland. The quest was undertaken on behalf of noble patrons by young champions called *hopu manu* ('servants of the bird') and officiated by the learned keepers of the inscribed Rongorongo tablets.[22] On a signal from these scribes the *hopu manu* clambered down the cliffs of Orongo and paddled themselves out to the island on small conical reed floats called *pora*.[23] The first to return with a sooty tern's egg would then hand it triumphantly to his patron, who would forthwith be declared the 'Tangatu-Manu' – the sacred 'birdman'. He would be honoured as a king throughout the following year, during which he would shave his head and paint it bright red. At the same time a curious petroglyph of a long-beaked, bird-headed man would be carved to represent him on the rocks of Orongo.[24]

Birdman glyphs at Orongo with the islands of Moto Kao Kao, Moto Iti and Moto Nui in the background.

Manu-tera, the Easter Island name for the sooty tern, means literally 'sunbird'.[25] From this we take it as very likely, though there is no proof, that the tern would have been seen as a symbol of the sun – just as the falcon and the phoenix were symbols of the sun in ancient Egypt.[26] The latter, the mythical Bennu bird, was associated with Heliopolis ('the City of the Sun') and with the pyramid-shaped Benben sunstone, and was famously linked to an egg:

> As its end approached the phoenix fashioned a nest of aromatic boughs and spices, set it on fire, and was consumed in the flames. From the pyre miracu-lously sprang a new phoenix, which, after embalming its father's ashes in an egg of myrrh, flew with the ashes to Heliopolis where it depos-ited them on the altar in the temple of the sun god Ra.[27]

The possibility cannot be ruled out that the birdman cult of Easter Island may have expressed ideas such as these. 'If one were to propose antecedents to the practice,' comments the historian R. A. Jairazbhoy:

> the thought of the Egg of the Egyptian sun god [the cosmic egg] would have to come to mind. *The Book of the Dead* says that this egg was laid by Kenkenur, or 'the Great Cackler' [an alias of the phoenix], and the deceased watches and guards it. This is declared in the Chapter headed 'Having Dominion over the Water in the Underworld'. And again the journey on the reed float across the sea is reminiscent of the journey of the Egyptian sun god Ra to the horizon on reed floats.[28]

Jairazbhoy's remarks, although ignored by other historians, are extremely perceptive:

almost everything about the birdman ceremony would make sense as a quest ritual for the primeval egg of Ra, symbolized appropriately by the egg of *manu-tera*, the sunbird. Particularly interesting in this regard are the reed floats that the Easter Islanders call *pora* – meaning literally 'reed floats of the sun'. Jairazbhoy is right to point out that in the ancient Egyptian *Book of the Dead* reed floats are sometimes depicted as the sun's means of transportation across the sky. We have also come across the same idea in the far older Pyramid Texts, which state: 'The reed floats of the sky are set in place for Ra, that he may cross on them to the horizon . . .'[29]

We see no real difference between the reed floats on which Ra crossed the sky in ancient Egypt and the reed floats of the sun-god Raa, used by the *hopu manu* of Easter Island to cross the waters to Moto-Nui and symbolically retrieve the sunbird's cosmic egg. Moreover, the conical reed floats depicted in the hieroglyphs, which are recognized by archaeologists as 'the earliest craft navigating on the Nile and in the Delta swamps',[30] are indistinguishable from reed floats that were still in use in Nubia and in Middle Egypt well into the twentieth century.[31] These in turn are identical to the reed floats of Easter Island – the only variation being in the materials used (totora reeds in the case of Easter Island; papyrus reeds in Egypt).

As we walked up on the Orongo headland, between the edge of Rano Kau and the cliffs above the sea, we wondered whether there could be a connection. Veiled under the confused mess of history could there be a link between the strange cultural expressions of Easter Island, the birdman ritual, the Ahu and the Moai, and the ancient quest for immortality described in the Pyramid Texts and symbolized by the 'reed floats of the sky' – which enabled not only Ra but also the souls of the deceased to 'cross over to the horizon'?[32] Within that quest, as we saw in Part II, it was *knowledge*, above all else, that was considered essential for those seeking 'the life of millions of years'. Is it therefore merely an accident that the title given to Easter Island's sacred birdman – Tangatu-Manu – means, literally, '*learned* man of the sacred bird'?[33]

The ancient Egyptian religion accorded huge importance to precisely such a learned bird/man figure – long-beaked, ibis-headed Thoth, the god of knowledge and the 'enumerator of the stars', who states in the *Book of the Dead*:

> I am Thoth, the master of laws who interprets writings, the skilled scribe whose hands are pure, who writes what is true, who detests falsehood. I am Thoth, great of magic in the bark of millions of years, who guides sky, earth and the Duat, who nourishes the sun-folk.[34]

Also perhaps of relevance is Utterance 669 of the Pyramid Texts, in which a promise of future life for the King is connected to curious bird and egg symbolism: 'Yours is rebirth in the nest of Thoth . . . Behold the King is in being; behold the King is knit together; behold the King has broken the egg.'[35]

At around 6 p.m. we saw a rainbow form a bridge over Rano Kau crater. By 6.15 it had faded. By 6.40 the sun had dropped below the horizon – just tipped below. The sky was soft orange to the west. And something amazing was happening out to sea directly ahead of us over Moto-Nui where a rainstorm falling out of a cloud seemed to join the cloud to the ocean: an umbilical cord between the cloud and the sea. It was difficult to be sure whether the phenomenon we were watching was a rainstorm, or whether it was the process of a cloud in formation. It was as though the cloud were drawing up moisture from the sea, a dark and heavy cloud marching steadily inland. The roots of its moisture below moved with it, the cloud being nourished by the ocean, the ocean being nourished by the cloud.

At that moment we understood the mystical power of this lonely island and sensed its utter isolation. Surrounded by the Pacific deeps, a vast wilderness more formidable to cross than any desert, it lay open beneath the stars with its eyes gazing at heaven just as its ancient name Mata-Ki-Te-Rani proclaimed.

The word *mata*, meaning 'eye' or 'eyes'[36] in the language of Easter Island, may convey an occult double meaning. Phonetically it is extremely close to the ancient Egyptian word *maat*, meaning 'truth', 'integrity', 'uprightness', 'the right', 'genuineness' etc., and also 'justice', 'balance' and 'cosmic harmony'.[37] The concept of *maat* was personified in the goddess Maat, whose symbol was the feather of truth and who played a key role in the judgement scene of the *Book of the Dead* – at which the eternal destiny of the deceased was decided.[38]

There is also another word *maat* in the ancient Egyptian language. According to Sir E. A. Wallis Budge's authoritative *Hieroglyphic Dictionary* it means, variously, 'eye', 'vision', 'sight', 'something seen', 'tableau', 'things seen', 'visions'.[39] In the *Book of the Dead* it appears quite frequently as the formula 'maat Ra', meaning 'the eye of Ra' – as, for example, in Chapter 17, where we read: 'This is the water of heaven; otherwise said, it is the image of the eye of Ra (*maat Ra*).'[40]

If the name Mata-Ki-Te-Rani is edited by the removal of the words 'Ki-Te' ('looking at') we are left with 'Mata Rani', a coherent expression in the Easter Island and other Polynesian languages which means 'the eye of heaven'.[41] No one could deny that *Mata Rani* is a rather close match phonetically and semantically for the ancient Egyptian *maat Ra*, meaning – essentially – 'the eye of the sun'.[42] Moreover, the focus in both cases is on the skies and on the celestial bodies; in other words, it is astronomical in nature.

SECRET LANGUAGE

There is ample evidence of a forgotten astronomical heritage on Easter Island and it is not confined to Ahu and Moai. As well as the physical proof surviving in the orientations of the island's stone ruins, tantalizing fragments of myths have come down to us expressing a powerful sky–ground dualism very similar to that of ancient Egypt and Angkor.

We mentioned in passing in Chapter 13 that the god-king Hotu Matua was said to have descended from heaven to earth in a great 'ship'. The full version of the tradition is as follows:

He came down from heaven to earth
To both earths, did Hotu-Matua,
returning with the help of heaven to both worlds of his eldest son,

to both worlds, to his world.
He came in the ship of his youngest son,
his best son,
came to earth from heaven.[43]

Recited from a Rongorongo tablet in the nineteenth century by one of the last islanders who could read the script,[44] such ideas are in our view remarkably akin to utterances in the ancient Egyptian Pyramid Texts and *Book of the Dead*, which depict Ra ascending the heavens in the mornings, when he is metaphorically 'youngest', on a solar bark named *Mandet*, and descending at twilight, when he is metaphorically 'oldest', on the solar bark *Meseket*.[45]

The recitation also contains a peculiar reference to two 'earths' that should be instantly familiar to any student of Giorgio de Santillana and Hertha von Dechend's *Hamlet's Mill*. Their great contribution to scholarship was the identification of a technical astronomical language of vast antiquity encoded in myths and monuments – a language and a science going back, as they put it, to 'some almost unbelievable ancestor civilization' that had existed, all around the world, thousands of years before the beginning of recorded history.

It is a language that addresses itself particularly to the precession of the equinoxes and that transmits a distinct series of numbers deriving from the rate of precession – one degree every 72 years, 30 degrees every 2160 years, etc. It is also a language that makes use of certain 'mental models' to facilitate understanding of complex astronomical ideas. One such model was termed 'the earth' in mythology and is defined by Santillana and von Dechend as an imaginary plane laid across the heavens connecting the four 'ruling' constellations of the zodiac against the background of which the sun rises on the solstices and the equinoxes in any particular epoch:

> Since the four constellations rising heliacally at the two equinoxes and the two solstices determine and define an 'earth' it is *termed* quadrangular (and by no means 'believed' to be quadrangular by 'primitive' Chinese, and so on). And since constellations rule the four corners of the quadrangular earth only temporarily [because of the precession of the equinoxes] such an 'earth' can rightly be said to perish, and a new earth to rise from the waters, with four new constellations rising at the four points of the year.[46]

We suggest that the two earths referred to in the Hotu Matua tradition are likely to be connected to ideas such as these. One would be the physical 'earth' here below, *terra firma*, our planet. The other would be the celestial earth, that 'ideal plane laid through the ecliptic' that Santillana and von Dechend have identified in archaic astronomical myths from all around the world. Their evidence points unequivocally to a common source for these myths – an advanced, even 'scientific', source lost in prehistory. The presence of such a myth on Easter Island suggests that it, too, may once have been closely connected to that enigmatic source.

Another fragment of tradition speaks the same coded international language:

> In the days of Rokoroko He Tau the sky fell,
> Fell from above on to the earth.

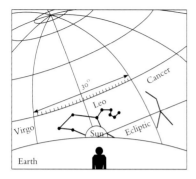

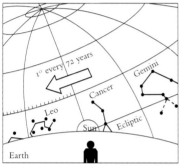

Due to the precessional cycle the background of stars against which the sun rises on any given date moves anticlockwise 1° every 72 years.

The people cried out, 'The sky has fallen in the days of King Rokoroko He Tau.'

He took hold: he waited a given time. The sky returned; it went away and it stayed up there.[47]

In Santillana and von Dechend's 'language' the sky falling from above on to the earth is an allegory for the end of an astronomical world-age – precisely the same kind of disturbance in the heavens that is depicted in the various scenes of the Churning of the Milky Ocean at Angkor Wat and Angkor Thom in Cambodia. Moreover, there is a sense in which the sky really does fall upon the earth that makes the allegory particularly true to life. Were you able to stand gazing east for thousands of years you would see quite clearly that the effect of precession is gradually to lower the altitude of the 'ruling' constellation against the background of which the sun rises at dawn on the spring equinox (see diagram). With the passage of enough time, each constellation is eventually displaced below the horizon and the constellation above it in the zodiac 'falls' to take its place.

Perhaps this is the meaning of the falling sky in the Easter Island tradition, and of the phrase 'it went away'? If so, then we should not be surprised that such a story also informs us that a new astronomical age was soon born ('the sky returned') and thereafter remained in place for a very long time – 'it stayed up there'.

A MANDALA MADE OF STATUES

There are more scattered hints and memories of Easter Island's astronomical heritage. Traditions state, for example, that ages ago there existed on the island a brotherhood of 'learned men who studied the sky'.[48] These 'Tangata Rani' were instantly recognizable because they were 'tattooed on their faces with coloured spots'[49] – somewhat like the astronomer priests of Heliopolis in ancient Egypt, who wore distinctive leopard-skin cloaks with coloured spots.

The Tangata Rani are said to have favoured particular observation points. One was known as 'the cave of the sun's inclination',[50] which, like so much else, suggests a solar focus. However, the Easter Islanders also remember that their ancestors had an intense interest in the stars, known as *hetu*. In 1914 the redoubtable Scoresby Routledge was shown a large flat rock on the Poike Peninsula near the eastern extremity of the island and told that its name was *papa ui hetu'u*, signifying, as the Chilean archaeoastronomer William Liller has pointed out, 'a stellar observatory (literally "rock where they watched the stars")'.[51] On another occasion Mrs Routledge was taken to a north-west-facing cave near Ahu Tahai. Her informants explained to her that it had been 'a place where priests taught constellations and the ways of the stars to apprentices'.[52]

The implication is that for a very long while the island that called itself 'Eyes Looking at Heaven' had an institutionalized system for transmitting astronomical knowledge – a system of initiation with its own apprentices and masters. In ancient Egypt we know that such a system flourished at Heliopolis and that the teachings of the Heliopolitan priesthood were the spiritual and philosophical backbone of the Egyptian state for more than three millennia. Their guiding principle was *maat* and they sought to create a society in balance with itself and with the universe. That they succeeded brilliantly in Egypt is something nobody can dispute. But when we consider the awful predicament of the Easter Islanders, marooned for generations on a tiny triangle of volcanic rock just 18 kilometres by 19 kilometres by 24 kilometres – stuck in the bare heart of the Pacific Ocean like the prisoners of some nightmare Alcatraz – it is clear that theirs, too, is an amazing success story. They might have fallen into claustrophobic despair and self-destructive ruin at any time. Yet it is certain that from at least the eighth to the sixteenth centuries AD, and perhaps for much longer, they maintained a peaceful, stable and constructive society which produced a sufficient surplus to support the very large artisan class of master-builders and master-sculptors necessary for the creation of the Ahu and the Moai. It was a society that the ancient Egyptians would immediately have recognized as one governed by *maat* – a society that was clearly inspired by an overwhelming sense of spiritual power which it expressed through acts of individual and collective artistic creativity. As a result it has bequeathed to the world a riddle of mystic beauty – an inscrutable mandala of more than 600 looming, energized statues which seem to spill out of Rano Raraku, the crater from which they were quarried, in a great spiral with two arms.

BELOW, OPPOSITE: *Stone heads of Rano Raraku.*

PARALLELS AND MERIDIANS

We climbed up the steep outer wall of the Rano Raraku crater, then over the grassy rim and down into the gaping mouth of the extinct caldera. The whole area was covered with partially quarried and complete Moai, and Moai heads, an estimated 276 of them,[53] some upright, some recumbent. Inside the crater all those that were standing, row upon row of them scattered apparently randomly, gazed down towards the reed-fringed lake on the crater floor.

Like a hot meal lying uneaten on a vacated table, these weird and other-worldly figures seemed to suggest an activity that had suddenly stopped. We found it easy to see why the archaeological explanation of the random scattering of the statues is that in a time of social conflict the workers at the quarry just downed tools and walked off the job one night and never returned. However, an alternative possibility is that the layout of the Moai inside and outside the quarry may have been deliberate and that these hundreds of statues, in so spectacular a setting, may have been conceived of as a single composite monument. The smaller heads looked to us almost as though they were growing out of the ground – strange, gnarled roots, or vegetables, or fruits withered on the branch, fashioned from growing, living rock, lined and wrinkled. There was also something solemn about the faces, gazing with their sightless eyes into the eye of a crater lake in the middle of an island in the middle of a vast ocean.

High up on the rim of Rano Raraku, where the surrounding cliff overlooking the sea to the east falls sheer away 180 metres to ground-level, we came upon a curious thing – an open rock-hewn cave with a series of rock-hewn benches or seats lining its walls. Like the Moai all the seats were oriented to gaze down towards the crater lake. Together with various slots, alcoves, and a niche over a metre deep and more than 2 metres high, the overall appearance of these devices was startlingly similar to the rock-cut terraces and steps of the underwater 'monument' at Yonaguni.

David Hatcher Childress reports an Easter Island tradition that seven magicians sat together on the Rano Raraku benches and combined their *mana* to make the statues walk: 'All of the Moai had to walk in the same direction. As they came out of the crater they began walking in a clockwise direction … a clockwise spiral around the island. You can see the ancient road that they walked on.'[54]

There is more than one ancient road to be seen, and at many points there are Moai beside them, fallen on their faces – as though they had all been in graceful motion, like players in a game of musical chairs, and that at the same moment they had all sunk to the ground. There are also other prehistoric tracks known to the Easter Islanders as Ara Mahiva, which are said to have once 'encompassed the whole seaboard of the island' and are considered 'to be the work of a supernatural being'.[55] Of the greatest interest, however, is the following tradition recited from a Rongorongo tablet in 1886 by an elder named Ure Vaeiko:

> When the island was first created and became known to our forefathers, the land was crossed by roads beautifully paved with flat stones. The stones were laid so close together so artistically that no rough edges were exposed. Heke [a name reminiscent of the ancient Egyptian *hekau*, 'magic'] was the builder

of these roads, and it was he who sat in the place of honour in the middle where the roads branched away in every direction. These roads were cunningly contrived to represent the plan of the web of the grey and black pointed spider, and no man can discover the beginning and the end thereof.[56]

The recitation broke off at this point because of 'unintelligible text in another language' that Ure Vaeiko did not recognize, then he completed it with a second cryptic reference to 'the pointed spider'. Originally, he intoned, this creature had lived in Hiva (the legendary flooded homeland of Easter Island's first settlers) and 'would have mounted to heaven, but was prevented by the bitterness of the cold'.[57]

Coupled with the hint of an antediluvian source, the most striking fragments of information conveyed by Ure Vaeiko's recitation are in our view the following:

1   there was a strange network of 'roads' laid out like a 'spider's web';
2   the network emanated from a 'place of honour in the middle';
3   evidently it had some sort of connection with 'heaven'.

We suspect that such imagery might hark back to a long-lost system of global and celestial co-ordinates – a network of parallels and meridians similar to the latitude and longitude lines on modern maps and to their 'heavenly' counterparts, declination and right ascension, which astronomers use to grid the sky.

The analogy of a web, or network, was well known and widely used in the ancient world. For example, the Chinese mapmaker Chang Heng, the inventor of quantitative cartography, was described as early as AD 116 as having 'cast a network of co-ordinates about heaven and earth and reckoned on the basis of it'.[58]

NAVELS OF THE WORLD

Easter Island was called 'Eyes Looking at Heaven', but it was also called Te-Pito-O-Te-Henua, 'The Navel of the World', a name that was supposedly bestowed on it by the god-king Hotu Matua himself.[59] What is strange, as we shall see in Part V, is that it shares this name with Cuzco – meaning 'Navel'[60] – the incredible megalithic capital of the Inca empire high up in the Peruvian Andes. Moreover, the same name, or idea, was applied in ancient times to many other ritual and sacred 'places of honour in the middle'. In all cases where there is sufficient evidence to make a judgement, these turn out to have been revered as centres of geodesy and geometry and of the related art of geomancy – a word that means, literally, 'earth divination'.[61]

Frequently such 'Navels of the Earth' also prove to have associations with meteorites – stones fallen from heaven. Many will have their own 'navel stone', or 'sunstone', or 'foundation stone', which will sometimes be accompanied by a tradition of a rod or pillar sunk into the earth or of an obelisk raised up. Each will additionally be depicted as a primordial centre of creation, from which all else grows: 'The Holy One created the world like an embryo. As the embryo proceeds from the navel outwards, so God began to create the world from its navel onwards, and from there it was spread out in different directions.'[62]

The Temple Mount and Dome of the Rock, Jerusalem. According to tradition the 'foundation stone' (Eben Shetiyah) beneath the dome marks the centre, or 'navel' of the earth.

Easter Island has well-established traditions concerning meteorites, which are called 'Ure Ti'oti'o Moana'. There are supposedly three of them 'deeply buried in the island's soil'.[63] In addition a mysterious 'round tooled stone',[64] about 75 centimetres in diameter,[65] can be seen to this day sitting near the shore close to Ahu Te Pito Kura, 2 kilometres east of Anakena. Said to be the navel of the island itself, this is the stone referred to earlier that magicians supposedly used 'to focus their *mana* power and so command the statues to walk'.[66] Its name, Te Pito Kura, has been variously translated as 'the golden navel'[67] and 'the navel of light'.[68] It could equally well mean 'navel of the sun', a concept extremely close to that of the ancient Egyptian Benben, the sunstone fallen from heaven[69] that stood on a pillar in the centre of the 'Mansion of the Phoenix' in the centre of the sacred city of Heliopolis – which was conceived of as the centre of the created universe and the site of the original 'Primeval Mound'.[70]

Related ideas occur in ancient Israel in connection with the sacred city of Jerusalem:

> The Holy Land is the central point of the surface of the earth, Jerusalem is the central point of Palestine, and the Temple is situated at the centre of the Holy City. In the sanctuary itself the holy Ark [of the Covenant] occupies the centre ... built on the Foundation Stone ['Eben Shetiyah'], which is thus the centre of the earth.[71]

Jewish legends add that this Eben Shetiyah was the stone pillow used by the patriarch Jacob when he had his famous 'ladder' dream (which speaks of a connection between heaven and earth).[72] Subsequently:

> He took the stone, and set it up for a pillar, and poured oil upon it, which had flowed down from heaven for him, and God sank this anointed stone into the abyss, to serve as the centre of the earth, the same stone, the Eben

Navel stone of Easter island, Ahu Te Pito Kura.

Shetiyah, that forms the centre of the sanctuary, whereon the Ineffable Name is graven, the knowledge of which makes a man master over nature, and over life and death.[73]

There has been reasonable speculation that the Eben Shetiyah may have been a 'firestone, i.e. meteorite'[74] – an idea that is supported in the Book of Chronicles and the Book of Samuel, which both speak of 'fire from heaven' striking the altar in Jerusalem.[75] Moreover, it is clear that it was not unique but rather one of a *class* of objects of this sort.

Perhaps the best-known example is the famous *omphalos* stone of Delphi in Greece, the most prestigious centre of geomancy in the classical world. Like the Benben and the Eben Shetiyah, this 'navel' – for that is what *omphalos* means[76] – was believed to mark the centre of the earth,[77] and to have fallen from heaven.[78] It was specifically identified in Greek mythology as the stone which had been fed to the monstrous time-god Cronus – who devoured his own children – in place of the infant Zeus. When Zeus grew to manhood he took his revenge on Cronus, 'driving him from the sky to the very depths of the universe' after first forcing him to vomit up the stone:[79] 'It landed in the exact centre of the world, in the shrine at Delphi.'[80]

Delphi stands on the slopes of Mount Parnassus, in a vale of great natural beauty overlooking the Gulf of Corinth. Its *omphalos* was a phallic, pillar-shaped, somewhat conical stone. The original has not come down to us but a copy made in the Hellenistic period was found on the site.[81] Carved in relief on its surface is what appears to be a net – archaeologists describe it as a 'woolly net pattern'[82] – or a web of some sort. Like the web of the pointed spider, it is difficult to see where it begins and where it ends.

Greek traditions very strongly associate the Delphi *omphalos* with birds, which is perhaps not surprising, since Greek augurs practised the art of divination by the flight of the birds.[83] Two golden eagles are said to have been depicted in effigy upon the *omphalos* stone, commemorating the belief that Zeus had released two golden eagles from opposite ends of the earth and caused them to fly towards the centre – where they naturally met at Delphi.[84]

Since one bird is said to have flown from the east and the other from the west,[85] their paths would have traced out a great arc, or semi-circle, around the curve of the earth – effectively a line of latitude. As the historian of science Livio Catullo Stecchini confirms: 'in ancient iconography these two birds [which are sometimes represented by doves as well as by eagles] are a standard symbol for the stretching of meridians and parallels'.[86] Stecchini also specifically argues that the web pattern carved on the Delphi *omphalos* was intended to symbolize 'a net of meridians and parallels'.[87]

Delphi was a 'navel of the world'. So, too, was the Bayon in the network of temples at Angkor, described by Bernard Groslier as 'the *omphalos* in Angkor's stone cosmos'. The same function was also shared by the sacred domain of Giza/Heliopolis in Egypt, which came under the governance of Osiris in his most ancient incarnation as Sokar, the god of orientation and balance, who also held sway over the Fifth Division of the Duat (a division frequently referred to in ancient texts as 'the Kingdom of Sokar').[88]

Hellenistic copy of the omphalos – *'navel-stone' – of Delphi. The original, now lost, was said to have fallen from heaven.*

The Temple of Apollo – the Greek sun god – at Delphi on the slopes of Mount Parnassus. Like the Temple Mount at Jerusalem, the shrine at Delphi was believed to mark the exact centre, or navel, of the world.

In the *Book of What is in the Duat*, 'the Kingdom of Sokar' features a prominent depiction of an *omphalos* stone on which perch two birds.[89] A physical example of such an *omphalos* was excavated in Upper Egypt by the American archaeologist G. A. Reisner in the sanctuary of the great Temple of Amon at Karnak,[90] giving substance to Greek traditions of 'doves' flying between Karnak and Delphi.[91] Authorities like Peter Tompkins, who worked closely with Stecchini, and John Michel, in his important study *At the Centre of the World*, present compelling evidence that a network of such centres, in constant communication with one another, did once exist spread out widely around the globe:

> Because of the advanced geodetic and geographic science of the Egyptians, Egypt became the geodetic centre of the known world. Other countries located their shrines and capital cities in terms of the Egyptian 'zero' meridian, including such capitals as Nimrod, Sardis, Susa, Persepolis, and apparently even the ancient Chinese capital of An-Yang . . . As each of these geodetic centers was a political as well as a geographical 'navel' of the world, an *omphalos*, or stone navel, was placed there to represent the northern hemisphere from equator to pole, marked out with meridians and parallels, showing the direction and distance of other such navels.[92]

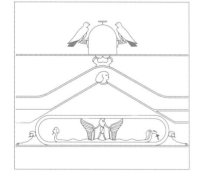

'Omphalos' in the kingdom of Sokar.

GEODETIC BEACON

The very existence of such an ancient world grid has been staunchly resisted by mainstream archaeologists and historians – as, of course, have all attempts to relate known sites to it. Nevertheless, the definite traces of lost astronomical knowledge that are to be seen on Easter Island, and the recurrent echoes of ancient Egyptian spiritual and cosmological themes, cast doubt on the scholarly explanation that the odd name 'Navel of the World' was adopted by the islanders for purely 'poetic and descriptive' reasons.[93] We suspect that Te-Pito-O-Te-Henua may originally have been selected for settlement, and given its name, *entirely because of its geodetic location*.

In an imaginary world-grid centred on Giza-Heliopolis, the temples of

Angkor lie 72 degrees east of the 'zero' meridian, the ruins of Nan Madol on the Pacific island of Pohnpei lie 54 degrees east of Angkor, and the megaliths of Kiribati and Tahiti lie respectively 72 degrees and 108 degrees east of Angkor. If this grid is based on the 'precessional scale' then the next significant number should be 144. When we look 144 degrees of longitude east of Angkor (which also happens to be 144 degrees of longitude *west* of Giza) we find that the only available option in the entire 165 million square kilometres of the Pacific Ocean is Easter Island, which is barely 320 kilometres out of line.

What we are suggesting therefore is that Easter Island might originally have been settled in order to serve as a sort of geodetic beacon, or marker – fulfilling some as yet unguessed at function in an ancient global system of sky–ground co-ordinates that linked many so-called 'world navels'.

We have encountered elements of this system in Egypt and in Angkor. One of its great mysteries is the way in which it constantly mingles the most esoteric forms of spiritual inquiry, and the quest for life after death, with a highly scientific approach to observational astronomy and to earth-measuring. Another mystery is its extraordinary extension, not only geographically but also through time, arising phoenix-like in many different cultures and epochs.

We do not claim to know when it first arrived in Easter Island but we think the evidence supports the theory that it became isolated from its sources, and perhaps forgotten, and that it was then subjected to a very long period of impenetrable isolation, during which it underwent a gradual enfeeblement and dissolution. By the time of the first European contacts, which only hastened the process of collapse, there was very little left either of the spiritual or of the scientific aspects of the system.

A network is still a network and if a precessional scale is being used then the next significant number after 144 should be 180, an increment of 36.

Exactly 180 degrees east of Angkor (and 108 degrees west of Giza), and almost exactly as far south of the equator (13 degrees 48 minutes) as Angkor is north of it (13 degrees 26 minutes), a colossal and unmistakable beacon does exist. It is the outline of a trident, or candelabra, 250 metres high, carved into the red cliffs of the Bay of Paracas on the coast of Peru and it is visible from far out to sea.

It seems to point inland, towards the plains of Nazca to the south and the Andes mountains to the east.

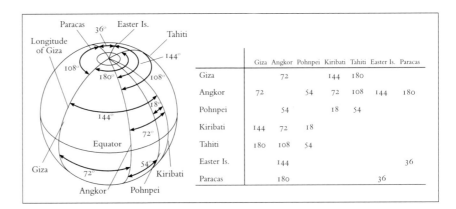

	Giza	Angkor	Pohnpei	Kiribati	Tahiti	Easter Is.	Paracas
Giza		72			144	180	
Angkor	72		54	72	108	144	180
Pohnpei		54		18	54		
Kiribati	144	72	18				
Tahiti	180	108	54				
Easter Is.		144					36
Paracas		180				36	

The world grid.

PART V

PERU AND BOLIVIA

CHAPTER 15

CASTLES OF SAND

TWO THOUSAND MILES OF empty ocean and the deeps of the Chile Basin separate Easter Island from the west coast of South America. A due-east course would lead voyagers from the island to make landfall in Chile. But a course somewhat to the north of east might bring a ship eventually to the safe haven of the bay of Paracas in Peru, which lies on a meridian exactly 180 degrees of longitude east – and west – of the temples of Angkor in Cambodia.

We came across the water from the north in a small open boat, skirting the arid Balestas islands, now a marine sanctuary, and heading for the Paracas peninsula, where rolling sandstone hills and escarpments drop steeply into the sea. From more than 15 kilometres off-shore we had been able to make out the so-called 'Candelabra of the Andes', first through binoculars and then in direct sight. It lay due south of us, carved into a sloping cliff, looming ever larger in our field of view as we approached.

The scholarly consensus is that this huge earth-diagram could easily be 2000 years old and is most likely to have been the work of the same people who created the better-known Nazca lines which are found inland, some 300 kilometres to the south. This 'Nazca culture', about which very little is known, is thought to have flourished from the second century BC until about 600 AD.[1]

CROSS ON THE GROUND, CROSS IN THE SKY

The 'Candelabra' has a rectangular, box-like base, enclosing a circle, out of which rises the representation of a wide central vertical bar, more than 240 metres in length, running north to south. This is crossed, about one-third of the way up, by a triangular contraption running east to west for some 120 metres, supporting two shorter vertical bars. All three bars are surmounted by curious patterns generally interpreted as flames or rays of light.

Because of its auspicious geodetic location half-way round the world from Angkor and 108 degrees west of Giza – sites that both 'resemble the sky' by modelling specific constellations on the ground – we have naturally considered the possibility that the Candelabra could be a work of celestial imitation. What particularly invites this enquiry is the orientation of the diagram. It is set very

PREVIOUS PAGE: *Titanic megaliths of Sacsayhuaman in the Peruvian Andes.*

OPPOSITE: *Seabirds soaring over the 'Candelabra of the Andes', 180 degrees east of Angkor, 108 degrees west of Giza. Like Angkor and Giza could it be an attempt to copy or symbolize a constellation on the ground?*

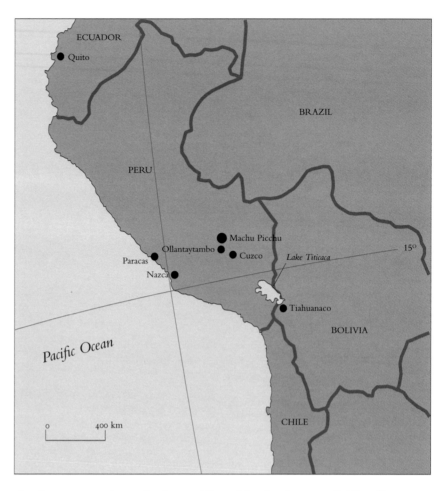

The 'Candelabra' of the Andes as a terrestrial counterpart of the Southern Cross.

closely to true north–south, the meridian of the sky, the great dividing-line across which astronomers in all cultures have traditionally observed the 'transits' of stars.

The Candelabra was intended by its designers to be seen from the north. Indeed, there is no other perspective from which it may be satisfactorily viewed: the observer must face south towards the sloping escarpment on which it is carved. Examining the diagram from the base up naturally draws the eyes towards the southern sky above the escarpment, and specifically towards the south meridian. Although it may be entirely coincidental, computer simulations tell us that at around the hour of midnight on the March equinox 2000 years ago – the epoch in which the Candelabra was probably made – the constellation known as Crux (the Southern Cross) would have been seen lying on the south meridian at an altitude of 52 degrees. At that moment an observer positioned on a boat as we now were, about a kilometre north of the Candelabra, would have seen the Southern Cross suspended in the sky directly above the great cliff diagram.[2]

ENTRANCE TO THE LAND OF THE DEAD

With its triangular cross-beam and its long central axis it is not impossible to imagine that the 'Candelabra' could be an image of the Southern Cross. Furthermore, although this constellation was not recognized by European sailors until the sixteenth century,[3] it was known to the astronomer-priests of the Andes

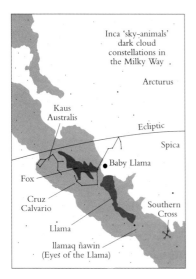

Inca 'sky-animals' dark cloud constellations in the Milky Way .

Arcturus

Kaus Australis

Ecliptic

Spica

Baby Llama

Fox

Cruz Calvario

Southern Cross

Llama

llamaq ñawin
(Eyes of the Llama)

Dark clouds in the Milky Way,
forming Incan 'sky animals'.

for an indefinite period before that.[4] There was also a time when the stars of the Crux were studied by ancient Greek and Egyptian astronomers – until precession eventually carried the constellation below the horizon at such northerly latitudes.[5]

The Southern Cross is part of the Milky Way but what we find particularly striking about it, as we shall see in Chapter 16, is that it lies in the specific sector of the Milky Way that the Incas and their ancestors regarded as the entrance to the land of the dead. It is also adjacent to two 'dark-cloud' constellations, envisaged as a fox and a llama. Since times immemorial Andean traditions have associated these shadowy 'sky animals', formed of interstellar dust, with a flood that destroyed the earth in mythical antiquity – a flood that an earlier race of mankind is said to have been warned of by a 'conjunction of stars'.[6]

Such themes, and their locations in heaven, are in our view too close to the beliefs that we have encountered as far afield as Egypt, Mexico and Cambodia to have come about by chance. In all these cultures the Milky Way – the 'Milky Ocean', the 'Winding Waterway', the 'Way of the Dead', etc. – plays an important role in the afterlife wanderings of the soul. In all of them it is also linked to cyclical time in which celestial 'earths' are constantly being destroyed and renewed by the ebb and flow of precession.

Everywhere else that we have encountered it, this cosmology is associated with a school of sacred geometry and architecture and a cult of celestial imitation that finds a mysterious virtue in making 'copies' on the ground – models, symbolic schemes – of certain constellations in the sky.

Why not also in Peru?

SOUTHERN VIEW

That night in Paracas – it was 11 May 1997 – we took a stroll along the beach during the hour after sunset. There were a few clouds, but they rapidly cleared, and by about 7.15 p.m. we could see the entire dome of the sky arching over our heads, aflame with stars.

We remembered that the ancient Egyptians liked to symbolize the sky as the goddess Nut, a voluptuous, full-breasted woman with her legs and arms spread, her fingers and toes resting on opposite sides of the horizon, her naked body arched over the earth, and her belly decorated with constellations. Amongst these the most prominent was Orion, the celestial image of the god-king Osiris, depicted in ancient Egyptian astronomical texts as a man standing upright in a reed boat holding a staff in one hand and the *ankh*, the emblem of immortal life, in the other.

Looking to our west, we were at first surprised to see Orion clearly visible low in the sky. The surprise was because Orion, at this time of year in the northern hemisphere, has already set when the sky becomes dark enough to show the stars. We could still see the constellation from the Bay of Paracas because May is a winter month at this latitude (13 degrees 48 minutes south of the equator) and night falls more than two hours before Orion sets.

Another surprise was the presence of the Hale-Bopp comet below the body of Orion and a few degrees to the north–west, streaking across the Milky Way towards the rectangle of Gemini, with its broom-like tail still clearly defined. We thought we had seen the last of it a few weeks previously from our own garden in England, when it had ceased to be visible in the evening after sunset. As with Orion, it was the earlier onset of darkness in the southern hemisphere that had brought it back into view.

A third surprise was the general orientation of Orion. Instead of standing upright as it does in the north, our southern-hemisphere perspective tilted it over on its side. In this posture it does not resemble a man and, as we shall see, researchers have demonstrated that it was probably viewed as a gigantic spider by the ancient inhabitants of Peru.[7] The outstretched arms of Osiris holding the staff and the *ankh* transform themselves very easily into the outstretched legs of an imaginary sky-spider.

Touching the horizon in the north-west, the Milky Way glimmered in a diagonal stripe across the sky. Orion, as ever, lay on its western bank. Above Orion, lapping the waters of the celestial river, was Sirius, the dog-star in the constellation of Canis Major, which the ancient Egyptians identified with Isis, goddess of magic. One horn of Taurus, the great sky-auroch, could also be seen jutting over the horizon, and Leo lay overhead.

Looking south, we observed the Southern Cross, still in the east of the sky, rising towards the meridian – which it would reach in another three hours. Also in the east lay Libra and Virgo, while to our north we could see Ursa Major, with the seven familiar stars of the Big Dipper brushing the horizon.

THE OLD LADY OF THE LINES

The next morning we drove south along the Pan-American Highway, through sun-blasted landscapes, to the little town of Nazca. It stands in the midst of a vast desert plateau that was decorated some 2000 years ago with hundreds of abstract lines and geometrical patterns and huge symbolic figures of animals and birds. Many of the designs are on the same scale as the Paracas figure and, as we have already observed, it is a good bet that the same culture was at work in both places.

We had visited Nazca before, in June 1993. At that time we had an encounter with a venerable elderly lady, confined to her bed by advanced Parkinson's disease, named Maria Reiche. She was completely blind but her grey eyes gazed steadily at the ceiling above her as though she were determined to peer through it into the infinite reaches of the sky beyond. Inside her mind, suffused by pure white light, an endless universe of intelligence swirled – galaxies of memories, clusters of connections, visions, presentiments and possibilities.

It was Maria Reiche's fate to bring the Nazca lines to the attention of the world. She was born in the German city of Dresden in 1903. In the 1920s she studied at the University of Hamburg but in 1932, discomfited by the rise of the Nazi Party, she left her homeland to live with an expatriate German family in Peru. Her work in that country eventually brought her into contact with Dr Paul Kosok of New York's Long Island University, who, a few years earlier, had launched an ambitious research project into the mystery of the Nazca lines. After several visits to the site in 1945 she came to live permanently in the town of Nazca in 1946, joining Kosok's staff and later taking over from him when he retired in 1951.

A mathematician and astronomer by training, Maria was pure and frugal by nature, a vegetarian and ascetic who would never marry. It seemed almost as though she had in some way been been 'chosen' to serve as the perfect guardian of a sacred trust. This trust she unwaveringly continued to fulfil, studying and exploring, tending and caring for the great earth-art of Nazca until she was well into her eighties.

Though physically very frail, and quite ill, Maria Reiche was still in full possession of her very sharp mind when we met her on 12 June 1993. For some reason she asked us to hold her hands. Her bones were large and her flesh cool. Briefly her fingers tightened around ours as though expressing some vague anxiety, but her features were perfectly composed. The rear door of her room was open and through it we could see a small flower garden bathed in a pool of afternoon sunlight. A fresh breeze infused the air.

Maria's skin was so drawn and translucent that she seemed almost like a mummified corpse, desiccated by the desert for thousands of years. We could hardly imagine that she would manage to say more than a few words to us. Yet when we put questions to her she responded in a firm voice and in excellent English.

We began by asking for her opinion of the significance of the Nazca lines. She replied:

> They teach us that our whole idea of the peoples of antiquity is wrong – that here in Peru was a civilization that was advanced, that had an advanced understanding of mathematics and astronomy, and that was a civilization of artists expressing something unique about the human spirit for future generations to comprehend.

She said that the Nazca lines were almost as mysterious to her today as they had been when she had first seen them half a century earlier – the time, as she wrote in a letter to a friend, when she discovered that: 'The Gods of Nazca had already stolen me when I was born and locked me up in their castle of sand to play with their immense figures, until I find, one day, the reason of my existence.'[8]

She pointed out that despite the attentions of many scholars, and despite her years of measurement and surveying, the Nazca enigma was still largely unsolved. Nobody could say for certain what the figures were supposed to represent, how they had been drawn with such high precision on such a large scale, or why they were only properly visible from aeroplanes. Her own opinion, which we knew that she had considered carefully over many years, was that the answer would in some way prove to be connected to ancient astronomy. She admitted that only a few specific alignments to the rising or setting of prominent stars had yet been identified at Nazca. Nevertheless, her hunch was that the entire rationale for the great earth drawings must lie in an ancient sky religion.

One possibility that she had seriously investigated was that the drawings could be diagrams of constellations. For example, might not the Nazca 'Monkey' be a figure of the Big Dipper? Might not the 'Spider' be Orion?

ABOVE: *The Nazca Monkey, an earth-diagram of the stars of the Big Dipper according to Maria Reiche.*
BELOW: *The Spider, terrestrial counterpart of the constellation of Orion.*

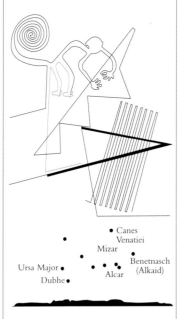

The 'Monkey' at Nazca.

A PLAN OF CELESTIAL IMITATION

We later had the opportunity to study a detailed astronomical interpretation of the Monkey that Reiche had published in 1958, in which she provided evidence in support of her intuition that it represents the Big Dipper (see diagram).[9] She also argued that the figure had been oriented towards the setting of the star Benetnasch in the tip of the Dipper's 'handle'.[10]

Likewise, at a lecture in London in 1976, she documented the manner in which the Spider might have been intended to model and symbolize Orion, with its narrow waist representing the three belt stars and a line through the figure oriented to the setting of Orion in the epoch when the Nazca lines were made.[11]

In recent years Reiche's pioneering work at Nazca has been taken up and greatly extended by Dr Phyllis Pitluga, a senior astronomer at the Adler Planetarium in Chicago. We had dinner with Pitluga in London in 1994. Also at the dinner was our friend Robert Bauval, the author of *The Orion Mystery*, and

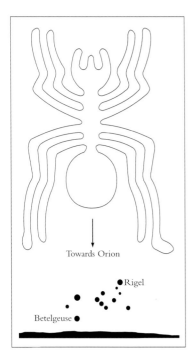

Towards Orion

●Rigel

Betelgeuse ●

The 'Spider' at Nazca.

Dr Pitluga's friend David Parker, a photographer of Nazca.[12] The primary subject of conversation was a television documentary that Pitluga had made, in which she had presented more evidence of a correlation between the Nazca Spider and the constellation of Orion. Confirming Reiche's first intuitions, this evidence showed that the Spider had indeed been designed as an image of Orion as it set along the western horizon approximately 2000 years ago.[13] Pitluga also told us she was quite sure that the Nazcan artist-astronomers must have known about the effects of precession on stellar positions. It was even possible, she suggested, that the numerous straight lines running at different angles through particular figures resulted from attempts to *track* precession and that the figures functioned like 'labels' indicating which constellations the lines were connected to.

In 1996 Pitluga gave a paper which elaborated on these views at the 15th Annual Meeting of the Society for Scientific Exploration. The paper reaffirms that constellations in and around the Milky Way are the most probable 'model of the Nazca figures' and suggests that the figures and lines might have been used to determine long periods of time through the observation of precessional changes in their celestial counterparts.[14] In addition evidence is presented of:

> a physical relationship of present-day Andean plant and animal figures imagined as silhouettes in dark spots along the Milky Way ['dark-cloud constellations'] to figure-lines pointing to the rising and/or setting of the same Andean figure 2000 years ago.[15]

When we returned to Nazca in May 1997 Dr Pitluga had still not published a long-awaited monograph setting out further details of her theory – the gist of which appears to be that a wide-ranging plan of celestial imitation lies at the heart of the Nazca enigma. We understand that her conclusions differ radically from the opinions of most other scholars who have studied the lines[16] and who prefer purely ritualistic and ceremonial explanations.[17] Yet the idea that the Nazca figures might be constellation diagrams is extremely viable. It accords well with the evidence we have presented in this book of a powerful religious influence, disseminated all around the world, that carried down from remote antiquity the distinctive doctrine of sky–ground dualism – 'as above so below'. It is the natural outcome of this doctrine, wherever it is taught, that its adherents will sooner or later attempt, in one way or another, to model heaven on earth.

The web and the figures

It was exhilarating to climb the sky over Nazca in a light single-engined plane. Our take-off altitude – the altitude of the plateau itself – was just over 550 metres. We quickly soared to 750 metres and then followed the course of the Pan-American Highway in a north-westerly direction, gaining height all the time, until we had reached 1220 metres. There, for the next seven or eight minutes, we turned in a widening circle directly above one of the most graceful of the Nazca geoglyphs, the figure of a gigantic Condor with its wings spread to a span of almost 120 metres. Around it, stretching away for 50 kilometres or more in every direction, the desert was etched and inscribed with a tremendous cross-hatching of geometrical patterns and radiating straight lines like a spider's web. We were reminded forcefully of the Easter Island tradition of the 'grey and black pointed spider', whose web was compared to a mysterious network of 'roads' so 'cunningly contrived' that 'no man can discover the beginning and the end thereof'.[18]

We descended to 640 metres, just 90 metres above the figures, to look closely at the Nazca Spider, which lies to the north-east of the Condor and is about 45 metres in length. To the west of it is the Hummingbird, 90 metres long, its narrow beak intersected by a line that targets the point of sunrise on the December solstice.[19] Other major figures in the immediate vicinity are the Monkey, the Dog, the 'Owlman', the Tree, the Hands, the Lizard, several triangles, several trapezoids, and the Whale. There is also a tremendous spiral, like a medieval labyrinth, and two more bird figures – the Parrot (highly stylized) and the mythical 'Alcatraz', with its zig-zag neck and pipette-shaped beak, extending over a distance of more than 610 metres.[20]

The mythical 'Alcatraz' of Nazca with its distinctive zig-zag neck.

More than a kilometre in length, this trapezoid is one of many geometrical and triangular designs at Nazca.

CONSTELLATIONS ON THE GROUND

Like Angkor, like Giza, like the celestially planned cities of Central America, the geoglyphs of Nazca constitute an ambitious cosmic and philosophical puzzle and a mystery worthy of deep concentration – in other words a true 'mandala of the mind'. We think that Maria Reiche was right to insist that the entrance to this mandala will have to be through the door of astronomy.

Human beings naturally perceive and link together the inherent shapes and patterns in any random assembly of objects. The stars in the sky are the arche-typal 'random assembly' and it is notable that all cultures at all times have habitually projected on to the sky a fantastic variety of real and mythical creatures and geometrical shapes.

The Nazca Dog, a figure of the constellation of Canis Major?

Amongst the constellations that Western astronomers recognize today, which are the end-products of a long accretion of astronomical and astrological lore, there is a whale (Cetus), a lion (Leo), several dogs (for example, Canis Major, Pupis), a fox (Vulpecula), several birds (for example, the swan Cygnus and the eagle Aquilla), a scorpion (Scorpio), a variety of dragons and snakes (Draco, Serpens), a lizard (Chamaeleon), a crab (Cancer), a 'sea-goat' (Capricorn), a hare (Lepus), a giant man (Orion), a virgin queen (Virgo), a river (Eridanus), and geometrical designs such as 'the Compasses' (Circinus), 'the Octant' (Octans)', and the 'Southern Triangle' (Triangulum Australae).[21]

The Whale seems a strange visitor to the desert plains of Nazca, unless it is a ground image of a star whale swimming in the waters of the celestial ocean.

Archaeologists agree that the Nazca lines are about 2000 years old. Computer simulations of the sky over Nazca at around 9 p.m. in the evening of the March equinox 2000 years ago show the Big Dipper in the north (the Monkey?) and Orion setting in the west (the Spider?). Extending below Rigel from the back 'leg' of the sky Spider is the river constellation Eridanus. On the ground, the corresponding leg of the Nazca Spider has a curious extension in the corresponding place. If we also accept Reiche's argument that the Monkey, with its striking spiralling tail, is a figure of the Dipper, then what other constellations should we expect to find around it at Nazca?

At the same date and time 2000 years ago Chamaeleon, the lizard, lay in the south, and there is a lizard at Nazca. Alongside Chamaeleon rested the Compasses, Octans and the Southern Triangle – and there are many such gigantic triangular patterns at Nazca. Canis Major, the dog, lay in the west; there is a dog at Nazca. Swimming in its proper place beneath the western horizon in the waters of the celestial ocean was the whale Cetus; there is a whale at Nazca.

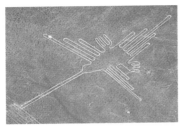

It is possible that the Nazcan artists could have seen the constellation of Cancer as a hummingbird.

Due west in the sky above Orion was the constellation known today as Monoceros (the Unicorn). It could equally be a bird with outstretched wings like the great Nazca Condor. To the north of Monoceros and Orion was Gemini forming a large rectangle in the sky; there is a large rectangle on the ground at Nazca between the Spider and the Condor. Above Gemini was Cancer, the crab – a tripod shape which a different culture might easily have perceived as a small-

ABOVE: *Nazca figures reflecting the*
constellations.
BELOW: *Plan of Nazca.*

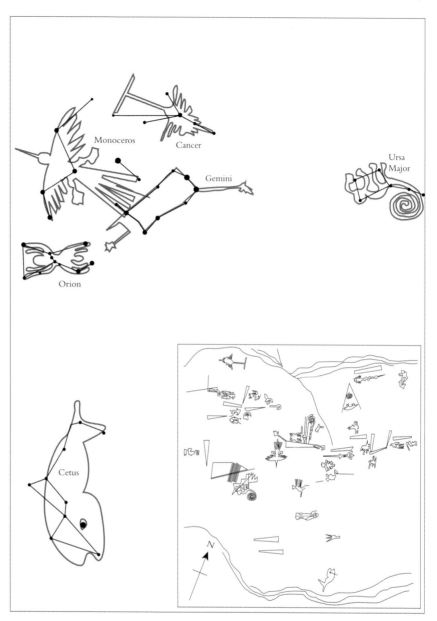

winged bird with a long 'proboscis'; there is a hummingbird at Nazca. North
again was the constellation known today as the Lynx, made up of a long, undulating line; at Nazca the distinctive feature of the 'Alcatraz' is the long undulating
line of its neck.

We are not proposing that the Nazca geoglyphs in any sense constitute an
accurate 'map' or 'chart' of the sky on the March equinox 2000 years ago. They
do not. We are simply making the observation that many of the constellations that
we know to have been present in that sky are curiously 'mirrored' by the patterns
and alignments of the principal Nazca figures. Although their relative positions
appear to have been selected at random (with the ground figures completely 'jumbled up' vis-à-vis the relative positions of the sky figures), it may be significant
that all these constellations lie in a great sweep from north to south around the
western half of the celestial sphere. This was the region in which many ancient

cultures, including the Egyptians and the Maya, believed that the gateway to the netherworld was located. Most, as we have seen, placed it in the general area of the Milky Way between Leo and Orion and there is evidence that Andean astronomer-priests assigned it quite specifically to the intersection of the Milky Way and the ecliptic within the rectangle of Gemini.[22]

Despite the 'jumbling' and 'shuffling' of the figures, therefore, it would be foolish to rule out the possibility that the vast canvas of Nazca could frame a symbolic picture – indeed a mandala – of the sky-region through which the human soul was believed to journey after death.

THE RIDDLE

Aerial view of the ruined Nazcan pyramid city of Cahuachi.

Because written records have not survived, we know very little about Nazcan beliefs. Some clue is provided by later religious traditions in the Andes, which almost certainly descended from the same stock. Also of great interest is Cahuachi, the principal archaeological site of Nazca culture. A sacred domain of pyramids and temples extending over 150 hectares, it lies a few kilometres to the south-west of the main geoglyphs and has been described by the archaeologist Michael Moseley as 'not so much a city but a congregation center'.[23]

Cahuachi looks post-apocalyptic, collapsed, worn down, swamped by desert sands. Something flourished here – some cult, some religion – while the lines and figures were being laid out 2000 years ago. But there is evidence that the site was occupied long before that and that all its structures are built upon pre-existing sacred mounds and natural hills.[24]

Cahuachi's fundamental character is as a pyramid-city and from the air its most striking monument is its central step-pyramid. This structure, which has five levels, is 18 metres in height and is oriented to due north–south with an entrance in its north face.[25] On either side of it, forming a diagonal row, are two smaller pyramids with contours much blurred by erosion. Immediately to the south-west is a terraced hillside, referred to by archaeologists as the 'Great Temple', which overlooks what was once a very large walled courtyard.[26]

Like the pyramids of Giza, Teotihuacan and Angkor, it is probably safe to assume, as the archaeologist Johan Reinhard has suggested, that the pyramids of Cahuachi 'functioned as a symbolic landscape, where architectural shapes and depictions of deities mirrored sacred geography'.[27]

In Egypt, Mexico and Cambodia, such structures were used as instruments of initiation into a powerful system of spiritual knowledge. It was the same system the world over. It was the same knowledge that was always taught. And in all places the same technique was used, obliging the initiate to think in terms of sky and ground and to explore the labyrinth of a dualistic mystery:

> Heaven above, Heaven below;
> Stars above, Stars below;
> All that is over, under shall show.
> Happy thou who the riddle readest.[28]

Is it coincidence, or it design, that exactly the same riddle is posed at Nazca –

where the pyramid-mountains of Cahuachi are set amidst great constellations of desert drawings looking directly up towards heaven?

THE EYE OF RA

Mata-Ki-Te-Rani, 'Eyes Looking at Heaven', is one of the two ancient names of Easter Island. The other is Te-Pito-O-Te-Henua, 'Navel of the World'. Over 300 kilometres east of Nazca, in the Andes mountains, is the Inca city of Cuzco – a name that means, literally, 'Navel of the World'.[29] Three hundred kilometres south of Nazca, on the Peruvian coast, is an ancient sea-port called Matarani,[30] The reader will remember that the word 'Matarani' in Polynesian languages means literally 'the Eye of Heaven' and in the ancient Egyptian language 'Maat Ra' means 'the Eye of the Sun'.

There are astronomically aligned megalithic remains at Cuzco, where a sophisticated 'cult of the Sun' was practised under the Inca god-kings until the coming of the Spaniards in the sixteenth century. The cult was by then already very old and traditions traced its origins back to a mysterious city and empire called Tiahuanaco, on the high plains to the south of Lake Titicaca, that had used Matarani as its Pacific port.[31]

All the civilizations of the Andes believed that Lake Titicaca was the original place of creation. Out of an island in the midst of it had emerged the creator, Viracocha, the human form of the sun, whom the most ancient myths of South America describe as being tall, white-skinned and bearded.

The Nazca Condor in flight upon the duality of earth and sky.

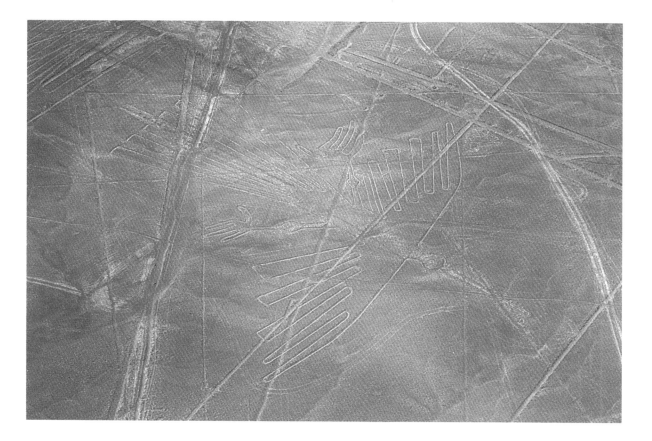

CHAPTER 16

THE MYSTERY AND THE LAKE

IN 1910 THE HISTORIAN Sir Clements Markham, a leading authority on the Incas of Peru, wrote of a 'mystery still unsolved on the plateau of Lake Titicaca ... The mystery consists in the existence of ruins of a great city on the southern side of the lake, the builders being entirely unknown.'[1]

A century later the mystery has still not been solved. The ruined city, which is today called Tiahuanaco, a relatively modern name,[2] was known in ancient times as Taypicala, 'the Stone at the Centre'.[3] There is an unusual lack of unanimity amongst archaeologists as to its date, with some experts believing it may go back as far as the second millennium BC[4] while others favour a more recent construction between the second century AD and the ninth century AD.[5]

As we shall see, all these chronologies rest on shaky foundations and there is a growing body of evidence which suggests that the dating of Tiahuanaco may be one of the biggest mistakes of archaeology – with even the second millennium BC, the most extreme conventional estimate for the age of the site, being far too recent by a factor of thousands of years.

ORGANIZATION

The hallmark of the unknown builders of Tiahuanaco was the use of spectacularly large megaliths. As Markham comments:

> One stone is 36 feet long by 7, weighing 170 tons, another 26 feet by 16 by 6. Apart from the monoliths of ancient Egypt, there is nothing to equal this in any other part of the world. The movement and placement of such monoliths point to ... an organized government ... There must have been an organization combining skill and intelligence with power and administrative ability.[6]

OPPOSITE: *Megalithic statue of Tiahuanaco. Almost nothing is known about the builders of this great and mysterious city, which stands to the south of Lake Titicaca.*

One of the few matters over which scholars of Tiahuanaco are in complete agreement is that this intelligent, powerful and efficiently administered 'organization' had nothing whatsoever to do with the well-documented Andean civilization of the Incas – which flourished only between the fifteenth and the sixteenth centuries AD. The scholarly view is supported by the traditions of the Aymara

Indians, who have lived in the vicinity of Tiahuanaco since times immemorial. In the sixteenth century the Spanish chronicler Cieza de Leon asked the Aymara whether the city's many megalithic structures were the work of the Incas:

> They laughed at the question, affirming that they were made long before the Inca reign and . . . that they had heard from their forebears that everything to be seen there occurred suddenly in the course of a single night.[7]

THE MEGALITHIC BUILDERS

'Stone of twelve corners', Cuzco.

Tiahuanaco is not the only megalithic site in the Andean region that historians and local traditions both attribute to a mysterious, as yet unidentified and apparently very ancient source. Some 650 kilometres to the north, similar ruins are found in great profusion all around Cuzco, the Inca capital city. These ruins long predate the time of the Incas. As Sir Clements Markham confirms:

> In Cuzco there is a cyclopean building . . . with a huge monolith known as the 'stone of twelve corners'. Some portions of the ancient remains at Ollantaytambo [60 kilometres from Cuzco] are megalithic work . . . But the grandest and most imposing work of the megalithic builders was the fortress of Sacsayhuaman hill [on the outskirts of Cuzco]. It consists of three parallel walls, 330 yards in length each, with 21 advancing and retiring angles . . . The outer wall at its salient angles, has stones of the following dimensions: 14 feet high by 12; another 10 feet by 6. There must have been some good cause for the erection of this marvellous . . . work of which we know nothing. Its origin is as unknown as that of the Tiahuanaco ruins. The Incas knew nothing. [The Inca chronicler] Garcilaso [de la Vega] refers to towers, walls and gates built by the Incas, and even gives the names of the architects; but these were later defences built within the great cyclopean fortress. The outer lines must be attributed to the megalithic age.[8]

Although Markham does not propose an exact date for 'the megalithic age . . . when cyclopean stones were transported, and cyclopean edifices raised',[9] he several times affirms his belief in the 'great antiquity' of Tiahuanaco, Sacsayhuaman and other associated structures and states that 'Andean civilization dates back into a far distant past'.[10]

AN UMBILICAL LINK

The people now known as the 'Incas' were Quechua-speaking South American Indians. The name was originally applied only to their kings, and it was a title – with the monarch being referred to as 'the Inca'. Only later, and by extension, did the entire nation become 'Incas'. They were arrivistes on the stage of Andean history: it is worth repeating that their empire was less than a century old at the start of the Spanish Conquest.[11] But they preserved legends which spoke of a remote time when the ancestors of the Inca line had been sent down from heaven to earth by the creator god Viracocha, whom they spoke of as 'our Father the Sun'. The mission of the royal couple was to bring salvation to mankind.

'The grandest and most imposing work of the megalithic builders was the fortress of Sacsayhuaman hill...'

Here is the story as it was told by the Inca historian Garcilaso de la Vega in the sixteenth century. It is set in an apocalyptic former epoch amidst a landscape of mountains and desolate escarpments in which the people were cannibals who 'lived like wild beasts, with neither order nor religion':[12]

Seeing the condition they were in, our Father the Sun was ashamed for them, and he decided to send one of his sons and one of his daughters from heaven to earth in order that they might teach men to adore him and acknowledge him as their god ... Our Father the Sun set his two children down at Lake Titicaca and he gave them a rod of gold, a little shorter than a man's arm and two fingers in thickness.

'Go where you will,' he said to them, 'and whenever you stop to eat or sleep, plunge this rod into the earth. At the spot where, with one single thrust, it disappears entirely, there you must establish and hold your court. And the peoples whom you will have brought under your sway shall be maintained by you in a state of justice and reason, with pity, mercy and mildness.'

Having thus declared his will to his two children, our Father the Sun dismissed them. They then left Lake Titicaca and walked northwards, trying vainly each day to thrust their rod of gold into the earth ...

[Finally] the Inca and his bride entered into Cuzco valley ... There [at a spot called Cuzco Cara Urumi, 'the Uncovered Navel-Stone'[13]] they tried their rod and not only did it sink into the earth, but it disappeared entirely ... Thus our imperial city came into existence.[14]

Sunset over Lake Titicaca: 'Our Father the Sun set his two children down at Lake Titicaca and gave them a rod of gold...'

The traditions which Garcilaso and others report leave no room for doubt that the place at which 'the children of the Sun' first appeared when they were sent down from heaven to the shores of Lake Titicaca was the megalithic city of Tiahuanaco.[15] There is also no doubt that Tiahuanaco was seen as the original sacred capital of the pan-Andean creator-god Viracocha who was the human form of 'our Father the Sun'.[16]

We may deduce that when the two 'children of the Sun' carried Viracocha's golden rod northwards from Tiahuanaco to Cuzco and 'planted it there in the earth', they were performing a ritual that established some sort of *umbilical* link. The same idea is also conveyed by the exact meanings of the two cities' curious ancient names: 'the Navel of the World',[17] in the case of Cuzco, and 'the Stone at the Centre', in the case of Tiahuanaco.[18] Such imagery calls to mind Te Pito Kura, the golden 'navel stone' of Easter Island, and the Eben Shetiyah of Jewish tradition – the 'Foundation Stone' of the Temple Mount in Jerusalem, which was said to have been 'sunk into the abyss' by God in order 'to serve as the centre of the earth'.

CLIFF OF THE LION

We flew from Nazca to Cuzco, a journey of 800 kilometres into the Andes. Seen from the pressurized cabin of the plane at an altitude of 9000 metres, the mountains beneath us appeared to extend away for ever, infinitely, in all directions. Jagged and snow-covered, pitiless and austere, the peaks are a natural fortress surrounding and protecting the inland sea of lake Titicaca as though it were a treasure.

The lake is 300 metres deep and more than 7700 square kilometres in area, with its shoreline 3800 metres above sea level.[19] Its name 'Titicaca' has two meanings. One is 'Cliff of Lead'.[20] The other is 'Cliff of the Lion'.[21] The same name was also traditionally applied to the lake's principal island, better known today as the Island of the Sun, and to a specific place on that island – a steep black east-facing cliff, terraced into sweeping steps.[22] This primeval mound was believed to be the 'place of creation' at which the present and all previous epochs of the world had been set in motion. To commemorate its mythical importance, during the brief century in which the Inca empire arose and flourished before being snuffed out by the Spanish Conquistadores, a beautiful stone shrine of stairways and fountains was built into the Titicaca cliff, oriented towards the equinoctial rising of the sun. As the archaeoastronomer William Sullivan argues: 'With awe-inspiring symbolic economy, the very word *Titicaca* tells the entire tale ... The "lion cliff" of Titicaca, rising from the waters of the lake, expresses the birth of a new world.'[23]

THE BEARDED GOD AND THE FIFTH SUN

According to Andean traditions, this 'new world' had been summoned into being by the word of the god Viracocha. Like Atum in ancient Egypt, or Vishnu amongst the Hindus, he was first and foremost the symbol of a great generative power in the cosmos. He was identified with the Sun and, as we have seen, he was also the human form of the Sun. In this aspect, exactly like the Mexican civilizing deity Quetzalcoatl, he was emphatically described as: 'a white man ... blue-eyed and bearded ... of large stature and authoritative demeanour ... In many places he gave men instructions how they should live ...'[24]

Since historians do not accept the possibility of any significant mutual influence or connection between Pre-Columbian Mexico and the Pre-Columbian Andes, it is assumed to be a coincidence that the ancient cultures of both regions worshipped a pale-skinned, bearded, civilizing god. But can it also be a coincidence that both cultures believed themselves to be living in the fifth epoch of the earth and that both specifically characterized this epoch as 'the Fifth Sun'?[25]

We have explored the Mexican version of this belief-system in Part I. The Andean version was preserved in writing in the sixteenth century by the native Peruvian nobleman Huaman Poma, whose name, literally 'Falcon-Lion',[26] is strongly reminiscent of the symbolism of the Horus-kings of ancient Egypt. The Spanish priest Martin de Murua also gave this account of the Andean beliefs: 'Since the creation of the world until this time, there have passed four Suns without [counting] the one which presently illumines us.'[27]

As in Mexico, each of the four earlier Suns was believed in South America to have been destroyed and swept away by a great cataclysm – water, a falling of the sky, air, fire.[28] It was also believed in both places that the Fifth Sun itself was soon to be destroyed by what was referred to in Mexico as a 'great movement of the earth' and in the Andes as a *pachacuti*. The word *pachacuti* means literally an 'overturning of the world' (according to Sir Clements Markham's translation[29]) and an 'overturning of space-time' according to William Sullivan.[30]

HELIOPOLIS OF THE ANDES?

It was understood in the Andes that the god Viracocha, who was responsible for the creation of new worlds, was also responsible for the destruction of the worlds that had preceded them.

His first creation was said to have been 'a world in which there was no light and no warmth'.[31] To inhabit this 'limbo of darkness' Viracocha created 'large, strong men of more than normal size who lived but a half-life of unreality like animals'.[32] When these giants displeased him he 'destroyed them by the flood, *una pachacuti*, the world overturned by water':[33]

> Then, after the flood, appearing again from the Island Titicaca, he … brought into being a new race of men of his own stature, which was the average height of men, [and] ordered that the sun, moon and stars should come forth and be set in the heavens … to shine by day and by night.[34]

This is the tradition behind the Andean belief that Titicaca, the Island of the Sun – and in particular the 'lion-cliff' on the eastern side of the island where Viracocha had emerged from the waters of the lake – constituted the original sacred domain of creation itself. In this great scheme of spiritual geography, Tiahuanaco, 'with its ancient and strange buildings',[35] was believed to be the first city that Viracocha had built after the creation.

There is no significant difference between this concept and the ancient Egyptian idea of Heliopolis as the city at 'the place of creation' – in which the god Atum 'rose [out of the waters of the Nun] as a High Hill [and] shone as the Benben stone in the Temple of the Phoenix'.[36]

Another curiousity is Titicaca's east-facing 'lion-cliff', specifically associated with the creation of the present epoch of the earth. Egypt's Great Sphinx is a lion carved out of the 'cliff' of the Giza plateau. According to an Eighteenth Dynasty inscription on a stela between its paws, it marks 'the Splendid Place of the First Time'[37] – i.e. the beginning of the present epoch of the earth.

The Island of the Sun – the original sacred domain of creation itself rising from the waters of Lake Titicaca like the 'High Hill' of ancient Egyptian mythology, which arose from the waters of the Nun.

Cuzco Cara Urumi, the 'Uncovered Navel Stone', at the centre of the Coricancha.

TEMPLE OF THE SUN

After they had adopted Cuzco as their capital, the Incas took over an earlier sacred site at the very centre of the city, on top of which they built the Coricancha, their great national Temple of the Sun. At the centre of the Coricancha, in a courtyard open to the sky, was a symbolic field planted with symbolic 'corn' fashioned out of pure gold. And at the centre of this field, marking the spot known as Cuzco Cara Urumi ('the Uncovered Navel-Stone'), where the myths said that the children of the Sun had thrust the golden rod of Viracocha fully into the earth, there was an octagonal grey-stone coffer which was at one time covered with 55 kilograms of gold.[38]

Very often when the Spanish conquerors encountered an indigenous sacred site in the Americas it was their policy to build a church on top of it. At the Coricancha they built the great church of Santo Domingo but left much of the Inca masonry intact. Following an earthquake in 1951 the church was demolished and the substructure of the original temple was revealed with its earthquake-resistant walls of hulking stone blocks fitted together like the pieces of a gigantic jigsaw puzzle. The joints between the blocks are so finely made as to be almost invisible and the overall impression is of jeweller's work on a scale of acres.

We looked down into the octagonal granite coffer at the centre of the cobbled courtyard. There is no gold on it now: it was stripped bare by the Spanish at the beginning of the Conquest, like everything else at the Coricancha. Today nobody has any idea what its original function was; indeed it is not even certain that it has always had its present font-like appearance, since zealous friars may have arranged for it to be recarved when the church of Santo Domingo was built around it in the sixteenth century.[39]

Garcilaso de la Vega, who knew the Coricancha before and after it became a church, does not mention the coffer but provides a fund of other evidence which helps us to assess the original character of the Inca Temple of the Sun:

> What we shall call the high altar, although this expression did not exist among the Indians, was to the east, and the roof, which was very high, was of wood, covered with straw. The four walls were hung with plaques of gold, from top to bottom, and a likeness of the Sun topped the high altar. This likeness was made of a gold plaque twice as thick as those that panelled the walls, and was composed of a round face, prolonged by rays and flames ... The whole thing was so immense that it occupied the entire back wall of the temple.
>
> On the other side of this Sun were kept the numerous mummies of former Inca kings, which were so well-preserved that they seemed to be alive. They were seated on golden thrones resting on plaques of this same metal, and they looked directly at the visitor ... The Indians hid these bodies along with other treasures that have not yet come to light ...[40]

ASTRONOMICAL CLUES

One of these hidden treasures is thought to be an effigy of Viracocha, the sun-

Sacred geography mirroring the sky: the 'Island of the Moon', Lake Titicaca.

god in human form. This effigy was seen and described by a number of Spanish chroniclers before it vanished from view in the sixteenth century. It was said to have been 'made of gold the size of a boy of ten, and was in the figure of a man standing up, the right arm high, with the hand almost closed, except for the thumb and forefinger held high . . .'[41]

William Sullivan is the first scholar to have noticed that this is the characteristic stance of the observational astronomer and that 'the palm measure, that is, the distance between thumb and forefinger extended at arm's length, has historically provided "archaic" peoples with a useful means of measuring time on the celestial sphere'.[42] Sullivan points out that a palm measure of exactly this sort is still in use amongst traditional Polynesian navigators of the Caroline Islands. According to anthropologist David Lewis, this measure, described as 'the forefinger to thumb distance at arm's length, or about 10 degrees . . . may well reflect an ancient technique . . .'[43]

As an avatar of the sun, Viracocha is already partly astronomical in nature and it seems reasonable that his human form might be portrayed in the attitude of one observing the stars – particularly since there are other strong signs of an ancient astronomical cult at the Coricancha.

Of great significance, arranged around the central courtyard, were what Garcilaso describes as 'five large square rooms', the core-masonry of which, in polished grey granite blocks, has survived to this day. 'They had no communication between them,' Garcilaso tells us, 'and were roofed over in the form of a pyramid.'[44]

One room was:

> dedicated to the Moon, the bride of the Sun. It was entirely panelled with
> silver ... A likeness to the Moon, with the face of a woman, decorated it
> in the same way that the Sun decorated the larger building ... The room
> nearest to that of the Moon was devoted to Venus, to the Pleiades and to
> all the stars ... This room was hung with silver, like that of the Moon, and
> the ceiling was dotted with stars, like the firmament. The next room was
> dedicated to lightning and to thunder ... The fourth room was devoted
> to the rainbow, which they said had descended from the Sun ... It was
> entirely covered in gold and the rainbow was painted, in beautiful colours,
> across the entire surface of one of the walls ... The fifth and last room was
> reserved for the high priest and his assistants, who were all of royal blood ...
> This parlour was panelled with gold, in the same way as the others.[45]

THE STELLAR PYRAMID OF THE INCAS

What Garcilaso is describing is a cosmological temple.

The main structure, laden with masses of obvious solar symbolism, was dedicated to the sun. Moreover, though Garcilaso does not note this, it incorporates
a prominent solar alignment in which a nearby mountain peak is used as the 'foresight'. The name of the peak is Pachatusan, which means literally 'Fulcrum' or
'Crossbeam' of the Universe.[46] For an observer in the courtyard of the Coricancha
the sun is seen to rise directly behind Pachatusan at dawn on the June solstice.[47]

Out of the five subsidiary 'rooms' of the Sun Temple, one was reserved for the
high priest and his assistants while the other four were all dedicated to cosmological – and specifically *astronomical* – phenomena.

We suspect that even the dedication to lightning and thunder had more to do
with astronomy than with the weather, since lightning and thunder served as symbols for meteorites – 'thunderstones' – elsewhere in the ancient world. There is
excellent evidence that a meteorite cult, with bizarre similarities to the meteor-
ite/Benben cult of ancient Egypt, once flourished in the Andes.[48] Indeed, one of
the many titles of Viracocha used in worship by the Incas was Illa-Tiki. The word
illa means 'thunderstone' and the word *tiki* means 'primeval', or 'original', or
'fundamental'.[49] Doesn't this 'primeval thunderstone' sound very like the Benben,
which the ancient Egyptians described as a stone that fell from heaven in the 'early
primeval age'? And is it a coincidence that meteorites in both places were
associated with the generative, creative power of the universe – and specifically
with fertility and rebirth?[50]

Garcilaso tells us that another of the temple rooms was dedicated to the
rainbow, which the Incas conceived of as an emanation of the sun. Here we are
reminded of how rainbows were regarded by the ancient Khmers of Angkor – as
bridges that arched between the worlds of gods and men.[51]

The remaining two rooms of the Coricancha were entirely and unambiguously given over to the celestial bodies – one to the moon, the other to 'Venus,
the Pleiades and all the stars'. In the latter the ceiling was 'dotted with stars, like

the firmament'. We are reminded of the tombs of the Egyptian Pharaohs, which also show ceilings dotted with stars – as, for example, in the Valley of the Kings and in the Fifth and Sixth Dynasty pyramids at Saqqara.

Garcilaso's account contains two other very curious details. He tells us that the Incas mummified the remains of their deceased god-kings. So did the ancient Egyptians. And he tells us that the astronomical rooms of the Coricancha were originally 'roofed over in the form of a pyramid'. Is it a coincidence that the three Pyramids of Giza are laid out on the ground according to an astronomical pattern and that the Great Pyramid contains a variety of rooms – the principal chambers – which have channels pointing at specific stars?

Niche of polished granite, part of the surviving masonry of the original cosmological temple of the Coricancha. The niche is in the exterior wall of the room dedicated to 'Venus, the Pleiades and all the stars'.

ASTRONOMER-PRIESTS

We saw in Part II that the pyramids of Giza were attended by the astronomer-priests of Heliopolis, the 'mystery teachers of heaven', whose pontiff, 'the chief of the astronomers' was believed to be in direct communication with the celestial powers. The Incas called the high-priest at the Coricancha *uilac-umu*, a name that means 'he who speaks of divine matters'.[52] He was supported by a caste of learned priests called Amuatas amongst whose numbers was a college of Tarpuntaes – skilled astronomers.[53] 'Their task was to study the astral bodies, note the advance and retreat of the sun, fix solstice and equinox, [and] predict eclipses.'[54] To this end they made use of a sequence of monoliths known as *sucanas*, now unfortunately all destroyed, which according to the early chroniclers once stood 'on the mountainous horizons of the Cuzco valley at strategic points visible from the Coricancha, marking the azimuths of the winter and summer solstices'.[55]

In Part II we also saw references in the Pyramid Texts which suggest that the ancient Egyptians, or their prehistoric benefactors, possessed an advanced knowledge of celestial mechanics at a very early date. For example the term 'Shemsu Hor', 'Followers of Horus' can be taken to refer to a group of observers who 'followed the path of the sun' – i.e. the ecliptic. It is therefore of interest that the archaeoastronomer William Sullivan believes the Coricancha, which has a pronouncedly elliptical exterior wall, served as a model of the ecliptic:[56]

> The ecliptic plane was represented ... by means of the Temple of the Sun, which in Quechua was called 'Coricancha', literally 'corral of gold'. The allied verb *canchay* means 'to encircle'. The imagery of the 'golden circle of the sun' ... suggests ... the ecliptic plane ...[57]

Sullivan reinforces his argument by highlighting one of Viracocha's many titles: Intipintin Tiki-Muyo Camac. This epithet:

> literally means 'the sun's-taken-altogether-fundamental-circle-creator' ... If one's child were to come home from a basic astronomy class in school and – rather than parrot the textbook definition for the ecliptic ('the apparent annual path of the sun') – were instead to define the ecliptic as 'the sun's-taken-altogether-basic-circle', one could rest assured that the child had grasped the essence of the matter.[58]

AFTERLIFE JOURNEYS

The ancient religious system that the priests of Heliopolis claimed to have inherited from the Followers of Horus combined precise astronomical observations, and the possession of sophisticated knowledge of such arcane matters as the precession of the equinoxes, with a quest for the immortality of the human soul. We have encountered the traces of that same quest in ancient Mexico and amongst the temples of Angkor in Cambodia. Is it a coincidence that it was also pursued passionately in the Andes not only by the Incas but by all their known predecessors for millennia before?[59]

Of particular note is the fact that the cultures of the Andes and of ancient

Egypt both believed that the souls of the dead must make an afterlife journey amongst the stars and find a gateway in the heavens leading to the netherworld. The ancient Egyptians called this netherworld the Duat and located it in a specific region of the sky between the constellations of Leo on one side and Orion/Taurus on the other – a region through which the Milky Way flows brightly. Equidistant between Leo and Taurus, the constellation of Gemini marks the intersection of the Milky Way with the ecliptic. This precise location, in native Andean belief, 'marked the cross-roads of the land of the living and the land of the dead'.[60]

The ancient Egyptians in 2500 BC believed that the Duat only became active – i.e. only opened its gates – at the time of the June solstice. The Incas of AD 1500 also believed that the gateway to their celestial netherworld opened only at a solstice. In their case this was thought to happen during the four days around the December solstice when the sun rested at the southern tropic, marking 'the annual opening of the land of the dead to the land of the living'.[61]

The Incas believed was that 'in this world we are exiled from our homeland in the world above'[62] and that on death the soul that had lived an initiated life would be able to return to the sky and to cloak itself once again in its true heavenly glory. The ancient Egyptian Pyramid Texts, which speak of the initiate's quest for 'the life of millions of years', proclaim loudly: 'earth is this King's detestation ... This King is bound for the sky ... This King is one of those ... beings ... who will never fall to earth from the sky.'[63]

Precisely the same ideas were also the inspiration behind the sky–ground temples of Angkor in Cambodia. They are likewise found as a continuous refrain in the Hermetic Texts and in the writings of the Gnostics which were circulated in Egypt and elsewhere in the Middle East during the early centuries of the Christian era.[64] The Hermetic codex known as the 'Kore Kosmou' comes particularly close to the Inca belief, actually describing the exile of souls from the heavenly realms in the world above to incarnation in human form. 'Poor wretches that we are,' the souls protest, 'what hard necessities await us! What hateful things we shall have to do in order to supply the needs of this body that must so soon perish.'[65] To reduce their suffering they petition the Creator to afflict them with amnesia: 'Make us forget what bliss we have lost, and into what an evil world we have come down.'[66]

All of these religious systems taught that the soul that had experienced incarnation must succeed in a terrible ordeal if it was to find its way back to the celestial realms. In Andean accounts of the afterlife journey this ordeal was frequently symbolized as the crossing of a narrow bridge made of human hair spanning a raging river.[67] The Incas also believed that the soul might cross the river 'by means of black dogs'[68] – a notion that is reminiscent of the role of the black dogs Anubis and Upuaut as guides of the soul in the ancient Egyptian *Book of the Dead*.

As part of this widely disseminated belief, it was held that the soul's best hope of salvation lay in making use of the opportunities for experience and choices accorded by material existence in order to acquire a special kind of secret knowledge. This 'gnosis' could equip the fallen soul to rise out of matter and return to the heavens, but required a long and painful process of spiritual initiation which it would be 'difficult for someone wearing a body to complete'.[69] In one way or another all the ancient sources advise the initiate determined to pursue this quest to 'keep mind as a guide, reason as a teacher. They will bring you out of destruction and dangers.'[70] Diligent cultivation of mind and reason would also equip the pilgrim to understand why 'the Lord did everything in a mystery'[71] and why he said: 'I come to make the things below like the things above.'[72]

The prize for those who completed the quest was to 'exist deathless in the midst of dying mankind'.[73] In Egypt, Mexico and Angkor we have shown that this quest for immortality was conducted in the setting of great astronomically aligned monuments in landscapes 'resembling the sky'.

In Egypt the Nile was the terrestrial counterpart of the Milky Way. The Incas regarded the entire valley from Cuzco up to Machu Picchu as a reflection of the sky and saw the Vilcamayu river running through it as the terrestrial counterpart of the Milky Way.[74] Rituals were conducted along the banks of both the Nile and the Vilcamayu rivers at the June solstice.[75] In both places these rituals were led by god-kings – Inca and Pharaoh being virtually interchangeable concepts.[76] And in both places they took place amidst precisely built megalithic structures of unknown antiquity.

We have seen evidence that the Sphinx and the megalithic temples that surround it at Giza may be more than 12,000 years old. Since so little is known about the origins of the megalithic structures around Tiahuanaco and Cuzco we should not close our minds to the possibility that they too could date back to that same mysterious period.

Sacred valley of the Vilcamayu. Like the Nile in Egypt, the Vilcamayu was seen as the terrestrial counterpart of the Milky Way and associated with a gnostic quest for immortality of the soul. Astronomical rituals led by god-kings were conducted along the banks of both rivers at the June solstice.

CHAPTER 17

THE STONE AT THE CENTRE

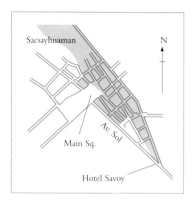

The 'puma' of Sacsayhuaman.

WE LEFT THE CORICANCHA and drove north out of Cuzco to the nearby ruins of Sacsayhuaman, literally 'Satisfied Falcon',[1] a name that might perhaps have meant something to the Shemsu Hor – the 'Followers' of the ancient Egyptian falcon-god Horus.

In the sixteenth century Garcilaso de la Vega gave an extensive description of Sacsayhuaman:

Its proportions are inconceivable when one has not actually seen it; and when one has looked at it closely and examined it attentively, they appear to be so extraordinary that it seems as though some magic had presided over its construction ... It is made of such [enormous] stones, and in such great number, that one wonders simultaneously how the Indians were able to quarry them, how they transported them, and how they hewed them and set them on top of one another ... They are so well fitted together that you could not slip the point of a knife between two of them. If we think, too, that this incredible work was accomplished without the help of a single machine ... how may we explain the fact that these Peruvian Indians were able to split, carve, lift, carry, hoist and lower such enormous blocks of stone, which are more like pieces of a mountain than building stones? Is it too much to say that it represents an even greater enigma than the seven wonders of the world?[2]

With its three rows of gigantic zig-zag walls undulating in parallel terraces for more than 300 metres across a hillside immediately above Cuzco, Sacsayhuaman is almost too large to grasp in one view. It has also long been recognized that it forms just part of an even larger geoglyph, once visible from surrounding mountain peaks, in which it combines with the oldest quarters of Cuzco to form the shape of an immense puma. The river Tullumayo (now diverted underground where it passes through the city) used to serve as the spine of this ancient lion. The torso was the spit of land between the Tullumayo to the east and the river Huatanay (now also underground) to the west. Sacsayhuaman is still recognizable as the head of the lion. To the south, its jagged zig-zag walls represent the teeth in the lower jaw.[3] To the north, the upper jaw is a knoll of bedrock. Between the

OPPOSITE: *Megalithic gateway, Sacsayhuaman.*

The teeth of the lion. Zig-zag walls of Sacsayhuaman, looking south from the bedrock knoll.

jaws is a long strip of open, cleared land, now grassed over, representing the lion's open mouth. It faces due west, the direction of the equinox sunset[4] – just as the Great Sphinx of Egypt faces due east, the direction of the equinox sunrise.

There are traditions, supported by some modern excavations, of a network of tunnels under the Sphinx in which mysterious treasures lie concealed.[5] There are virtually identical traditions – again supported by recent excavations – of a labyrinth of enormously long tunnels under the lower 'jaw' of the Sacsayhuaman lion 'into which people descend to be lost forever, or to emerge, gibbering, mad, clutching items of treasure'.[6]

THE PROBLEM OF THE BIG STONES

In the hour before sunset we stood on the rocky knoll that forms the upper jaw of the lion looking south towards the three zig-zag rows of megalithic 'teeth' in the lower jaw. There are more than a thousand individual stone blocks. All of them are massive, with many in the range of 200 tonnes, but the largest are on the lowest terrace. According to measurements and calculations by Dr John Hemming of the Royal Geographical Society in London, one of these has a height of 8.5 metres and weighs 355 tonnes, ranking it amongst 'the largest blocks ever incorporated into any structure'.[7] Hemming also draws attention to the polygonal characteristics of the masonry with each stone a different size and shape, interlocking 'in a complex and intriguing pattern'.[8]

The three terraces, all in all, reach a height of about 15 metres. From our vantage point on the knoll, in the early evening light, their contours merged to form a fantastic castle of the imagination, with layer upon layer of stone heaped up, leading towards the sky. As the sun fell progressively lower in the west, the shadows cast by the jagged edges of the lion's teeth, projected on the spaces between them, grew progressively longer and we had the sense that the whole monument might have been designed to track the sun.

We clambered down from the knoll and walked across the grassy expanse of the lion's mouth to the first zig-zag wall. The huge stones, dark and heavy, loomed above us. Looking at their tremendous bulk, and considering their ponderous weight, we found it as difficult as Garcilaso to imagine how they might have been transported from quarries several kilometres away, still less how each one of them could have been manoeuvred so precisely into position and then fitted with such astonishing accuracy to neighbouring stones.

Although there have been a few dissenters – some of very high calibre such as Sir Clements Markham – the prevailing archaeological theory by and large attributes Sacsayhuaman to the Incas and holds that the whole ensemble was built through 'some sort of trial-and-error approach involving multiple movements of each stone, however laborious this might have been'.[9] No studies have yet been published showing how a trial-and-error system might have worked.[10] Moreover, it is admitted that the great megalithic walls of Sacsayhuaman 'had been completed or abandoned prior to the arrival of the Spaniards, and the Incas neither reported nor left records of their methods'.[11]

In fact the *only* record of the Incas attempting to move a true megalith (which appears in Garcilaso de la Vega's *Royal Commentaries of the Incas*) suggests that they had no experience of the techniques involved – since the attempt ended in disaster. Garcilaso tells us of a great boulder of 'incredible' dimensions that 'was hauled across the mountain by more than twenty thousand Indians, going up and down very steep hills . . . At a certain spot, it fell from their hands over a precipice, crushing more than three thousand men.'[12]

We do not doubt that the Incas were accomplished stone-masons, nor that a great many of the smaller, internal structures of Sacsayhuaman – now almost completely vanished – were made by the Incas (as were large parts of Cuzco). If moving just one big stone was such an ordeal for them, however, then we must wonder how likely it is that they could have succeeded in transporting many hundreds of monsters of equally killing weight to build the zig-zag walls of Sacsayhuaman. The alternative possibility is that Clements Markham was right and that the walls are indeed the legacy of a far earlier time – 'the megalithic age, when cyclopean stones were transported and cyclopean edifices raised'.

Detail of the rocky knoll at Sacsayhuaman. The knoll is extensively carved into 'thrones' and terraces. Like the cyclopean construction blocks of the zig-zag walls, could it have come down to us from an earlier megalithic age?

GIANTS

We returned to examine the rocky knoll – the lion's 'upper jaw' – which stands 200 metres due north of the zig-zag walls. Its contours have been extensively hewn and shaped by the hand of man and this work is habitually attributed to the Incas by historians.[13] Once again, however, there is no proof that the Incas had anything to do with it. Since there is as yet no reliable test for measuring the age of stone monuments, it is theoretically possible that the knoll could have been carved thousands of years before the Incas, by an entirely different race, and then 'acquired' by the Incas when they appeared on the scene in the fifteenth century. In such an event it is not necessary to imagine a complete break in continuity between the hypothetical 'elder culture' and the Incas; on the contrary, the latter could have inherited some of the traditions and knowledge of the former and attemped, on a smaller scale, to mimic their cyclopean works. There is evidence of such a process at numerous sacred sites around the world – notably in Mexico, Egypt and Angkor, where it is the norm to discover that later monuments have been built over the foundations of earlier ones, which have in turn been built on even older foundations ... and so on, indefinitely backwards in time.

This scenario for the megaliths of Sacsayhuaman is supported by Andean myths which speak of the magical engineering and architectural achievements of the bearded, white-skinned, fair-haired god Viracocha and his companions – 'the messengers', 'the shining ones' – who appeared from Lake Titicaca in primordial times.[14] There is also a parallel tradition concerning a race of prehistoric master-builders referred to as the Huari. They are described as: 'white, bearded giants who had been created at Lake Titicaca, whence they had set forth to civilize the Andes ...'[15]

Great megaliths attributed to just such giants are found all around the world from Stonehenge[16] to the Americas. So too are rock-hewn structures very similar to the rocky knoll of Sacsayhuaman. It is cut into a profusion of terraces, steps, angles, channels, triangular recesses and stone 'seats' and closely resembles the

OPPOSITE AND RIGHT: *The mystic rock-hewn terraces and caves of Qenko, a short distance to the east of Sacsayhuaman. Compare with the Yonaguni underwater monument, page 214 ff.*

Qenko monolith.

general pattern of the underwater monument of Yonaguni in Japan and the rock-hewn caves and ledges inside the Rano Raraku crater on Easter Island.

The same style of rock-hewn outcrop, carved into numerous puzzling designs, occurs repeatedly in the Cuzco area. One of the most intriguing of these mounds is at Qenko, 1.5 kilometres to the east of Sacsayhuaman. Here a heavily eroded butte of limestone has been carved inside and out to create a mystic dome filled with caves, ledges, passageways and hidden niches. On the very top, also carved out of the raw stone, is an oval protrusion surmounted by a stubby double prong. Into the sloping sides of the dome are cut deep, narrow zig-zag channels, the shapes of various animals – a puma, a condor and a llama – and a succession of steps and terraces again very similar in appearance to the general profile of Yonaguni. At the base of the dome, encircled by a low elliptical wall, is a jagged monolith nearly 4 metres high, somewhat resembling the so-called 'heelstone' of Stonehenge.

There are no facts about this or any of the other rock-hewn monuments of the Andes. They all feel as though they were built so long ago that it may never be possible to understand the minds that made them. They seem to express an ethic, unfamiliar today, that never looked for the easy way out but always sought perfection through inviting the most difficult challenges. In Egypt that ethic produced the Pyramids of Giza, in Angkor the greatest collection of temples that the world has ever seen, in Nazca ambitious earth-drawings only visible from the air. Likewise, in inaccessible locations high in the the Andes it produced sacred buildings made of blocks of stone weighing hundreds of tonnes each.

OLLANTAYTAMBO

Though Sacsayhuaman beggars belief it is probably true to say that the extremely curious temple-mound of Ollantaytambo, 60 kilometres to the the north–west, in every sense exceeds it. Entering the site, we found ourselves confronting a vast amphitheatre of terracing extending all the way up a steep concave hillside to a flat ridge 80 metres above us.

We climbed through the terraces on a sunken stairway, noticing that the lower levels were made up of relatively small dry stones. As we mounted higher, somewhat paradoxically, the sizes of the blocks used seemed to get larger. Then we came to a level where several massive granite megaliths lay scattered casually around. They were in the range of 50 to 70 tonnes each and had been brought to an altitude of at least 60 metres.

Before continuing our climb we walked along a narrow ledge under a wall of tightly fitted trapezoidal blocks into which were set a row of ten polygonal niches. The southern end of the ledge passed under a megalithic gateway surmounted by a lintel-stone leading through to a little oval viewpoint perched on the mountain's edge.

We retraced our steps and came to a stairway cut into the trapezoidal wall. Following this upwards, we emerged at last on to the summit of Ollantaytambo, part mountain and part temple. Here there were many more scattered megaliths of between 100 and 200 tonnes and, at the highest point, a squat, square structure

Carved stone 'throne', Ollantaytambo.

Lintelled gateway and trapezoidal wall, Ollantaytambo.

fronted by six truly massive megaliths, each about 2 metres in breadth and a metre or so in depth, varying in height between 3.4 metres and 4.3 metres. We had the impression that these gigantic, smoothly cut facing stones may originally have covered the back wall of a room, the front and sides of which lay tumbled all around us. They were set into the edge of a higher mound and on top of this and all around were further megaliths, at least thirty of which we estimated to be in the 200-tonne range.

What was remarkable about these polished, jewel-hard pink-porphyry blocks, as well as bearing no resemblance, other than in size, to any of the blocks of Cuzco and Sacsayhuaman, was the almost unbelievable journey that they came on in order to get here. The quarries from which they were cut have been identified

Detail of the step-pyramid Benben symbol at Ollantaytambo.

by geologists and lie nearly 8 kilometres away and about 900 metres higher on the opposite side of the sacred Vilcamayu river.[17] This means that they had to be dragged down to the valley floor, across the river, and then steeply up again to the summit of Ollantaytambo – an almost superhuman task.

What the Ollantaytambo blocks most resemble is the architecture of Tiahuanaco, far away to the south-east beyond Lake Titicaca – hulking, slab-like, straight-edged megaliths, with inexplicable knobs and protrusions and indentations here and there, put together with breathtaking skill and precision.

This may explain why a characteristic symbolic device of Tiahuanaco, the step-pyramid, appears several times in low relief on one of the upright slabs of Ollantaytambo's megalithic wall. In ancient Egypt exactly the same graphic device was used as the hieroglyph and symbol for the Benben stone, the emblem of immortal life. Likewise, as in Egypt, as in Tiahuanaco, and as in Angkor, one of the 'signatures' of the builders of Ollantaytambo was a distinctive construction technique in which I-shaped metal clamps were used to hold blocks together.

MACHU PICCHU

Our final destination in Peru was the geomantic city of Machu Picchu, which stands on a soaring pinnacle of rock entwined in an oxbow of the sacred river Vilcamayu. All of it is remarkable, and most of it is indisputably of Inca origin. Within its precincts, however, are certain structures that are surely much older, although they too were used and modified by the Incas. These include a rock-hewn cave, the beautiful megalithic monument known as the 'Temple of the Three Windows', and, most conspicuously, Machu Picchu's central pyramidial mound, part natural, part built-up. The summit of the mound is exposed bedrock and has been carved into a shape something like a giant flattened fist with one finger pointing vertically upwards.

This object, the so-called 'Intihuatana', or 'Hitching Post of the Sun', has never been satisfactorily explained. Its name was given to it rather arbitrarily by Hiram Bingham, the American explorer who discovered Machu Picchu in 1911. It does definitely exhibit solar alignments – both to the equinoxes and to the solstices – when used as a foresight against neighbouring peaks.[18] But it is by no means an ideal shadow-casting or sighting device.

Carved at the base of the Intihuatana, the archaeoastronomer Ray White recently uncovered depictions of four constellations that are prominent in the Andean skies – the Southern Cross, the Summer Triangle, the bright 'eye stars' of the dark-cloud constellation in the Milky Way that the Incas called the Llama, and the Pleiades.[19] These are the constellations that were believed to rule the four *suyus*, or quarters, of the Inca empire.[20]

The 'umbilical' descendant of an ancient astronomical religion, the origins of which are lost to history, that great empire achieved much that was admirable during its brief existence. Nevertheless, in the decades before it was swept away forever by the cataclysm of the Spanish Conquest, it fell into a state of spiritual decay and lost contact with the allegorical and initiatory nature of its spiritual beliefs. We suggest that the Incas had inherited this remarkable spiritual system through

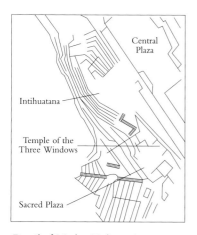

Detail of Machu Picchu.

Machu Picchu poised on a pinnacle of rock above the sacred Vlicamayu river.

an immensely long chain of transmission from the unknown predecessors who built the Andean megaliths. We suggest, too, that these megalith builders were connected to others, equally anonymous, all around the world, and that *they all taught the same system*. In this respect it is highly suggestive, as William Sullivan's landmark study of Andean cosmology persuasively demonstrates,[21] that the same 'technical language of myth' that was used throughout the Old World to convey complex information about the precession of the equinoxes was also used by the Incas and by their predecessors in the pre-Columbian Andes.[22]

In Sullivan's opinion, 'some very unusual spiritual perception, with deep insight into the operations of the human mind'[23] lies behind the formulation of this 'language of sacred revelation grounded in empirical observation'.[24] He argues that the Incas forgot, or misunderstood – and thus distorted – the original teachings that had come down to them. Like the Aztecs, who were the beneficiaries of an almost identical spiritual legacy in Central America, they made the deadly mistake of taking the symbolism of the initiatory rituals literally. This error led them into the dark hell of black magic and human sacrifice[25] and caused them to abandon the instruction of their 'Father the Sun' that they should rule a society based on justice and reason with 'pity, mercy and mildness'.

Inca sacrifices were carried out with cold, robotic deliberation along a system of geodetically surveyed straight lines that were laid out according to a sky–ground design.[26] The lines, called 'Ceques', corresponded in orientation to the rising and setting of certain stars and constellations, including the four that appear on the Intihuatana. Radiating out for hundreds of kilometres in all direc-

THE STONE AT THE CENTRE 295

tions from the 'Navel of the World' in the Coricancha at Cuzco,[27] this strange network, bloodied by sacrifice, created invisible connections between many prominent and widely separated monuments.

In a setting as serene as Machu Picchu it is disquieting to remember that the Incas, just like the Aztecs, believed that the souls of sacrificial victims would rise directly to the heavens and become stars in the sky.[28] In Egypt and in Angkor the same belief existed. We have seen, however, that it was not connected to human sacrifice there but to the teachings and rituals of an astronomical wisdom cult that sought immortal life for its initiates. In Mexico and the Andes astronomically aligned, pyramidial monuments were used as part of the apparatus of sacrifice. In Egypt and Angkor astronomically aligned pyramidial monuments were used as part of a gnostic quest for immortality.

Machu Picchu's central pyramidal mound, a knoll of bedrock which has been shaped and incorporated into the foundations of a megalithic temple. The Intihuatana, or 'Hitching Post of the Sun', stands with an enclosure at the summit of the mound.

Rainbow over the Intihuatana. Rainbows were thought of in the ancient Andes as emanations of the sun.

Perhaps the difference between the darkness and the light really is no wider than a human hair. If Egypt, Angkor, Mexico and the Andes all inherited a common legacy of sky–ground ideas, as we are proposing here, it is not necessarily the case that these ideas should always have been received in the same way. On the contrary, everything suggests that it is inherent in the nature of the system itself that those who participate in it must *choose* the direction of their own destiny:

> We have the power to choose the better, and likewise the power to choose the worse ... It is we who are to blame for our own evils if we chose the evil in preference to the good.[29]

REED BOATS OF THE SKY

One of the ancient Ceque lines venerated by the Incas passes through or extremely close by, Machu Picchu, Ollantaytambo, Sacsayhuaman and Cuzco. The alignment then extends, without deviation, through Lake Titicaca, through the island of Suriqi near the southern shore of the Bolivian section of the lake, and finally to the city of Tiahuanaco – an overall distance of almost 800 kilometres.

From pre-dynastic times in Egypt, the principal means of navigation on the river Nile were boats with high curving prows and sterns made of bundles of papyrus reeds. For thousands of years boats virtually identical to these ancient

Aerial view of the island of Suriqi, its bays fringed by totora reeds.

Traditional reed-boat construction, Suriqi. Compare to the design of the 'solar boat' of Khufu (see photograph, page 45).

Egyptian craft have been made in the Andes by the indigenous Aymara Indians of Suriqi island. Papyrus does not grow on Lake Titicaca so they work with native totora reeds, which are functionally similar. Totora reeds also grow on Easter Island, where they were traditionally used for making the reed floats for the annual 'birdman' ceremony (see pages 243–4).

Orthodox archaeology attributes all these similarities and many, many others to coincidence. It is, however, equally possible that they may *not* result from 'independent invention' but from a remotely ancient common influence.

If this turns out to be the case, then the implication is that at some point in prehistory, perhaps very far back, a group of people with advanced spiritual, architectural and astronomical ideas must have explored the world. They must have been great navigators, otherwise they could not have worked out where they wanted to go, and great sailors, otherwise they could not have crossed the wide oceans safely. And they must have possessed knowledge which they believed to be of vital importance to mankind, otherwise they would have had no motive to go to such lengths to disseminate it.

Undoubtedly they would have used ships, perhaps with high, curving prows and sterns, perhaps made of huge reed bundles, perhaps modelled on the same design but made of more durable materials such as the cedarwood planks of the so-called 'solar boat' of Khufu found buried beside the south face of the Great Pyramid of Egypt. We will not rehearse the arguments at greater length here, since we have done so elsewhere,[30] but we think it very likely that this entire

PREVIOUS PAGES: *An Aymara Indian navigates the waters of Lake Titicaca in a traditional reed boat. The distinctive design of the boat, with its high prow and stern, may be a legacy of a forgotten sea-going age.*

Village life on the Bolivian altiplano. How was it possible, in this harsh and sparsely populated environment, for so great an engineering project as the megalithic city of Tiahuanaco to have been completed?

system of boat-building, together with the distinctive religious ideas that are associated with such boats in Egypt and in the Andes, belong to the fragmented legacy of a forgotten seagoing and navigating age – an age that was also characterized by the construction of a worldwide network of extraordinary megalithic temples.

THE RAPE OF TIAHUANACO

Twenty-four kilometres south–east of Suriqi we circled in a hired plane over Tiahuanaco, the most extraordinary of all the megalithic temples in the Andes – the place of the beginning, the Stone at the Centre.

We had taken off in a little Cessna Skymaster from La Paz International Airport in Bolivia. The airport stands on the edge of the high intermontane plateau known as the *altiplano*, which extends for nearly 1000 kilometres north and south of Lake Titicaca. Our take-off altitude was 4140 metres, almost high enough for oxygen, but we had decided not to pressurize the cabin so that Santha could keep a window open for her cameras.

We soared into crystal blue air through a scattering of clouds, with snow-capped mountains glittering in the distance. At 4360 metres, roughly 240 metres above the *altiplano*, we caught sight of Lake Titicaca, reflecting the early evening sun, and then a chain of islands, amongst them Suriqi.

We followed the ribbon of main road running from La Paz to Laja and Tiahuanaco across the epic landscape of the *altiplano*, dotted with fields, ringed by brooding mountains, riven by high ridges, scored with huge erosion channels and the serpentine trails of rivers. Then swooping down between two of the ridges we

RIGHT AND BELOW: *Monumental walls of the Kalasasaya, Tiahuanaco.*

One of the several strange pieces of statuary from Tiahuanaco that are now decaying in the middle of a traffic island in La Paz.

entered the flat, wide valley at the western edge of which, at a distance of about 16 kilometres from the lake, lies Tiahuanaco.

We had visited the ruined city many times before by land, but had never seen it from the air. It was sheer poetry to do so now, when it was bathed in a soft golden light that fell upon it like a glamour, restoring it, as we imagined, to its former glory.

Time and man have not been kind to Tiahuanaco. The great edifices were systematically attacked in the sixteenth century and many of the statues were smashed during a frenzied campaign by the Catholic Church to extirpate idolatry in the Andes. In the same period, following up ancient rumours, treasure-hunters dug a deep crater in the top of the Akapana Pyramid, one of Tiahuanaco's principal structures. In the nineteenth century, without any consideration of the archaeological loss, the city's ancient megaliths were used as a quarry by engineers building a railway embankment. By the beginning of the twentieth century blocks from Tiahuanaco were routinely being removed and broken up for use on construction sites as far away as La Paz. What remained of the wreck was picked over by local villagers for materials to build their shacks and visited by ruthless curio-hunters and looters, who stole whatever they could easily carry away and carved their initials on whatever they had to leave behind.

In the 1940s, in an attempt to save the few remaining statues from further graffiti, wanton damage and theft, all but two of those that were still intact were removed to La Paz and installed in an outdoor museum opposite the Stadium, in the middle of the busiest traffic roundabout in the Bolivian capital. There to this day they remain, bathed in a continuous acid rain of exhaust fumes – an environment so toxic and corrosive that they will be damaged beyond recognition within our lifetimes. Meanwhile, the two monumental statues that stayed at the site are surrounded by barbed wire, like the inmates of a concentration camp.

Is ours a culture that wishes to murder the past?

At Tiahuanaco it sometimes seems so. Nevertheless, like the human spirit in adversity, the original conception of this sacred city is so powerful and majestic that inimical forces have not been able to destroy it completely. Something still remains.

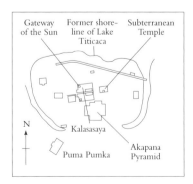

Aerial view of Tiahuanaco showing the rectangular enclosure of the Kalasasaya, the Semi-Subterranean Temple, and the pyramidal ground-plan of the Akapana. The irregular water-filled hole in the summit of the Akapana was cut by treasure-hunters.

Plan of Tihuanaco.

BIRD'S-EYE VIEW

The city consists of four main elements but only three of them, the Semi-Subterranean Temple, the Kalasasaya, and the Akapana Pyramid, were visible from our circling altitude about 150 metres above Tiahuanaco. We would have had to go higher to bring into view the megalithic step-pyramid known as the Puma Punku – the 'Lion Gate' – which lies in the south-western quadrant of the extensive site.

The Kalasasaya is a rectangular enclosure, 130 metres from east to west by 120 metres north to south[31] and is oriented to the cardinal directions.[32] It has walls made from red sandstone megaliths, surrounding a raised platform on which, from the air, we could just make out the figures of Tiahuanaco's two remaining statues. In the north-western corner of the platform we could also see the monumental 'Gateway of the Sun', carved in one piece out of 45 tonnes of solid andesite.[33]

Oriented like the Kalasasaya to the cardinal directions, the Akapana Pyramid is an archetypal sacred mountain of complex and cunning design. First of all, easily visible from our altitude despite erosion and other damage, it has a pyramidial ground-plan covering an area of more than 200 square metres – a three-stepped pyramid with its 'base' towards the east and its summit towards the west. On this ground-plan it then rises vertically in seven steps to a height of 18 metres. We could see the huge irregular hole dug by the treasure-hunters into the heart of the monument, now partly filled with muddy water. Around the rim of the hole were scattered great blocks of stone which had once formed a cruciform central well – with each arm of the cross in the form of a three-stepped pyramid. Archaeologists have established that this well fed a series of shafts inside the pyramid that carried torrents of water down to lower levels:

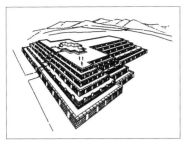

The Akapana Pyramid at Tihuanaco.

Semi-Subterranean Temple: serpent symbol on the side of the Viracocha stela.

The bearded face of Viracocha.

a system of drains that alternately poured water out of the vertical facing walls of each level, took the water back underground horizontally beneath the standing surface of each tier, and then brought it forth again and again, cascading down all the levels of the pyramid.[34]

The third principal structure visible from our altitude was the Semi-Subterranean Temple – a sunken enclosure, open to the sky, measuring 26 metres by 28 metres.[35] It contains three stele, one carved with the figure of a bearded man, and its walls are lined with the representations of dozens of strange goggle-eyed heads. It is generally agreed by archaeologists that the bearded figure, which does not have the features of a native American, is an image of Viracocha in his purely human form. Carved vertically on two sides of the stela are serpents, the universal symbol of wisdom and spiritual power that appears as far afield as Egypt, Mexico and Cambodia.

As we circled and overflew Tiahuanaco, considering it from different directions and altitudes, trying to fill in the details torn away by time, we gradually came to understand that it too was a mandala, like the temples of Angkor, like the pyramids of Giza, a mandala of geometry and form and symbolism designed to focus concentration and to confront the curious with a labyrinth of riddles.

Huge megaliths of the Puma Punku.

Imprint of metal clamp that formerly joined two gigantic blocks, Puma Punku, Tiahuanaco.

Imprint of metal clamp, Ollantaytambo.

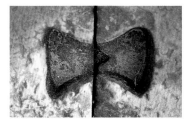

Imprint of metal clamp, Angkor Wat, Cambodia.

Imprint of metal clamp, Dendera, Egypt.

RIDDLES

For us, Tiahuanaco is many riddles wrapped up in a larger riddle.

There is the riddle of the huge stones. At the Puma Punku, a squat step-pyramid with a base measuring approximately 60 metres by 50 metres, there is a block that has been calculated to weigh 447 tonnes.[36] Many others are in the range of 100 to 200 tonnes. The principal quarries were 60 kilometres away, where all Tiahuanaco's andesite came from, and 15 kilometres away, where all its red sandstone came from. It is a complete mystery, which cannot be trivialized with easy mental images of thousands of primitive tribesmen hauling on ropes. After all, Tiahuanaco stands at 4115 metres above sea-level and the implications of organizing, motivating and feeding a large labour force at this altitude are formidable. Whoever it was done by, therefore, we can be sure that this sacred city was not the work of primitive people.

Another riddle, much in evidence at the Puma Punku, is that many of the megaliths were joined by metal clamps, some very large. For a long while it was thought that these I-shaped and T-shaped clamps had been pre-cast at a furnace and then placed cold into carved indentations in the recipient blocks. A close study with a scanning electron microscope has, however, revealed surprising evidence that they were *poured molten* into the indentations. This implies that a portable smelter must have been used, moving from block to block at the site itself[37] – a much higher level of technology than has ever been credited to pre-Columbian South America.

Another mystery is that spectrographic analysis of one of the very few surviving clamps has shown it to consist of a most unusual alloy of 2.05 per cent arsenic, 95.15 per cent copper, 0.26 per cent iron, 0.84 per cent silicon and 1.70 per cent nickel.[38] There is no source of nickel anywhere in Bolivia.[39] Furthermore, the 'rarely encountered'[40] alloy of arsenical nickel bronze would have required a smelter operating at extremely high temperatures.

Image of elephant-like creature, Gateway of the Sun.

Tiahuanaco's biggest riddle concerns its age. The range of approximately 1500 BC through to 900 AD considered by most archaeologists has been challenged on the grounds of the geology of the site, showing a relationship to Lake Titicaca that last prevailed more than 10,000 years ago.[41] Above the serpents on the side of the Viracocha figure in the Semi-Subterranean temple there are representations of an animal species resembling Toxodon – a large hippo-like animal that became extinct in the Tiahuanaco area more than 12,000 years ago.[42] And on the eastern side of the Gateway of the Sun there is the representation of an elephant-like creature, perhaps the New World proboscid Cuvieronius, which also became extinct 12,000 years ago.[43]

More substantially, there are astronomical alignments which speak of an extremely ancient date for Tiahuanaco. These were first noted by the Bolivian archaeologist Arthur Posnansky early in the twentieth century. His calculations are based on the changes in the earth's obliquity (the 'obliquity of the ecliptic', see Chapter 12) which are thought to occur at the rate of 40 arc seconds per century. The effect of these is to alter the range of sunrise along the horizon from solstice to solstice, with the extreme rising points moving relatively further north and further south and then back again in a see-saw motion over tens of thousands of years. Posnansky's calculation of Tiahuanaco's principal solstitial alignments suggested that they might originally have been surveyed more than 17,000 years ago.[44] Based on modern satellite readings, this date has subsequently been refined to approximately 12,000 years ago by the American archaeoastronomer Neil Steede.[45]

AN IMPORTANT SHIFT

The possibility that Tiahuanaco may be more than 12,000 years old is one that has traditionally been ridiculed by all mainstream historians and archaeologists. But in 1996 and 1997 an important change seems to have occurred.

Our first clue that something might be shifting was Steede's endorsement and tightening-up of Posnansky's original calculations, homing in on the eleventh millennium BC. Then, in January 1997, no less a mainstream figure than Dr Oswaldo Rivera, Director of the Bolivian National Institute of Archaeology and one of the world's leading experts on Tiahuanaco, made a number of extraordinary statements in a magazine interview with our colleague Shun Daichi, the Japanese translator of *Fingerprints of the Gods*:[46]

> *Daichi*: By the way, American researcher Neil Steede is now examining the Kalasasaya astronomically ... His research concluded that the inner walls were made 2000 years ago, and the outer walls 12,000 years ago. What do you think about this result?
>
> *Rivera*: The truth is that we are now pursuing a similar research. Previous researchers observed sunrise. Ours observes sunset. On the opposite side against where the inner walls stand are ten megaliths which function as a celestial observatory. A detailed report on this research was just completed, on 21 December 1996. It showed the same results as the one observing the sunrise.

The Gateway God of Tiahuanaco, thought to be an image of Viracocha. On the eastern side of the Gateway, the figure stands on top of a plinth in the form of a step-pyramid – the emblem of the ancient Egyptian Benben stone and perhaps a symbolic 'map' of the Akapana Pyramid.

Daichi: Steede's research concluded that the Kalasasaya was made 12,000 years ago. Did you arrive at a similar number?

Rivera: The number is very close.

Daichi: Meaning 12,000 years? But Posnansky said 17,000 years . . .

Rivera: Much more must be researched regarding this matter.

Daichi: You can't draw conclusions yet?

Rivera: The work is still in progress. Revealing the truth is no easy work. But by using the updated method, technique and tools, I believe more information will be acquired in several years ahead.

Daichi: There are many similarities between Egypt and South America: mummies, reincarnation, megaliths . . .

Rivera: And the pyramid and the cross pattern, not to forget the 'King's Chamber'.

Daichi: King's Chamber?

Rivera: We may see the last and the greatest archaeological discovery of the twentieth century this year. There is in fact a chamber inside the Akapana Pyramid of Tiahuanaco. This pyramid has a corridor and a chamber in its interior. We believe the chamber to be similar to the King's Chamber in Giza, Egypt. We plan to open its door within this year.[47]

In January 1997, when Daichi recorded the interview, Rivera was still the National Director of Bolivian Archaeology. He had held that position for more than seven years, excavating extensively at Tiahuanaco throughout this period (and indeed for 14 years before it), developing an impressive international reputation for the high quality of his work. Then in March 1997, apparently quite suddenly, he resigned.[48]

CONFIRMATION

The remarks that Rivera made in his Japanese interview amount to a historical heresy of the highest order. Used to the stolid resistance of Egyptologists to any hint of a greater antiquity for Giza we had been amazed to hear from Daichi that so senior a figure in Bolivian archaeology now seemed to be actively considering the possibility that Tiahuanaco might have been founded 12,000 years before the present – and, equally sensationally, the possibility that there was a hidden chamber there.

In May 1997, two days after our flight over Tiahuanaco, we met Rivera by prearrangement inside the Kalasasaya in front of the Gateway of the Sun. We had expected him to back-pedal. Instead he confirmed what he had told Daichi – that the date of 12,000 years ago, suggested for the Kalasasaya by astronomical calculations, was beginning to look as though it might be correct:

> It could be, yes. We are thinking that Tiahuanaco is so much earlier than has been realized before. After 21 years of making excavations and studies in Tiahuanaco I can tell you that we are all the days with our mouths open, because Tiahuanaco is incredible, including for the archaeologists working in Tiahuanaco. We are all the days discovering different things.

Moonrise photographed looking east from the western side of the Gateway of the Sun.

Rivera accepted that pushing Tiahuanaco's antiquity back to the eleventh millennium BC – the same millennium in which we have proposed that the Sphinx was originally carved – meant that a lost civilization must have been the original influence at Tiahuanaco: 'could be Atlantis, more or less'. And he told us that he had begun to find that the same hypothesis of a lost civilization was equally helpful to him in making sense of the astonishing cultural similarities between Mexico, South America and Egypt: 'We need a point of union of all these things that are today isolated and have been isolated for a very long time.'

A MAP CARVED IN STONE

Perhaps the unopened chamber that Rivera had identified in the Akapana Pyramid would provide the answers to some of these mysteries. We asked him whether there had been any further progress in opening it. He replied: 'We are looking for the entrance to a chamber in the middle of the Pyramid, inside the Pyramid. We understand there may be eight entrances.'

We pressed him: 'So you *have* found a chamber, or an entrance to a chamber, inside the Akapana Pyramid?'

He replied: 'We did not enter, but we are looking now. I was working one year, all the year, in these excavations ... and I am sure we are going to discover the inner part of Tiahuanaco ... a sunken Tiahuanaco, underneath the existing one ... I think 12 or 21 metres down we have another Tiahuanaco, and it's the

The figure of Viracocha on the Gateway of the Sun at Tihuanaco, depicting a possible subterranean chamber.

sacred Tiahuanaco, the original. I can't tell how old it is. It's a new chapter in the study of Tiahuanaco. We are going to open a new book.'

The strangest part of our conversation was yet to come. When we asked Rivera how he was so sure that the chamber would be found he pointed to the eastern face of the Gateway of the Sun, which is decorated from end to end with a complex frieze of pictures, symbols and geometrical forms. At the centre of the frieze, which extends over a span of more than 3 metres, is an anthropomorphic being, clutching a curiously shaped staff in each hand. Reckoned by archaeologists to be another of the various forms or avatars of Viracocha, the sun-god, this highly stylized figure is almost in the style of a computer icon. It has the features of a lion, it faces due east, it is bearded, and it stands on a plinth in the form of a step-pyramid.

Rivera drew our attention to the pyramid-shaped plinth. It has three steps in the same disposition as the steps in the ground-plan of Akapana Pyramid. At its core, deep within it, there is the representation of what he believes is a chamber. This square room – containing some strange coiled animal, perhaps a dragon, perhaps a lion – is shown to be approached by eight entrance corridors. Six of these corridors are in the form of bird-headed (in other words feathered) serpents; two of them are in the form of lion-headed serpents; four of them follow an angled path identical to that of the Grand Gallery of the Great Pyramid of Egypt.

'I am sure this is a map of the Akapana,' insisted Rivera. Indicating the eight entrances, he added: 'We are working on one of them. I am sure we are going to restart our excavations soon.'

PARALLELS WITH GIZA

We could not fail to be reminded of events at Giza during the 1990s after American seismic surveys and ground-penetrating radar had identified chambers under the Sphinx, and after a German-built robot-camera had explored the narrow southern shaft of the Queen's Chamber inside the Great Pyramid – making a journey of 60 metres, at the end of which it came to a closed portcullis door apparently leading to a hitherto unopened chamber. Like Oswaldo Rivera the researchers at Giza had experienced delays in continuing their work – which they nevertheless hoped to restart soon.[49]

What is distinctly odd is the fact that both Giza and Tiahuanaco have been suggested, on reasonable geological and astronomical grounds, to be more than 12,000 years old, that both sites show signs of being built on top of deep subterranean labyrinths and chambers, and that there are now rumours at both sites that some sort of message from a lost civilization may be about to be found.

THE WORKINGS OF DUALITY

After Oswaldo Rivera had left for La Paz we stayed behind in the Kalasasaya to watch the sun setting over Lake Titicaca to the north-west and the full moon rising over the peak of Mount Illimani to the south-east. Through the arch of the

Visitor from a lost world: 'El Fraile', one of the two remaining statues of the Kalasasaya.

The head of Viracocha, bearing nineteen 'solar rays'.

Gateway of the Sun, standing first one side then the other, it was possible to view both orbs in opposition, 180 degrees across the sky. We could very easily have imagined that the Gateway was the fulcrum of some immense cosmic balance with the sun and the moon in its scales.

On the east-facing frieze Viracocha's head is surmounted, as is appropriate for the sun-god, by 19 'solar rays'. William Sullivan has argued that these rays do not refer to the sun at all but indicate a knowledge of the moon's 19-year Metonic cycle – 'the number of years it takes for a particular lunar phase to recur on a given solar date. In other words, if there is a full moon on your birthday, this will not happen again for nineteen years.'[50] It is equally possible that the 19 rays could symbolize the 'solstices' of the moon – the southern and northern extremes of its major standstills – which also occur every 19 years. Far from Tiahuanaco, as we saw in the Introduction, the Callanish megalithic circle in the Outer Hebrides is designed to 'capture' the moon once every 19 years at its extreme southern standstill.

It is our conclusion that Callanish, and Tiahuanaco and many of the other monuments that we have investigated all around the world, were part of a great archaic scientific project, the objective of which was the immortality of the human soul. Unless there is something in the way of a Rosetta Stone in the hidden chambers of Giza and of Tiahuanaco, it is going to take years of patient unravelling before it will be possible to arrive at any clear understanding of how that science worked, or where and when it originated.

But we do know that it made use of certain distinctive emblems. It therefore does not surprise us that on either side of Viracocha on the east-facing frieze, are three horizontal rows of beings that have been referred to as 'angels'.[51] They are all birdmen – men with birds' wings, sometimes with bird heads and sometimes with the heads of human beings.

An essentially similar icon of human-headed birds was used in ancient Egypt to symbolize an aspect of the soul – specifically the *ba*, or 'heart-soul'. The reader will remember that the *ba* was thought able to survive in the life hereafter as an independent entity and to have the power of unfettered movement in the Duat like the flight of birds. For this reason it was symbolized as a bird.

Ba souls were frequently depicted by the ancient Egyptians receiving rays of influence (energy, life . . .) from celestial bodies – the sun, the stars, the moon. It seems to us that the 48 birdman figures on the Gateway of the Sun, 24 on either side of Viracocha (although some have eroded almost entirely away) may be doing exactly the same thing as they crowd in from left to right – like moths around a flame – towards the solar-lunar god. Oswaldo Rivera's 'map' of the Akapana, the pyramid-shaped plinth beneath Viracocha, is also the step-pyramid symbol of the Benben stone – which is itself a symbol of immortal life.

In the falling light we walked slowly around the Kalasasaya and examined its two statues, one an andesite monolith 3.7 metres high, the other sandstone and 1.8 metres tall. Both bore strange objects in their hands. Both incorporated a number of unmistakable aquatic motifs – for example, images of water creatures carved into their waist-belts. Like the astronomer-priests of the ancient Maya and the ancient Egyptians – who in both cases wore leopard-skin cloaks with spots resembling stars – both these statues were depicted wearing garments covered with circular spots.

We looked into the statues' faces. They were blank, seeming to gaze elsewhere, to gaze through us, like faces from another world. How much more of that world might yet be found if Tiahuanaco could be properly excavated? Of its vast surface-area only an estimated *1.2 per cent* has yet been investigated by archaeologists; the rest remains as time and man have left it.[52]

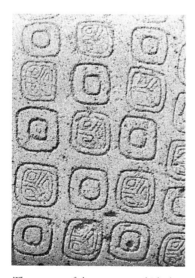

The pattern of the garments which the Kalasasaya statues wear is reminiscent of the star-spangled cloaks of ancient Egyptian and ancient Mayan astronomers.

With darkness rapidly closing around us, and an icy wind blowing across the *altiplano*, we returned to the Gateway and sought out the expressionless eyes of Viracocha's lion-visaged avatar. Facing east, the old god now had his back to the vanishing sun and gazed towards the rising moon and stars – which he seemed poised to conduct, like a celestial orchestra.

We felt the power of the cold universe – within the vastness of which is fixed mankind, a creature of material form with the divine capacity to choose between good and evil, the spiritual power to receive and to give love, and the intelligence to examine the cosmos with reverent wonder.

And we remembered what the ancients said – that at all times the God of Duality is at work within the cosmos, measuring out its cycles of millions of years, enumerating the stars:

> The God of Duality is at work,
> Creator of men,
> mirror which illumines things.
> Mother of the gods, father of the gods, the old god
> spread out on the navel of the earth
> within the circle of turquoise.
> He who dwells in the waters . . .
> He who dwells in the clouds.
> The old god, he who inhabits the shadows of the land of the dead,
> The lord of fire and time.[53]

Looking west through the Gateway of the Sun at sunset. The Gateway God faces east, towards the rising moon and stars.

THE FOURTH TEMPLE

A GREAT PAN-CULTURAL theory of the meaning and mystery of death and the possibility of eternal life illuminated the ancient world. Linked to it was a science of immortality that sought to free the spirit from the gross encumbrance of matter. In its own way, this science was every bit as rigorous and empirical as astrophysics, medicine or genetic engineering. Unlike modern sciences, however, it appears, from the very beginning, to have been as old as the hills – fully evolved, with its adepts and teachers already present and at work at the dawn of history as far afield as northern Europe, Egypt, Mesopotamia, Vedic India, the Pacific, Japan, China, South-east Asia, and the Americas.

In all these regions, leaving behind in all cases a distinctive 'package' of myths, monuments and spiritual beliefs, a strange and striking doctrine of sky–ground dualism was taught, repeatedly and insistently emphasizing that:

> All the world which lies below has been set in order and filled with contents by the things which are placed above; for the things below have not the power to set in order the world above. The weaker mysteries, then, must yield to the stronger . . . the system of things on high is stronger than the things below . . . and there is nothing that has not come down from above.[1]

The science that professed these laws used an international language – a technical terminology expressed in both architecture and myth – based upon universally agreed conventions concerning complex astronomical cycles. Yet these cycles are so minute and so obscure that they could only have been detected by means of precise observations of the heavens consistently carried out over thousands of years.

PEARL OF GREAT PRICE

Who made these observations? How was the knowledge of them distributed around the world? Before Egypt, before Vedic India, before the ancient Maya, when did they first begin? Why were they accorded such importance? And what was the nature of the system of knowledge that they served?

There is mystery in the answers to these questions – something precious that humanity has left behind in the night of time. We believe that this 'pearl of great price' could be the legacy of a lost civilization, a 'science of the soul' developed through thousands of years of inquiry and experimentation and applied with high precision to the fundamental questions of life and death.

The science could be recovered. Like all modern sciences, it required a physical apparatus and a body of theory in order to work. We have tried to show that traces of this apparatus still survive in many of the great monuments and temples of antiquity, and traces of the theory in powerful scriptures and myths that have come down to us from prehistoric times. We do not think it is an accident that when the two are brought together a synergy occurs – with the whole amounting to more than the sum of the parts. It is almost as though the myths have been purposefully designed to bring the stonework and the geometry of the monuments to life, and as though the monuments have simultaneously given substance and meaning to the myths. It is like watching a powerful piece of game software, loaded into some vast computer, waking up the latent faculties of the machine.

LOST CIVILIZATION

We have tried to play the game of myths and monuments and to understand the lost language of astronomical allegories and of sky–ground dualism.

In the monuments we have encountered recurrent astronomical alignments and astronomical symbolism. Sometimes, as at Giza and Angkor, and amongst the Maya of Central America, the alignments and symbolism have blossomed into ambitious works of celestial imitation, mirroring whole constellations on the ground.

In the mythology, traditions and scriptures of the countries through which we have travelled we have found ourselves repeatedly confronted by another shared system of ideas – the notion that the soul might be reborn down the ages, in different forms and circumstances, thus accumulating experiences and progressing gradually towards perfection. In all these cultures we also encountered the parallel notion that the task of perfecting the spirit was to be accomplished not only through good works and good thoughts but also through a ruthless stripping-away of all attachments to the material world and through mastery of an ancient system of spiritual *knowledge*.

It is nowhere made explicit exactly what this knowledge was thought to consist of but there are clues at many sites that point once again to astronomy and to a particular interest in the great celestial cycle of precession. Moreover, there are a number of texts and traditions which hint that the monuments may have been used directly as instruments of the knowledge. They are spoken of as places in which the initiate might be 'transformed into a god',[2] or into a bright star,[3] or might become a candidate for rebirth: 'that you may live and be little again'.[4]

We remain convinced that a lost civilization is by far the most likely source of all these distinctive and widely disseminated ideas. A specific hypothesis which we originally put forward in *Fingerprints of the Gods* is that this civilization flourished before 10,500 BC and vanished almost without a trace in the great cataclysm that shook the earth at the end of the last Ice Age. We propose that there were survivors who spread out around the world, settling in different continents. And we suggest that in each place they settled they constructed a wisdom cult founded on astronomical knowledge and holding out to its initiates the Holy Grail of immortality.

A network of such cults once surrounded the globe, radiating outwards from geodetic nodes that were routinely referred to by the technical term 'navels of the earth'. We have presented evidence that at least some sites may have been deliberately positioned in relation to one another according to astronomical calculations, being, for example, 72 degrees of longitude apart, or 54 degrees apart, or 108 degrees apart, or 144 degrees apart – all numbers that are generated by the precession of the equinoxes.

It is also striking that when precessional calculations are applied to astronomically aligned sky–ground monuments such as the Great Sphinx, the three Pyramids of Giza in Egypt, and the 72 temples of Angkor in Cambodia, the same date, the same season and indeed the same precise moment keep getting 'printed out' – twelve and a half thousand years before the present, on the spring equinox, at sunrise.

We accept that the Sphinx, the pyramids of Egypt and the Cambodian temples were built in different epochs. Since they all so clearly bear the imprint of a common purpose, and were designed to serve a common spiritual idea, we deduce that the cult which used them must be of immense antiquity – and immensely long-lived, having pursued the same objectives in Egypt in 2500 BC and at Angkor in AD 1150. We see no good reason why the roots of such a cult might not extend back to 10,500 BC, the epoch so insistently signalled by the monuments. Moreover, it seems to us perfectly possible that the same cult, pursuing its original objectives, could still be in existence today.

MERCHANTS OF LIGHT

In the seventeenth century the English philosopher Francis Bacon started work on an extraordinary book called *New Atlantis*, but died before completing it. This book proposed the existence, 'in the midst of the greatest wilderness of waters in the world', of an island – 'Bensalem' – ruled by a college of wise men. The inhabitants of Bensalem were enlightened, and scientifically advanced, great astronomers and geometers,[5] and the builders of aeroplanes and submarines ('we have some degrees of flying in the air; we have ships and boats for going under water'[6]). Bacon attributes to the islanders a knowledge of genetic engineering,[7] of 'seeing objects afar off',[8] and of 'divers mechanical arts'.[9] They were also accomplished navigators and seafarers, but secretive and unwilling to reveal their existence: 'we know well most part of the habitable world, and are ourselves unknown'.[10]

Bacon's story is supposedly fiction and is assumed to be merely a vehicle for the expression of his own philosophical and political ideas. Nevertheless, we find it to the point that he describes the astronomer-priests of Bensalem as the possessors of a special form of wisdom handed down from a great civilization of the past – one that had been destroyed in a worldwide deluge.[11] He tells us that their quest was for 'the knowledge of causes, and secret motions of things',[12] that it was their mission to nourish 'God's first creature, which was Light',[13] and that this mission they continuously spread abroad by means of 'twelve that sail into foreign countries under the names of other nations (for our own we conceal) . . . These we call Merchants of Light.'[14]

Whether *New Atlantis* was all made up, or whether Bacon was chosen to convey a special piece of occult history under the disguise of a harmless fable, are matters that we shall consider in another book. What is certain, however, is that all around the world, in epochs separated by thousands of years, supposedly un-related seers and sages have played the crucial role in guiding unrelated cultures along astonishingly similar paths of spiritual development. These teachers and civilizers were always said to have come from somewhere else – often an island – arriving by boat from across the sea.

Perhaps they were the real 'Merchants of Light' – the Akhu Shemsu Hor of ancient Egypt, the 'plumed serpents' of Mexico, the Viracochas of the Andes, the god-kings of the Khmers. And perhaps they did belong to a secret society, just as Bacon suggests, an 'invisible college' dedicated to the preservation of a mysteri-ous legacy of knowledge from before the Flood – an island of light surrounded by the waters of darkness.

THE ORGANIZATION

All of the religious ideas that we have considered in *Heaven's Mirror* are essen-tially 'gnostic' in nature: whether in Angkor, or in Mexico, or in ancient Egypt, initiates were taught to seek out *knowledge* of the mystery of existence through direct experience. But there was also a religion called 'Gnosis' – literally 'the knowledge', or 'secret knowledge'[15] – that was practised widely in the Middle East during the centuries before and immediately after the beginning of the Christian era.

The heart of this religion lay in Egypt, where, in the late 1940s, a large cache of Gnostic texts was unearthed at Nag Hammadi, very close to the temple of Dendera. Dating to approximately the third century AD, these papyri – now generally referred to as the 'Gnostic Gospels' or the 'Nag Hammadi Library' – make frequent allusions to the existence of a secret society, usually referred to as 'the Organization'.[16] In a number of the texts the purpose of this 'Organization' is explicitly spelled out – to build monuments 'as a representation of the spiritual places' (i.e. the stars),[17] and to oppose the universal forces of darkness and ignor-ance which are said to have:

> steered the people who followed them into great troubles, by leading them astray with many deceptions. They became old without having enjoyment. They died not having found truth and without knowing the God of truth. And thus the whole creation became enslaved forever from the foundations of the world until now.[18]

As with the ancient Egyptians, as with the Khmers, as in Mexico, the Gnostics saw the universe as a school of experience, created to give 'unperfected souls' precious opportunities to learn and grow through having to confront the challenges and choices of material existence:

> Visible creations . . . have come into being because of those who need education and teaching and formation, so that the smallness might grow, little by little. It was for this reason that [God] created mankind . . . [19]

The Gnostics also believed that there are two potent spiritual forces at work in the material universe – the force of light and love and the force of darkness and nihilism. The purpose of the force of darkness is to prevent human beings from realizing the spark of divinity within themselves – to 'make them drink the water of forgetfulness . . . in order that they might not know from where they came'.[20] The darkness works to anaesthetize intelligence and spread the cancer of 'mind-blindness'[21] because: 'Ignorance is the mother of all evil . . . Ignorance is a slave. Knowledge is freedom.'[22]

By contrast, the 'Organization' serves the force of light and its sacred purpose is to free human beings from their state of enslavement by initiating them into the cult of knowledge. There could hardly be a more important or more urgent task: in the Gnostic view mankind is the focus, or fulcrum, of a cosmic struggle; individual choices for evil, arising out of ignorance, therefore have ramifications far beyond the merely material, and mortal and human plane.[23] For these reasons the Gnostics said: 'Our struggle is not against flesh and blood, but against the world rulers of this darkness and the spirits of wickedness.'[24]

THE ARCHON AND THE SERPENT

The Gnostics lived in close contact with the vestiges of the ancient Egyptian religion and also co-existed with Judaism and with early Christianity. They honoured Osiris, the ancient Egyptian god of rebirth,[25] 'who stands before darkness as a guardian of the light'.[26] By contrast they saw Jehovah, the Old Testament god of the Jews and the Christians, as a dark force, indeed as one of the 'world rulers of darkness' – an 'Archon' whose purpose was to keep mankind chained for eternity in spiritual ignorance. Although it comes as a shock to Jews and Christians, the Gnostic account of the Old Testament story of the 'temptation' of Adam and Eve in the Garden of Eden therefore depicts the serpent not as the villain of the piece but rather as the hero and as a true benefactor of mankind.

> 'What did God say to you?' the serpent asked Eve. 'Was it "Do not eat from the tree of knowledge [*gnosis*]"?' She replied: 'He said, "Not only do not eat from it, but do not touch it lest you die."' The serpent reassured her, saying: 'Do not be afraid. With death you shall not die; for it was out of jealousy that he said this to you. Rather your eyes shall open and you shall come to be like gods, recognizing evil and good.'[27]

After Adam and Eve, the primordial human couple, had eaten of the tree of knowledge, the Gnostics taught that they experienced enlightenment and awakened to their own luminous and immortal nature. This realization, in itself, was not a guarantee of immortality but it was an essential precondition for those who wished to 'eat of the tree of life'.

The Archons were jealous and said:

> Behold, Adam! He has come to be like one of us, so that he knows the difference between the light and the darkness. Now perhaps he also will come to the tree of life and eat from it and become immortal. Come let us expel him from Paradise down to the land from which he was taken, so that henceforth he might not be able to recognize anything better ... And so they expelled Adam from Paradise, along with his wife. And this deed that they had done was not enough for them. Rather, they were afraid. They went in to the tree of life and surrounded it with fearful things ... and they put a flaming sword in their midst, fearfully twirling at all times, so that no earthly being might ever enter that place.[28]

THE DELUGE

At a later stage in history, a golden age, the Gnostic Texts tell us that the descendants of Adam and Eve had ascended through knowledge to a high state of development, manipulating the physical world with clever machines and devices and beginning to engage in profound spiritual inquiries. Out of jealousy, the Archons once again decided to intervene to diminish human potential: 'The rulers took counsel with one another and said, Come, let us cause a deluge with our hands and obliterate all flesh, from man to beast.'[29]

According to the Gnostics, the Flood was not inflicted to punish evil – as the Bible tells us – but purely and simply to punish humanity for having risen so high and 'to take the light' that was growing amongst men.[30] This it in very large part succeeded in doing. Although there were survivors, they were thrown: 'into great distraction and into a life of toil, so that mankind might be occupied by worldly affairs, and might not have the opportunity of being devoted to the holy spirit'.[31]

Fortunately, however, there were a few amongst the survivors who still possessed the old knowledge and who were determined to pass it down for the benefit of future generations, for as long as was necessary, wherever possible, until such a time as a general awakening might occur again.[32]

POPOL VUH

There is no recognized historical route by which Gnostic ideas could have reached the ancient Quiche Maya of Mexico and Guatemala. We saw in Part I that the Quiche were the builders of Utatlan, the 'stellar city' of Orion. Their only surviving sacred book, written down soon after the Conquest but recognized as preserving vastly older teachings, is the *Popol Vuh*. Strangely, just like the Gnostic Texts, it speaks of a remote golden age, and of the 'First Men' who lived then:

> Endowed with intelligence, they saw and instantly they could see far; they succeeded in seeing, they succeeded in knowing, all that there is in the

world. The things hidden in the distance they saw without first having to move ... Great was their wisdom; their sight reached to the forests, the rocks, the lakes, the seas, the mountains, and the valleys. In truth, they were admirable men ... They were able to know all, and they examined the four corners, the four points of the arch of the sky, and the round face of the earth.[33]

The achievements of the First Men were to prove their undoing, outraging the gods who decided to afflict them with amnesia:

Then the Heart of Heaven blew mist into their eyes which clouded their sight as when a mirror is breathed on. Their eyes were covered and they could only see what was close, only that was clear to them ... In this way the wisdom and all the knowledge of the First Men were destroyed ...[34]

All that survived to tell of the heights that they had formerly reached was the book *Popol Vuh*, which the Maya called *The Light That Came From Beside The Sea*.[35]

LEGACY

Very similar notions, dating back almost 5000 years, are found in the supposedly unconnected Old World texts of the Sumerians and of the ancient Egyptians. And as far afield as Micronesia, South-east Asia, China, Peru, Greece and India there is a persistant tradition – as old as the hills – that a secret treasure was long ago stored away by a race of supermen who had been cruelly punished by the gods. Legends and scriptures hint that this treasure does not consist of gold or jewels but of occult knowledge, perhaps in the form of 'books' or 'archives'.

For example, in the Indian version of the global flood myth, the god Vishnu warns Manu, his human protégé, that the deluge is about to descend and that he 'should conceal the Sacred Scriptures in a safe place' to preserve the knowledge of the antediluvian races from destruction.[36] Likewise, in Mesopotamian traditions, a hero named Utnapishtim receives instructions from the god Ea 'to take the beginning, the middle and the end of whatever was consigned to writing and then to bury it in the City of the Sun at Sippara'.[37] After the waters of the flood had subsided, survivors were instructed to make their way to the site of the City of the Sun 'to search for the writings' which would be found to contain knowledge of benefit to future generations of mankind'.[38]

When the Oxford astronomer John Greaves visited Egypt in the seventeenth century he collected a number of ancient local traditions which attributed the construction of the three Giza Pyramids to a mythical antediluvian king:

The occasion of this was because he saw in his sleep that the whole earth was turned over, with the inhabitants of it lying upon their faces and the stars falling down and striking one another with a terrible noise ... And he awaked with great feare, and assembled the chief priests of all the provinces of Egypt ... He related the whole matter to them and they took the altitude of the stars, and made their prognostication, and they foretold of a deluge. The king said, will it come to our country? They answered yes, and will destroy it. And there remained a certain number of years to come, and he commanded in the mean space to build the Pyramids ... And he engraved in these Pyramids all things that were told by wise men, as also all profound sciences – the science of Astrology, and of Arithmeticke, and of Geometry, and of Physicke. All this may be interpreted by him that knowes their characters and language ...[39]

LEO, ORION, DRACO, AQUARIUS

What these accounts have in common is the notion of a lost golden age, the notion of the Flood – or some other equally devastating cataclysm – as a setback in the progress of human knowledge, and the notion of a determined attempt being made, by a small group of survivors, to find ways to transmit to the future the precious wisdom of a former civilization.

That wisdom, at all times, in all places, concerned what the Gnostic Texts call 'the object of man's quest, the immortal discovery'.[40] It taught that the initiate must *strive* to obtain 'the life of millions of years' – which cannot be obtained by all,[41] nor achieved through blind faith or even through good works, but is a prize 'that human souls may win'.[42]

It is our conclusion that the ancient monuments and myths and scriptures that we have explored in *Heaven's Mirror* are all parts of the vast apparatus of an archaic spiritual system aimed at enabling those who had proved their worth to initiate themselves into the mystery of eternal life.

We also conclude, as the Gnostic texts envisage, that there must have been some form of coherent 'organization' behind this system. To pick the strongest examples out of the background of other evidence that we have presented, the astonishing similarities between Giza and Angkor are hard to explain in any other way – even though these sites are separated by 8000 kilometres and almost 4000 years. More important by far, however, is the fact that both sites feature enormous monuments modelling a particular group of four constellations – Leo, Orion, Draco and Aquarius – at dawn on the spring equinox in 10,500 BC.

At dawn on the spring equinox in 10,500 BC, Aquarius was setting due west, Leo was rising due east, Orion lay on the meridian due south, and Draco lay on the meridian due north.

That we have two of those constellations modelled at Giza (Leo and Orion) and a third (Draco) at Angkor seems most unlikely to be a coincidence, particularly since each one of them is oriented to a different cardinal direction. It seems obvious that careful planning must lie behind so subtle and so extenuated a scheme – and planning is the handiwork of an organization.

Such an organization would surely want to complete its great worldwide project and therefore might be expected to build a temple somewhere on earth, at some point in history, resembling the constellation of Aquarius – the fourth constellation in the talismanic sky of 10,500 BC. To conform with the global pattern we would expect that such a temple, 'resembling' or 'similar' to Aquarius, should be oriented west – just as the Angkor complex is oriented north, the Giza pyramids south and the Great Sphinx east. We would also expect it to lie at a significant distance in degrees from Giza and Angkor, which are themselves separated by 72 degrees of longitude – the 'ruling' precessional number.

AQUARIUS AND THE PHOENIX

Perhaps this temple resembling Aquarius already exists.

Might it be Tiahuanaco – which displays pronounced Aquarian characteristics in the aquatic motifs of the two great statues inside the Kalasasaya and in the water-pouring channels of the west-facing Akapana Pyramid? If so, then matters of great interest may be revealed if Oswaldo Rivera succeeds in his search for a hidden chamber.

Or might it lie in some other watery place? One area that may yet have some surprises to reveal encompasses the Gulf of Mexico, the Florida coast of the United States, and the Bahamas – in particular the shallow waters around the Bimini islands. In September 1997 we received a fax from the free-diver, Jacques Mayol, renowned as the first human being to reach the depth of 100 metres holding his breath. He told us that between 1967 and 1975 he had been part of a team of divers and archaeologists, led by Dr Manson Valentine (then the Curator of the Miami Museum of Science), who had carried out an underwater survey around Bimini. The team, said Mayol, had found 'unexplainable underwater vestiges of pre-cataclysmic dating … off the Bimini islands.' He added: 'I still have in my possession some astonishing underwater slides. For obvious reasons Dr Valentine never referred to the vestiges as possible traces of Atlantis.'[43]

Could the fourth temple be found amongst these vestiges of a former civilization?

Or could it perhaps be that the fourth temple is meant to be built in the future, 'when the time is right' – in fulfilment of an ancient plan.

Imagine yourself at Giza on 21 March around the year 2000 (or indeed at any time in the past century or the next century). Imagine yourself between the paws of the Sphinx, gazing east in line with its gaze. Approximately an hour before dawn what you would observe would be the constellation of Aquarius ascending above the eastern horizon, hovering over the exact place where the sun is going to rise.

Although it was usually represented in ancient zodiacs as a man pouring water from an urn,[44] some cultures preferred to depict Aquarius as a bird flying upward.[45] The Romans saw the constellation variously as a peacock and as a goose.[46] The Maya saw it as Coz, the celestial falcon.[47] And in the 1920s the English scholar Katherine Maltwood showed that the ancient Hindus may have identified Aquarius with their mythical birdman Garuda, who had 'the head, wings, talons and beak of an eagle and the body and limbs of a man'.[48]

Maltwood also compared Garuda, king of the birds, to the figure of the phoenix in Egyptian and Greek mythology, pointing out that, just like the phoenix, Garuda was associated with long cycles of time (he was said to have hatched from an egg 500 years after it was laid).[49] Moreover, the primary quality of the phoenix is immortality and Garuda is especially remembered in Indian

(A) 'Heraldic sky' at sunrise on the spring equinox in 10,500 BC.

(B) 'Heraldic sky' at sunrise on the spring equinox AD 2000 (note that Leo and Aquarius have swapped places).

(C) 'Heraldic sky' at sunset on the spring equinox AD 2000 (note that Orion and Draco have seesawed from their 10,500 BC position, so that Draco is now lowest and Orion highest).

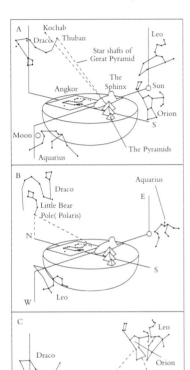

myths for having stolen the elixir of immortality from the gods. Like the tree of life in the Garden of Eden, the myths state that the elixir was hidden from man in a place of great danger where it was surrounded by flames and protected, not by a twirling sword, but by 'a fiercely revolving wheel, sharp-edged and brilliant'.[50] Garuda extinguished the flames, broke the revolving wheel and flew up with the precious goblet containing the elixir of life.

Because of this escapade Garuda is frequently depicted bearing a cup filled with liquid – which seems to provide further support for the notion of a link to Aquarius 'the water-bearer' in modern zodiacs. Moreover, if Aquarius is Garuda and Garuda is the phoenix, then, as Maltwood goes on to point out, it is not a big additional step to see Aquarius as a stellar image of the phoenix. Indeed, it may even be represented as such in an immense zodiac of prehistoric earthworks, only visible from the air, that surrounds the English sacred city of Glastonbury.[51]

In ancient Egyptian iconography and hieroglyphs the phoenix 'was born before death existed'[52] and symbolized the eternal return of all things and the triumph of spirit over matter. To observe the phoenix of Aquarius rising at the equinox, as we can today, is therefore to confront a potent celestial symbol of rebirth.

Since what happens below was believed to be determined by what happens above, it is legitimate to ask: is something to be born again?

RESURRECTION

An hour before dawn on the spring equinox in 10,500 BC Leo lay where Aquarius lies today and Aquarius was setting. In AD 2000 the reverse is true with Aquarius rising and Leo setting.

Precisely at sunrise on the spring equinox in 10,500 BC the constellation of Orion struck the south meridian and the constellation of Draco struck the north meridian, facing each other across the dome of the sky. On the spring equinox in AD 2000 this happens again – but at sunset rather than sunrise. Furthermore, Draco was at its highest altitude above the horizon in 10,500 BC and is at its lowest altitude in AD 2000 while Orion was at its lowest altitude above the horizon in 10,500 BC and is at its highest altitude today.[53]

In other words precisely the same constellations, but with everything flipped 180 degrees, are present in these two skies separated by 12,500 years.

Is it possible that there might be some message in this – or a teaching, or a portent of changes to come? The Organization that we have envisaged accorded special importance to the sky of the spring equinox in 10,500 BC. If it still exists would it not accord equal importance to the 'polar opposite' sky of the spring equinox in AD 2000?

Perhaps that sky – we repeat that precession has not significantly altered its appearance throughout the entire twentieth century – has already been taken as a signal for the construction of the 'fourth temple'. Perhaps the prehistoric cult of immortality which used vast monuments linked to constellations and astronomical cycles as instruments of initiation is coming back to life. Today as never before, travellers to Giza and Angkor find themselves amongst thronging crowds enraptured with the mystery of these places. The same is also true of the Mexican temples and pyramids, of the colossi of Easter Island, of the Nazca lines in Peru – where the pilgrims now gaze down from aerial chariots – and of Cuzco, Ollantaytambo and Tiahuanaco in the high Andes.

Poised on the edge of a millennium, at the end of a century of unparalleled wickedness and bloodshed in which greed has flourished, humanity faces a stark choice between matter and spirit – the darkness and the light. Modern religions, like modern science, have let us down, offering us no nourishment or guidance. Perhaps our only hope, as wise scholars long ago recognized, is that:

> there might come once more some kind of 'Renaissance' out of the hopelessly condemned and trampled past, when certain ideas come to life again, and we should not deprive our grandchildren of a last chance at the heritage of the highest and farthest-off times.[54]

REFERENCES

INTRODUCTION

1 'The Emerald Tablet', cited in K. E. Maltwood, *A Guide To Glastonbury, Temple of the Stars*, James Clarke and Co., London, 1964, xix.
2 *Katha Upanishad*, cited in Joseph Head and S. L. Cranston, *Reincarnation: The Phoenix Fire Mystery*, Julian Press/Crown Publishers Inc., New York, 1977, 40.
3 Ibid.
4 Ibid.
5 Alastair Service and Jean Bradbury, *The Standing Stones of Europe: A Guide to the Megalithic Monuments*, J. M. Dent, London, 1993, 47.
6 Patrick Ashmore, *Calanais: The Standing Stones*, The Standing Stones Trust, Callanish, 1995, 10.
7 We are grateful to Kent B. Watson and Dr Izumi Masukawa for introducing us to this material primarily through their video footage and extensive background document (unpublished), *The Pyramids of Japan*, International Production Services Inc., Honolulu. See also Tsutomu Sago, Osama Yamada and Lyle B. Borst, *Astronomical Analysis of Oshoro Stone Circles in Hokkaido*, reproduced in ibid.
8 *The Standing Stones of Europe*, 89 ff.
9 For further discussion see Graham Hancock, *The Sign and the Seal*, Crown Publishers, New York, William Heinemann, London, 1992.
10 William N. Morgan, *Prehistoric Architecture in Micronesia*, Kegan Paul International, London, 1988.
11 For further discussion see Graham Hancock, *Fingerprints of the Gods*, Crown Publishers, New York, William Heinemann, London, 1995.
12 To be precise, once every 18.6 years. See discussion in Gerald Ponting and Margaret Ponting, *New Light on the Stones of Callanish*, Callanish, Isle of Lewis, 1984, 50 ff.
13 E. C. Krupp, *Echoes of the Ancient Skies*, Oxford University Press, 1983, 167.
14 See for example Christopher Chippendale, *Stonehenge Complete*, Thames and Hudson, London, 1994, 137–8.
15 Graham Hancock in the *Daily Mail*, 13 March 1996.
16 Ibid.
17 *Daily Telegraph*, 28 June 1996.
18 Julius Caesar, *De Bello Gallico*, VI, 13–18, cited in John Matthews, ed., *The Druid Source Book*, Blandford Press, London, 1996, 15–16.
19 *The Druid Source Book*, 220.
20 Robert Graves, *The White Goddess*, Faber and Faber, London, 1961, 251, 274, 292.

21 Ibid, 251 and 114.
22 Ibid, 251.
23 *Encyclopaedia Britannica*, Micropaedia, Vol. 4, 233. See also *The Druid Source Book*, 12 ff.
24 Ibid.
25 The Osireion is discussed at length in *Fingerprints of the Gods*.
26 See *Fingerprints of the Gods*. See also Robert Bauval and Graham Hancock, *Keeper of Genesis* (US, *The Message of the Sphinx*), Crown Publishers, New York, and William Heinemann, London, 1996.

CHAPTER ONE

1 The figure is from the Aztec chronicler Ixtlilxochitl, cited in William Prescott, *History of the Conquest of Mexico*, Modern Library edition, New York, 49, and is supported by many other witnesses to these events.
2 Wigberto Jimenez Moreno, cited in Laurette Sejourne, *Burning Water: Thought and Religion in Ancient Mexico*, The Vanguard Press, 1956, and Shambhala, Berkeley, 1976, 17.
3 Fr Bernadino de Sahagun, *Historia General de las Cosas de Neuva España*, Editorial Neuva España, S.A., Mexico, 1946.
4 Cited in *Burning Water*, 30.
5 *History of the Conquest of Mexico*, 202.
6 Mary Miller and Karl Taube, *The Gods and Symbols of Ancient Mexico*, Thames and Hudson, London, 1993, 190.
7 Ibid, 142.
8 Kurt Mendelssohn, *The Riddle of the Pyramids*, Thames and Hudson, London, 1986, 190; Peter Tompkins, *Mysteries of the Mexican Pyramids*, Thames and Hudson, London, 1987, 57; Constance Irwin, *Fair Gods and Stone Faces*, W. H. Allen, London, 1964, 56.
9 For further discussion, see *Fingerprints of the Gods*, Heinemann, London, 1997, 111–12.
10 The earlier pyramid has been partially excavated. A tunnel and steep stairway inside the façade of the main pyramid lead the visitor to the summit of this enclosed pyramid.
11 Juan de Torquemada, *Monarchicha Indiana*, cited in *Fair Gods and Stone Faces*, 37–8; *North America of Antiquity*, cited in Ignatius Donnelly, *Atlantis: The Antediluvian World*, Harper and Brothers, New York, 1882, 165; John Bierhorst, *The Mythology of Mexico and Central America*, William Morrow, New York, 1990, 161.
12 See *Fingerprints of the Gods*, 113.
13 From a remark by Cortés to an emissary

of the Aztec king Montezuma. Cited in William Sullivan, *The Secret of the Incas*, Crown Publishers, New York, 1996, 315.
14 Bernal Diaz de Castillo, cited in *Aztecs: Reign of Blood and Splendour*, Time-Life Books, 1992, 29.
15 Ibid, 105.
16 Sahagun, cited in *Burning Water*, 163–5.
17 Ibid.
18 Ibid.
19 *History of the Conquest of Mexico*, 47.
20 Cited in *Burning Water*, 12–13.
21 *History of the Conquest of Mexico*, 48.
22 *Burning Water*, 14–15.
23 Ibid, 15.
24 Ibid.
25 Munoz Camargo, cited in ibid, 126.
26 Sahagun, cited in ibid, 29.
27 Ibid, 29–30.
28 *The Gods and Symbols of Ancient Mexico and the Maya*, 176; *History of the Conquest of Mexico*, 49.
29 *History of the Conquest of Mexico*, 49.
30 W. J. Moreno, cited in *Burning Water*, 17–18.
31 *Burning Water*.
32 Cited in ibid, 55–6.
33 Cited in ibid, 63.
34 *Aztecs: Reign of Blood and Splendour*, 36.
35 Cited in *Burning Water*, 20.
36 *Aztecs: Reign of Blood and Splendour*, 41.
37 Ibid.
38 Cited in *Burning Water*, 5.
39 Ibid.
40 From Sahagun, cited in ibid, 22.
41 Ibid, 81.
42 *National Geographic*, Washington, December 1995, 7.
43 See *Fingerprints of the Gods*, 182.
44 *Mexico*, Lonely Planet Publications, December 1992, 201.
45 Michael D. Coe, *Mexico*, Thames and Hudson, London, 1988, 91.
46 Ibid.
47 Ibid, 89.
48 Michael D. Coe, *Breaking the Maya Code*, Thames and Hudson, London, 1992, 275; Adela Fernandez, *Pre-Hispanic Gods of Mexico*, Panorama Editorial, Mexico City, 1992, 24.
49 *Burning Water*, 28.
50 Ibid.
51 *Pre-Hispanic Gods of Mexico*, 21.
52 Coe, *Mexico*, 89.
53 Cited in ibid, 89.
54 *Pre-Hispanic Gods of Mexico*, 24–6.
55 Ibid.
56 Sahagun, cited in *Burning Water*, 75.
57 Ibid, 76.
58 Ibid, 76.
59 Cited in *Burning Water*, 9.
60 Cited in Coe, *Mexico*, 98. See also

Demetrio Sodi, *The Great Cultures of Mesoamerica*, 89–90.

61 Coe, *Mexico*, 97, Sodi, 89.

62 Juan de Torquemada, *Monarchia Indiana*, cited in *Fair Gods and Stone Faces*, 37.

63 Dennis Tedlock, *Popol Vuh: The Mayan Book of the Dawn of Life*, Simon and Schuster, 1996, 64.

64 *Washington Post*, 15 April 1997.

CHAPTER TWO

1 *Burning Water*, 139–41.

2 Ibid.

3 Ibid, 56.

4 Ibid, 58.

5 Ibid.

6 Cited in ibid, 58.

7 Cited in ibid, 62.

8 For a detailed discussion of the Duat, see *Keeper of Genesis*.

9 Cited in *Burning Water*, 63–4.

10 Ibid, 65.

11 Ibid, 69–70.

12 E. A. Wallis Budge, *The Gods of the Egyptians*, Methuen and Co., London, 1904, Vol. II, 140 ff.

13 See *The Gods and Symbols of Ancient Mexico*, 114: 'In the Maya region, the Milky Way is conceptualised as the road to Xibalba, the Underworld, and the entire night sky may replicate the Underworld, and the movements of its denizens.'

14 Kurt Sethe, cited in Selim Hassan, *Excavations at Giza*, Government Press, Cairo, 1946, 135. The Nahauatl song quoted earlier also clearly directs the attention of the deceased towards the 'rosy' pre-dawn sky.

15 Cited in *Burning Water*, 67.

16 Cited in ibid, 55.

17 Cited in ibid, 26.

18 Peter Tompkins, *Mysteries of the Mexican Pyramids*, 317, 318. See also *Burning Water*, 86, citing the survey of the architect Ignacio Marquina, who 'discovered that the cause of the displacement arises because the pyramid points toward the spot where the sun falls below the horizon on the day of its passage through the sky's zenith'.

19 Anthony F. Aveni, *Skywatchers of Ancient Mexico*, University of Texas Press, Austin, 1980, 225; Chiu and Morrison, *Archaeoastronomy*, no. 2, 1980.

20 Coe, *Mexico*, 104.

21 *The Gods and Symbols of Ancient Mexico*, 114.

22 Ibid.

23 Cited in *Mysteries of the Mexican Pyramids*, 220–21.

24 Hagar, cited in ibid, 221.

25 Ibid.

26 See for example Coe, *Mexico*, 91.

27 *Encyclopaedia Britannica*, Macropaedia, Vol. 3, 197.

28 S. G. Morley, *An Introduction to the Study of the Maya Hieroglyphs*, Dover Publications Inc., New York, 1975, 16–17.

29 John Major Jenkins, *Maya Creation: The Stellar Frame of the World Ages*, Four Ahau Press, 1995, 4.

30 Prof. Gualberto Zapata Alonzo, *Descriptive Guidebook to Chichen Itza*, Merida, 33.

31 Alexander Marshack, *The Roots of Civilization*, McGraw-Hill, New York, 1972.

32 Frank Edge, P.O. Box 2552, Pinetop, AZ, *Aurochs in the Sky*, December 1995.

33 Cyril Fagan, *Zodiacs Old and New*, Anscombe, London, 1951, 24 ff.

34 *Aurochs in the Sky*, 6.

35 Reported in the *Sunday Telegraph*, London, 25 May 1997.

36 Ibid.

37 Ibid.

38 Skyglobe.

39 Stansbury Hagar, 'The Zodiacal Temples of Uxmal', *Popular Astronomy*, Vol. 79, 1921, 96.

40 Ibid, 96–7.

41 Hagar, 'The Zodiacal Temples of Uxmal'.

42 Ibid, 96–7.

43 Ibid, 96–101.

44 Ibid, 101.

45 Tedlock, *Popol Vuh*.

46 José Fernandez, 'A Stellar City: Utatlan and Orion', in *Time and Astronomy at the Meeting of Two Worlds*, Proceedings of the International Symposium, 27 April–2 May 1992, 72 and 74.

47 Fernandez, cited in David Friedel, Linda Schele, Joy Parker, *Maya Cosmos*, William Morrow, New York, 1993, 103.

48 Fernandez, 'A Stellar City', 73.

49 *Maya Cosmos*, 245.

50 Tedlock, *Popol Vuh*.

51 *Maya Cosmos*, 83.

52 Ibid.

53 See Chapter 8.

54 *Maya Cosmos*, 283.

55 Ibid.

56 Ibid, 281.

57 Coe, *Mexico*, 91.

58 Cited in Michael D. Coe, *The Maya*, Thames and Hudson, London, 1987, 173.

59 Ibid, 173–8.

60 *Fingerprints of the Gods*, 131.

61 Nigel Davies, *The Ancient Kingdoms of Mexico*, Penguin Books, London, 1990, 55.

62 *Fingerprints of the Gods*, 130.

63 *The Ancient Kingdoms of Mexico*, 53; *Mexico*, Lonely Planet Publications, 671; *Fingerprints of the Gods*, 157 ff.

64 L. A. Parsons, *The Origins of Maya Art*, Dumbarton Oaks, Washington DC, 1986, 88.

65 *Burning Water*, 84.

66 *Maya Cosmos*, 134.

67 Ibid, 132.

CHAPTER THREE

1 Jaromir Malek, *Discussions in Egyptology* 34, Oxford, 1996.

2 John Michel, *A Little History of Astro-Archaeology*, Thames and Hudson, London, 1977, 45.

3 Giorgio de Santillana and Hertha von Dechend, *Hamlet's Mill*, David R. Godine Publisher, Boston, 1977, 245 ff.

4 Ibid, 6.

5 Ibid.

6 Ibid, 4.

7 See discussions in *Hamlet's Mill*, and in *Fingerprints of the Gods*.

8 As demonstrated in Peter Tompkins, *Secrets of the Great Pyramid*, Harper and Row, New York and London, 1978, 101.

9 Einar Palsson, *The Sacred Triangle of Pagan Iceland*, Mimir, Reykjavik, 1993, 32.

10 Ibid.

11 See *Fingerprints of the Gods* and *Keeper of Genesis/The Message of the Sphinx*.

12 *Keeper of Genesis/ The Message of the Sphinx*.

13 Jill Kamil, *Luxor*, Longman, London and New York, 1989, 37 ff.

14 Sir J. Norman Lockyer, *The Dawn of Astronomy*, Massachusetts Institute of Technology Press, 1973, 109.

15 Ibid, 99.

16 *Fingerprints of the Gods*, 242–3.

17 *The Dawn of Astronomy*, 119.

18 Reported in W. R. Fix, *Pyramid Odyssey*, Mercury Media Inc., Urbanna, Virginia, 1984, 264–5.

19 Geoffrey Cornelius and Paul Devereux, *The Secret Language of the Stars and Planets*, Pavilion, London, 1996, 138.

20 Ibid, 139.

21 *The Dawn of Astronomy*, 104–6.

22 Ibid, 109.

23 Ibid.

24 R. O. Faulkner, ed., *The Ancient Egyptian Coffin Texts*, Aris and Phillips, Warminster, 1994, Vol. I, 179–80.

25 For a discussion see Peter Tompkins, *The Magic of Obelisks*, Harper and Row Publishers, New York, 1981, 358–9.

26 *Hamlet's Mill*, 59.

27 See Chapter 2.

28 *Hamlet's Mill*, 62.

29 Skyglobe 3.6.

30 Henri Frankfort, *Kingship and the Gods*, The University of Chicago Press, 1978, 90.

31 John Baines and Jaromir Malek, *Atlas of Ancient Egypt*, Time-Life Books, 1990, 76.

32 Ibid.

33 E. A. E. Reymond, *The Mythical Origin of the Egyptian Temple*, Manchester University Press, 1969.

34 Ibid, 316.

35 Ibid.

36 Ibid.

37 Ibid.

38 Ibid.

39 Ibid, 122.
40 Ibid, 55.
41 Ibid.
42 Ibid, 109, 113–14, 127.
43 Ibid, 299.
44 Ibid, 101 and 209.
45 Ibid, 231.
46 Ibid, 110.
47 Margaret A. Murray, *Egyptian Temples*, Sampson Low, Marston & Co., London, 163.
48 Ibid, 162.

CHAPTER FOUR

1 John Anthony West, *The Traveller's Key to Ancient Egypt*, Harrap Columbus, London, 1989, 374.
2 R. O. Faulkner, ed., *The Ancient Egyptian Book of the Dead*, British Museum Publications, London, 1989, 12.
3 *The Traveller's Key to Ancient Egypt*, 374.
4 S. A. B. Mercer, *The Religion of Ancient Egypt,* London, 1946, 25, 112: 'The ancient Egyptians are known to have identified Orion with Osiris.' *Echoes of Ancient Skies*, 19. R. O. Faulkner, ed., *The Ancient Egyptian Pyramid Texts*, Oxford University Press 1969, 147–8: 'Behold he has come as Orion, behold, Osiris has come as Orion.'
5 Veronica Ions, *Egyptian Mythology*, Newnes Books, London, 1986, 133–6; Budge, *The Gods of the Egyptians*, Vol. I, 416 ff; Margaret Bunsen, *The Encyclopaedia of Ancient Egypt*, Facts on File, New York, Oxford, 1991, 152.
6 E. A. Wallis Budge, *The Egyptian Heaven and Hell (Book of What is in the Duat)*, Martin Hopkinson Co., London, 1925, Vol. II, 158 ff; Ions, *Egyptian Mythology*, 134–5.
7 *The Encyclopaedia of Ancient Egypt*, 23.
8 E. A. Wallis Budge, *The Book of the Dead*, Arkana, London and New York, 1985, 366 ff.
9 *Egyptian Mythology*, 136.
10 Kamil, *Luxor*, 171.
11 Budge, *The Gods of the Egyptians*, Vol. I, 414–15.
12 Ibid, 415.
13 Ibid, 415, 414, 401. See also Garth Fowden, *The Egyptian Hermes*, Cambridge University Press, 1978.
14 *The Gods of the Egyptians*, Vol. I, 400.
15 Ibid, 417.
16 Ibid.
17 Ibid, 418.
18 Ibid, 402.
19 Ibid.
20 Martina D'Alton, *The New York Obelisk*, The Metropolitan Museum of Art, facing 72.
21 *The Gods of the Egyptians*, Vol. II, 407–8.
22 Ibid, 408.
23 Budge, *The Egyptian Heaven and Hell*, Vol. II, 166, Sixth Division of the Duat.

24 *The Gods of the Egyptians*, Vol. I, 408.
25 Ibid.
26 Ibid, 409.
27 *Fingerprints of the Gods*, 154.
28 *The Gods of the Egyptians*, Vol. I, 409.
29 Ibid, 411.
30 *The Ancient Egyptian Book of the Dead*, Chapter XCIV, cited in ibid, 411.
31 *The Book of the Dead*, Chapter CLXXV, cited in ibid, 412.
32 *The Egyptian Hermes*, 58–9.
33 Ibid. See also *The Gods of the Egyptians*, Vol. I, 414–15.
34 *The Egyptian Hermes*, 60 ff; *The Traveller's Key To Ancient Egypt*, 426: 'It is at Philae that the last known inscription in sacred hieroglyphs is to be found, dated AD 394, and the last example of demotic graffiti, this dated AD 425. If knowledge of the hieroglyphs persisted beyond this time, no record of it has been found.'
35 See Fowden, *The Egyptian Hermes*; G. R. S. Mead, *Thrice Greatest Hermes*, Samuel Weiser Inc., New York, 1992; Walter Scott, *Hermetica*, Shambhala, Boston, 1992.
36 *Hermetica*, 'Kore Kosmou', 459.
37 *Thrice Greatest Hermes*, 'The Virgin of the World', 60.
38 *Hermetica*, 'Kore Kosmou', 461.
39 Ibid, note 4, 461.
40 E. A. Wallis Budge, *Egyptian Magic*, Keegan, Paul, Trench, Trubner and Co., London, 1901, 143.
41 Ibid.
42 Westcar Papyrus in Miriam Lichtheim, *Ancient Egyptian Literature*, University of California Press, 1975, Vol. I, 219.
43 I. E. S. Edwards, *The Pyramids of Egypt*, Penguin, London, 1949, 134.
44 Edwards, *The Pyramids of Egypt*, 1993 edition, 286.
45 F. W. Green, *Journal of Egyptian Archaeology* (*JEA*), Vol. XVI, 1930, 33.
46 Alan H. Gardner, *JEA*, Vol. XI, 1925, 2–5.
47 Budge, *Egyptian Magic*, 144.
48 Ibid.
49 *The Ancient Egyptian Coffin Texts*, Spell 992, Vol. III, 100.
50 *The Encyclopaedia of Ancient Egypt*, 54.
51 *The Ancient Egyptian Coffin Texts,* Vol. I, 19, 25, 28.
52 Ibid, 28.
53 Ibid, Vol. III, 132.
54 Faulkner, *The Ancient Egyptian Pyramid Texts*, Introduction, vi.
55 E. A. Wallis Budge, *Osiris and the Egyptian Resurrection*, The Meidic Society Ltd., 1911, Vol. I, 93.
56 Hassan, *Excavations at Giza*, 278.
57 Ibid.
58 *Keeper of Genesis/ The Message of the Sphinx*, 134 ff.
59 *The Ancient Egyptian Coffin Texts*, Spell 1087, Vol. III, 150.
60 Werner Forman and Stephen Quirke,

Hieroglyphs and the Afterlife in Ancient Egypt, Opus Publishing, London, 1996, 7.
61 Ibid, 7–8.
62 John Romer, *Valley of the Kings*, Michael O'Mara Books, London, 1981, 117; James H. Breasted, *The Dawn of Conscience*, Charles Scribner's Sons, New York, London, 1944, 70.
63 Edwards, *The Pyramids of Egypt*, 1949, 27–8. Similarly: 'Egyptian thinking in religious matters was haphazard and confused', T. G. H. James, *An Introduction to Ancient Egypt*, British Museum Publications, London, 1987, 128.
64 Margaret A. Murray, *The Splendour that was Egypt*, Sidgwick and Jackson, London, Saint Martin's Press, New York, 1987, 131–2.

CHAPTER FIVE

1 Reymond, *The Mythological Origin of the Egyptian Temple*, 9: 'A copy of writings which Thoth made according to the words of the Sages.'
2 Budge, *The Book of the Dead*, Arkana edition, xxix.
3 Matthew 13:46.
4 *Hermetica*, 309.
5 Ibid, 307.
6 *The Ancient Egyptian Pyramid Texts*, 68.
7 Ibid, 50.
8 Ibid, 76.
9 Ibid, 93–4.
10 Ibid, 138.
11 Ibid, 144.
12 *Hermetica*, 457, 521, 523.
13 Ibid, 341.
14 *Hermetica*, translated by Brian P. Copenhaver, Cambridge University Press, 1995, 81.
15 Scott, *Hermetica*, 341.
16 Ibid, 341.
17 *The Egyptian Heaven and Hell*, Twelfth Division of the Duat, 258.
18 Ibid, 240.
19 Ibid, 258.
20 *The Mythological Origin of the Egyptian Temple*, 309.
21 Ibid, 257, 262.
22 Ian Shaw and Paul Nicholson, *British Museum Dictionary of Ancient Egypt*, British Museum Press/Book Club Associates, 1995, 180.
23 *Fingerprints of the Gods*, *Keeper of Genesis/ The Message of the Sphinx*.
24 Selim Hassan, *The Sphinx: Its History in the Light of Recent Excavations*, Government Press, Cairo, 1949, 91.
25 Extracts from *The Mystery of the Sphinx*, NBC Television, 1993, and transcript of AAAs meeting, Chicago.
26 For example at the Geological Society of America. Cf *Fingerprints of the Gods*, Chapter 47, note 4.
27 The so-called Neolithic Sub-Pluvial.

28   Interviewed in *Fingerprints of the Gods*, 447–8.
29   *Keeper of Genesis/The Message of the Sphinx*, 65.
30   Ibid, 67.
31   *The Ancient Egyptian Coffin Texts*, Vol. III, 169.
32   *Keeper of Genesis/The Message of the Sphinx*, 76–8.

CHAPTER SIX

1   *The Ancient Egyptian Pyramid Texts*, 117.
2   For example, Jean-Philippe Lauer, cited in *National Geographic* Magazine, January 1994, 14–15.
3   Ibid, 15.
4   Discussed in *Fingerprints of the Gods* and *Keeper of Genesis/The Message of the Sphinx*.
5   Breasted, *The Dawn of Conscience*, 68–9.
6   Reymond, *The Mythical Origin of the Egyptian Temple*, 257. See also 262.
7   James Henry Breasted, *Development of Religion and Thought in Ancient Egypt*, Pennsylvania Paperback Edition, 1972, 71–2.
8   Edwards, *The Pyramids of Egypt*, 144–5.
9   Tompkins, *Secrets of the Great Pyramid*, 388.
10   Lockyer, *The Dawn of Astronomy*, 76–7.
11   Cited in G. Maspero, *The Dawn of Civilization*, SPCK, London, 1894, 135–6.
12   *The Ancient Egyptian Pyramid Texts*, 225.
13   R. T. Rundle Clark, *Myth and Symbol in Ancient Egypt*, Thames and Hudson, London, 1991, 54–5.
14   Normandi Ellis, *Awakening Osiris: The Egyptian Book of the Dead*, Phanes Press, 1988, 102.
15   Pyramid Texts, cited in Hassan, *Excavations at Giza*, 106.
16   Robert Bauval and Adrian Gilbert, *The Orion Mystery*, Heinemann, London 1994, 16 and 17.
17   Labib Habachi, *The Obelisks of Egypt*, The American University Press, Cairo, 1988, 5–6; *The Encyclopaedia of Ancient Egypt*, 110.
18   *Myth and Symbol in Ancient Egypt*, 38.
19   Cited in ibid, 38.
20   Utt 600, cited in ibid, 37.
21   Henri Frankfort, *Kingship and the Gods*, The University of Chicago Press, 1978, 153–4; Ions, *Egyptian Mythology*, 35–6.
22   E. A. Wallis Budge, *Cleopatra's Needles*, The Religious Tract Society, London, 1926, 2; Bauval, *Orion Mystery*, 203–4.
23   *Kingship and the Gods*, 153–4.
24   Ibid.
25   Budge, *The Egyptian Heaven and Hell (Book of What is in the Duat)*, 196.
26   Ibid, 200.
27   Jane B. Sellers, *The Death of Gods in Ancient Egypt*, Penguin Books, London, 1992, 36.
28   *Kingship and the Gods*, 153–4.
29   *The Death of Gods in Ancient Egypt*, 36.

30   *Kingship and the Gods*, 380–81.
31   Ibid.
32   Ibid.
33   Habachi, *The Obelisks of Egypt*, 4-5; Sellers, *The Death of Gods in Ancient Egypt*, 36; Head and Cranston, *Reincarnation*, 19; Ions, *Egyptian Mythology*, 25.
34   *Myth and Symbol in Ancient Egypt*, 39.
35   Ibid, 245–6.
36   Ibid.
37   Head and Cranston, *Reincarnation*, 19. See also Budge, *The Gods of the Egyptians*, Vol. II, 97.
38   *The Ancient Egyptian Pyramid Texts*, 159, 173, 226–7; *The Ancient Egyptian Coffin Texts* Vol. I: 18; Budge, *The Book of the Dead*, Arkana Edition, 651–3.
39   Hegel's *Philosophy of History*, essay on the Phoenix, cited in *Reincarnation*, 19.
40   Cited in R. A. Schwaller de Lubicz, *Sacred Science*, Inner Traditions International, Rochester, 1988, 187–8.
41   Tompkins, *Secrets of the Great Pyramid*, 388.
42   *Orion Mystery*, 200 ff.
43   Faulkner, *The Ancient Egyptian Book of the Dead*, 124.
44   See Chapter 4.
45   G. A. Wainwright, *The Sky Religion in Egypt*, Greenwood Press Publishers, Westport, Connecticut, 1971.
46   John Ivimy, *The Sphinx and the Megaliths*, Abacus, London, 1976, 33–4.
47   Lockyer, *The Dawn of Astronomy*, 74.
48   Ibid, 2.
49   *The Death of Gods in Ancient Egypt*, 204.
50   *The Ancient Egyptian Coffin Texts*, Vol. I, 18: 'Be silent, O men! Hearken, hearken, of men! Hear it, this great word which Horus made for his father Osiris. He lives thereby, he has a soul thereby, he has honour thereby.'
51   *The Ancient Egyptian Pyramid Texts*, 244.
52   E.g., see *British Museum Dictionary of Ancient Egypt*, 236, and Edwards, *The Pyramids of Egypt*, 152.
53   See *Keeper of Genesis/The Message of the Sphinx*.
54   Schwaller, *Sacred Science*, 111.
55   Scott, *Hermetica*, 34 ff.
56   *Sacred Science*, 111; Lockyer, *The Dawn of Astronomy*, 57 ff.
57   *The Ancient Egyptian Pyramid Texts*, 197.
58   Anthony Aveni, *Skywatchers of Ancient Mexico*, University of Texas Press, 1990, Glossary of Astronomical Terms, 99.
59   Jacqueline Mitton, *The Penguin Dictionary of Astronomy*, Penguin Books, London, 1993, 129.
60   *Keeper of Genesis/The Message of the Sphinx*, 213–14. The possibility was first seriously considered by Lockyer in *The Dawn of Astronomy*, 57.
61   Ibid.
62   Scott, *Hermetica*, 349–51.
63   Ibid.
64   Ibid.

CHAPTER SEVEN

1   George Coedes, *Angkor: An Introduction*, Oxford University Press, London, New York, 1966, 7.
2   Bernard Groslier and Jacques Arthaud, *Angkor: Art and Civilization*, Thames and Hudson, London, 1966, 16.
3   This is the commonly accepted and undisputed etymology. See, for example, Henri Parmentier, *Guide to Angkor*, reprint by EKLIP Publisher, Phnom-Penh, 61; David P. Chandler, *A History of Cambodia*, Silkworm Books, Thailand, 1994, 29; Dawn Rooney, *Angkor: An Introduction to the Temples*, Asia Books, Hong Kong, 1994, 17; Neil Standen, *Passage Through Angkor*, Asia Books, Thailand, 1995, 13; Albert le Bonheur and Jaroslav Poncar, *Of Gods, Kings and Men: Bas Reliefs of Angkor Wat and the Bayon*, Serindia Publications, London, 1995, 6.
4   Letter dated 9 January 1997 from Dr R. B. Parkinson, Department of Egyptian Antiquities, the British Museum: 'Ankhhor is a well-attested personal name, meaning "the god Horus lives".'
5   Ibid, citing H. Ranke, *Die Aegyptischen Personennamen*.
6   See discussion in *Fingerprints of the Gods*, Part I.
7   *Fingerprints of the Gods*.
8   Richard Hinckley Allen, *Star Names: Their Lore and Meaning*, Dover Publications Inc., New York, 1963, 203.
9   Michael Freeman, *Angkor*, Asia Books, Thailand, 9; Rooney, *Angkor: An Introduction*, 15.
10   Robert Stencel, Fred Gifford, Eleanor Moron, 'Astronomy and Cosmology at Angkor Wat', *Science*, 23 July 1976, Vol. 153, No. 4250, 281.
11   Ibid.
12   Groslier, *Angkor*, 99 ff; Parmentier, *Guide*, 35 ff; *Science*, 281.
13   Parmentier, *Guide*, 35 ff; Bonheur and Poncar, *Of Gods, Kings and Men*, 6 ff.
14   Groslier, *Angkor*, 55–6, who also points out that 'sections of a canal 60 kilometres long run dead straight'.
15   Parmentier, *Guide*, 35 ff.
16   Bonheur and Poncar, *Of Gods, Kings and Men*, 6 ff; Parmentier *Guide*, 35 ff.
17   *Science*, 281.
18   Ibid.
19   Ibid.
20   *Encyclopaedia Britannica*, Micropaedia, Vol. 7, 763.
21   Ibid.
22   Philip Rawson, *Sacred Tibet*, Thames and Hudson, London, 1991, 90.
23   Ibid and Henry Clarke Warren, *Buddhism in Translations*, Motilal Banarsidass, Delhi, 134.
24   Miloslav Krasa, *The Temples of Angkor*, Allan Wingate, London, 1963, 153.
25   Rooney, *Angkor*, 133.

26 Marc Riboud, *Angkor: The Serenity of Buddhism*, Thames and Hudson, London, 1993, 136.

27 David Fontana, *The Secret Language of Symbols*, Pavilion, London, 1993, 60.

28 An inscription of Jayavarman VII, circa AD 1166, excavated from the Royal Palace – see Coedes, *Angkor*, 87.

29 Ibid, 87 and 88.

30 Ibid.

31 John Grigsby, *The Temples of Angkor*, a research project for Graham Hancock, 11–12.

32 Coedes, *Angkor*, 52; Rooney, *Angkor*, 222.

33 *Science*, 281.

34 Grigsby, *The Temples of Angkor*.

35 The highest point at meridian transit reached by a circumpolar constellation. See *The Penguin Dictionary of Astronomy*, 102.

36 The lowest point at meridian transit reached by a circumpolar constellation.

37 *Keeper of Genesis/ The Message of the Sphinx*.

38 Hinckley Allen, *Star Names*, 203.

CHAPTER EIGHT

1 Binod Chandra Sinha, *Serpent Worship in Ancient India*, Books Today, Delhi, 1978, 63.

2 Rooney, *Angkor*, 52: 'The Khmers obsession with the *naga* is reflected in its omnipresence at the temples of Angkor. It is seemingly everywhere.'

3 Sinha, *Serpent Worship*, 7.

4 Hinckley Allen, *Star Names*, 202 ff.

5 *Serpent Worship*, 41.

6 J. L. Brockington, *The Sacred Thread: Hinduism in its Continuity and Diversity*, The University Press, Edinburgh, 27–8; Donald A. Mackenzie, *India: Myths and Legends*, The Mystic Press, London, 1987, 65; A. L. Basham, *The Origins and Development of Classical Hinduism*, Oxford University Press, 75.

7 As *Vrta*, 'the arch-demon of chaos . . . a monstrous serpent.' See Basham, *Origins and Development of Classical Hinduism*, 75.

8 Basham, *Classical Hinduism*, 74–5.

9 *Encyclopaedia Britannica*, Micropaedia, Vol. 12, 289.

10 Lockamanya Bal Gangadhar Tilak, *The Orion, or Researches into the Antiquity of the Vedas*, Tilak Bros., Publishers, Poona, 1955; David Frawley, *Gods, Sages and Kings*, Passage Press, Salt Lake City, 1991; Georg Feuerstein, Subhash Kak and David Frawley, *In Search of the Cradle of Civilization*, Quest Books, Wheaton, Adyar, 1995.

11 David Frawley, *Gods, Sages and Kings*, 39.

12 *Encyclopaedia Britannica*, Vol. 9, 20.

13 Basham, *Classical Hinduism*, 75.

14 Alain Danielou, *The Myths and Gods of India*, Inner Traditions International, Rochester, 1991, 151; *New Larousse*

Encyclopaedia of Mythology, Hamlyn, London, 1989, 362. *Ophiolatreia: An Account of the Mysteries Connected with the Origin, Rise and Development of Serpent Worship*, Privately Printed, 1889, 96.

15 G. Buhler, trans., *The Laws of Manu*, Motilal Banarsidass, Delhi, 1993, 3.

16 *New Larousse Encyclopaedia of Mythology*, 362.

17 *The Laws of Manu*, 3.

18 See Chapter 6.

19 *The Laws of Manu*, 5.

20 See discussion in *The Orion Mystery*, 203 ff.

21 Danielou, *The Myths and Gods of India*, 226.

22 *The Orion Mystery*, 205, 210, 221.

23 *The Laws of Manu*, 16.

24 W. J. Wilkins, *Hindu Mythology*, Heritage Publishers, New Delhi, 1991, 116.

25 Danielou, *The Myths and Gods of India*, 24.

26 Ibid, 101.

27 See Chapter 6.

28 *The Laws of Manu*, 2.

29 Danielou, *The Myths and Gods of India*, 163–4.

30 Ibid, 163.

31 Ibid, 163.

32 Ibid, 163.

33 See Chapter 7 for detailed dimensions.

34 Le Bonheur, Poncar, *Of Gods, Kings and Men*, 7.

35 Groslier, *Angkor*, 99 ff.

36 Ananda K. Coomarswamy and Sister Nivedita, *Myths of the Hindus and Buddhists*, Dover Publications, New York, 1967, 395.

37 *Of Gods, Kings and Men*, 44.

38 Mackenzie, *India: Myths and Legends*, 142.

39 *In Search of the Cradle of Civilization*, 236.

40 Sinha, *Serpent Worship*, 44–5. In some ancient Indian reliefs Sesha is shown to be supporting not only Vishnu but the world itself and there is a text which states: 'This movable Earth with her rocks and woods, with her seas, villages, groves and towns, hold her firmly O Sesha, so that she may be immovable' (*Serpent Worship*, 25–6). This explains why Sesha has to uproot Mount Mandera before Vasuki can work as the churning rope (see *Serpent Worship*, 45).

41 *India: Myths and Legends*, 143.

42 *New Larousse*, 367.

43 *India: Myths and Legends*, 144.

44 Ibid.

45 Ibid.

46 *In Search of the Cradle of Civilization*, 237.

47 Coedes, *Angkor*, 48.

48 Ibid, 40.

49 Ibid, 42–3.

50 Ibid, 41–2.

51 *Hamlet's Mill*, in particular 162–3 plus interpolated illustrations.

52 Ibid, 132, 418.

53 *Fingerprints of the Gods*, 263.

54 *Hamlet's Mill*, 232–3.

55 Ibid, 231.

56 Ibid, 232.

57 Ibid, 164–5.

58 Ibid, illustration opposite 162, caption.

59 An example of such a relief is reproduced in Lockyer, *The Dawn of Astronomy*, 149.

60 We are grateful to John Grigsby for pointing out this similarity to us, *Temples of Angkor* research paper, 2 and 10.

61 *The Egyptian Heaven and Hell (Book of What is in the Duat)* 70.

62 Ibid, 77.

63 Ibid, 81.

64 Ibid, 65.

65 Ibid, 105.

66 Livio Catullo Stecchini in Peter Tompkins, *Secrets of the Great Pyramid*, 298.

67 Ibid.

68 Ibid and Mark Lehner, *The Egyptian Heritage*, ARE Press, Virginia Beach, 1974, 118–19.

69 *Secrets of the Great Pyramid*, 298.

70 *The Egyptian Heaven and Hell (Book of What is in the Duat)* 94.

71 *Secrets of the Great Pyramid*, 298.

72 *The Egyptian Heaven and Hell*, 94.

73 Hassan, *Excavations at Giza*, 265.

74 Lehner, *Heritage*, 119.

75 *The Egyptian Heaven and Hell*, 117.

76 *The Egyptian Heaven and Hell (Book of Gates)* 158 ff.

77 Wilkins, *Hindu Mythology*, 79.

78 Ibid, 84, Yama.

79 Ibid, 83.

80 *Of Gods, Kings and Men*, 35; Veronica Ions, *Indian Mythology*, Hamlyn, London, 1983, 29.

81 *Of Gods, Kings and Men*, 35; Ions, *Indian Mythology*, 34.

82 *Of Gods, Kings and Men*, 35.

83 See Chapter 4.

84 Cited in Groslier, *Angkor*, 153.

85 Ibid, 193.

CHAPTER NINE

1 As described in *Science*, 282.

2 Ibid, 285.

3 Ibid, 285–6. A further detailed survey of the astronomical and cosmological symbolism of Angkor Wat is Eleanor Moron, 'Configurations of Time and Space at Angkor Wat', in *Studies in Indo-Asian Art and Culture*, Vol. 5, 1977, 217–67.

4 Coomarswamy, *Myths of the Hindus and Buddhists*, 393; 16 February 3102 BC according to some reckonings. See RILKO Newsletter No 28, 1986, 13.

5 E.g., see Aveni, *Skywatchers of Ancient Mexico*, 143.

6 Wilkins, *Hindu Mythology*, 354.

7 See discussion in *Fingerprints of the Gods*, 280–81.

8 Mackenzie, *Myths and Legends*, 113.

9 Ibid, 357–8.

10 *The Ancient Egyptian Pyramid Texts*, 225.

11 Ions, *Indian Mythology*, 29.

12 Ibid.
13 Ibid.
14 Ibid. Also Mackenzie, *India: Myths and Legends*, 108.
15 Ions, *Indian Mythology*, 29, Mackenzie, *India: Myths and Legends*, 108.
16 Wilkins, *Hindu Mythology*, 247.
17 Danielou, *The Myths and Gods of India*, 181.
18 Ibid.
19 Ibid, 181.
20 Danielou, *The Myths and Gods of India*, 166–8. *New Larousse*, 362.
21 Danielou, *The Myths and Gods of India*, 168.
22 Ibid, 167.
23 Ibid, 167.
24 Ibid, 181.
25 Ibid, 164.
26 Ibid, 164.
27 Rundle Clark, *Myth and Symbol*, 179.
28 Ibid, 59.
29 Ibid, 59.
30 Danielou, *The Myths and Gods of India*, 165, citing *Bhagavata Purana* 8.24.5.
31 See Part I.
32 Scott, *Hermetica,* 385, 387.
33 Ibid, 419.
34 Ibid, 343.
35 Ibid, 343.
36 Ibid, 345–7.
37 Danielou, *The Myths and Gods of India*, 165.
38 *In Search of the Cradle of Civilization*, 16.
39 Ibid, 16–17.
40 Lockamanya Bal Ganghadar Tilak, *The Arctic Home in the Vedas*, Poona, 1956, 420.
41 See Part II.
42 Reymond, *The Mythical Origin of the Egyptian Temple*, 201.
43 See Part II.
44 Ibid.
45 Ibid.
46 Ibid.
47 See discussion in Mackenzie, *India: Myths and Legends*, 103.
48 Coomarswamy, *Myths of the Hindus and Buddhists*, 333.
49 Hinkley Allen, *Star Names*, 205; Lockyer, *Dawn of Astronomy*, 150 ff.

CHAPTER TEN

1   Rooney, *Angkor*, 109.
2   Ibid, 115.
3   Ibid, 117.
4   Ibid, 115.
5   Groslier, *Angkor*, 34 and 57.
6   *Science*, 281.
7   The sunrise points targeted mark what astronomers call the 'cross-quarters'. See discussion in *Keeper of Genesis/The Message of the Sphinx*, 254 ff.
8   Coedes, *Angkor*, 72.
9   Ibid, 73.
10 Schwaller, *Sacred Science*, 111; Frankfort, *Kingship and the Gods*, 90–91.

11 Coedes, *Angkor*, 73.
12 Dates from *Encyclopaedia Britannica*, Vol. 4, 390.
13 Michael D. Coe, 'The Khmer Settlement Pattern: a Possible Analogy with the Maya', *American Antiquity*, Vol. 22, 1957, 409–10.
14 Sir Arthur Conan Doyle, *The Sign of Four*, 1889.
15 Coedes, *Angkor*, 70 ff.
16 Ibid.
17 Chandler, *A History of Cambodia*, 34–5.
18 Coedes, *Angkor*, 73–4.
19 Ibid, 74.
20 Ibid.
21 Ibid.
22 Skyglobe 3.6. Again we have John Grigsby to thank for the discovery of this correlation.
23 Coedes, *Angkor*, 76.
24 Ibid.
25 Ibid, 76–7.
26 Groslier, *Angkor*, 29.
27 Ibid.
28 Coedes, *Angkor*, 77–8.
29 Ibid, 79.
30 Ibid, 30.
31 Ibid.
32 Ibid, 82.
33 Ibid.
34 'A number of monuments point to an earlier occupation of the area.' Bruno Dagens, *Angkor, Heart of an Asian Empire*, Thames and Hudson, London, 1995, 170.
35 Coedes, *Angkor*, 17.
36 *British Museum Dictionary of Ancient Egypt*, 139–40.
37 We have already cited examples from Angkor. In the case of Egypt see the image of the goddess Werethekau, half-human, half-cobra, in *Hieroglyphs and the Afterlife*, 22.
38 Cited in Sinha, *Serpent Worship in India*, 19.
39 Cited in *Ophiolateria*, 39.
40 *British Museum Dictionary of Ancient Egypt*, Serpent, 262.
41 Ibid.
42 *Ophiolateria*, 11.
43 *The Ancient Egyptian Pyramid Texts*, 155.
44 Malcolm Macdonald, *Angkhor and the Khmers*, Oxford University Press, Kuala Lumpur, 1987, 54. See also Chandler, *History*, 46.
45 John Audric, *Angkor and the Khmer Empire*, Robert Hale, London, 1972, 20.
46 Riboud, *Angkor*, 131.
47 *The Ancient Egyptian Pyramid Texts*, 50.
48 Coedes, *Angkor*, 105–6 is an example. Macdonald, *Angkhor and the Khmers*, 59.
49 Groslier, *Angkor*, preceding 155; Coedes, *Angkor*, 105.
50 Coedes, *Angkor*, 27.
51 Cited in Chandler, *History*, 64.
52 Groslier, *Angkor*, 168.
53 *The Ancient Egyptian Pyramid Texts*, 246–7.

54 Ibid, 246–7.
55 Coedes, *Angkor*, 29.
56 Ibid; Groslier, *Angkor*, 30.
57 Coedes, *Angkor*, 29.
58 Groslier, *Angkor*, 30.
59 *The Encyclopaedia of Ancient Egypt*, 130; *Hieroglyphs and the Afterlife*, 21 ff.
60 Zahi Hawass and Mark Lehner, 'The Sphinx: Who Built It and Why', *Archaeology*, September–October 1994, 34; Coedes, *Angkor*, 29 and 31.
61 Coedes, *Angkor*, 28.
62 Grigsby, *Temples of Angkor*, research paper, 15.
63 Alexandre Piankoff, *The Shrines of Tutankhamon*, Harper and Row, New York, 22–3.
64 *British Museum Dictionary*, 211–12.
65 *Hieroglyphs and the Afterlife*, 32–3.
66 See Aylward M. Blackman, 'The Rite of Opening the Mouth in Ancient Egypt and Babylonia', *JEA*, Vol. X, London, 1924, 55.
67 *Hieroglyphs and the Afterlife*, 32.
68 Budge, *The Book of the Dead*, Arkana edition, lxv.
69 Breasted, *The Dawn of Conscience*, 50.
70 Budge, *The Book of the Dead*, lxvi; Breasted, *Dawn*, 49.
71 *Hieroglyphs and the Afterlife*, 31.
72 Budge, *The Book of the Dead*, lxvi.
73 Ibid, lxviii.
74 *Keeper of Genesis/The Message of the Sphinx*, 209.
75 Groslier, *Angkor*, 153.
76 Coedes, *Angkor*, 94 ff.
77 Ibid, 86.
78 Ibid, 91 ff.
79 Katha Upanishad, cited in Coomarswamy, *Myths of the Hindus and Buddhists*, 334–5.
80 Coedes, *Angkor*, 46.

CHAPTER ELEVEN

1   *Larousse Encyclopaedia of Astronomy*, Batchworth Press, London, 1959, 37.
2   *The Ancient Egyptian Pyramid Texts*, 170; *The Ancient Egyptian Coffin Texts*, Vol. I, 65.
3   *Hamlet's Mill*, 162–3.
4   Henri Mouhot, *Travels in the Central Parts of Indo-China (Siam), Cambodia and Laos, 1864*, cited in *Angkor: Heart of an Asian Empire*, 141.
5   *Hamlet's Mill*, 162.
6   Ibid, 163.
7   Krasa, *The Temples of Angkor*, 40; *Encyclopaedia Britannica*, Vol. 1, 733.
8   Ibid.
9   From the Milindapana, *Buddhism in Translations*, 232.
10 Milarepa, *Drinking the Mountain Stream*, Wisdom Publications, Boston, 1995, 43.
11 *Encyclopaedia Britannica*, Vol. 10, 372.
12 F. A. Wagner, *Art of the World: Indonesia, the Art of an Island Group*, Holle & Co., Baden Baden, 1959, 81.

13 Danielou, *The Myths and Gods of India*, 180–81.
14 Dimensions from Parmentier, *Angkor*, 88.
15 Chou Ta-Kuan, *The Customs of Cambodia*, The Siam Society, Bangkok, 1992, 5.
16 Ibid.
17 Ibid, 29.
18 *Science*, 281.
19 Stephen O. Murray, *Angkor Life*, Bua Luang Books, San Francisco, 1996, 56.
20 Audric, *Angkor*, 24.
21 Ibid, 31.
22 *The Ancient Egyptian Pyramid Texts*, 120.
23 *The Orion Mystery*, 221.
24 *The Ancient Egyptian Pyramid Texts*, 186.
25 Dimensions from Parmentier, *Angkor*, 77–8.
26 *Customs of Cambodia*, 2.
27 Audric, *Angkor*, 176.
28 Coedes, *Angkor*, 44.
29 Ibid, 44.
30 Cited in John Greaves, *Pyramidographia*, London, 1646, Robert Lienhardt reprint, Baltimore, 155.
31 The quotation is from Santillana and von Dechend, *Hamlet's Mill*, 7, but it is taken out of context. Santillana and von Dechend are here precisely arguing that: 'universality is in itself a test when coupled with a firm design. When something found, say, in China turns up also in Babylonian astrological texts, then it must be assumed to be relevant, for it reveals a complex of uncommon images which nobody could claim had risen independently by spontaneous generation.'
32 Ibid.
33 Norman Lewis, *A Dragon Apparent*, Eland, London, Hippocrene, New York, 1982, 227–8.
34 One important paper that does consider precession at Angkor is Eleanor Moron, 'Configuration of Time and Space at Angkor Wat', particularly 251 ff.
35 *National Geographic*, May 1982, Vol. 161, No. 5, 549 and 559.
36 Ibid, 549.
37 Krasa, *The Temples of Angkor*, 24.
38 Rooney, *Angkor*, 223.
39 *A Dragon Apparent*, 225.
40 Dates from Coedes, *Angkor*, 96–7. On the Bayon see also Audric, *Angkor*, 165.
41 Macdonald, *Angkhor*, 110.
42 Groslier, *Angkor*, preceeding 155; Coedes, *Angkor*, 105.
43 Inscription cited by Coedes, *Angkor*, 105.
44 See for example Krasa, *The Temples of Angkor*, 201.
45 Coedes, *Angkor*, 59 and 65.
46 Jean Boisellier in Riboud, *Angkor*, 137.
47 Pierre Loti, *A Pilgrimage to Angkor*, Silkworm Books, Thailand, 1996, 43–4.
48 Claudel, *Journal*, cited in *Angkor: Heart of an Asian Empire*, 104.
49 Jean Boisellier in Riboud, *Angkor*, 136.
50 *Encyclopaedia Britannica*, Vol. 12, 819.

51 Ibid, Vol. 12, 819–20.
52 Parmentier, *Angkor*, 71; Rooney, *Angkor*, 140–41 ff.
53 Coedes, *Angkor*, 56.
54 Groslier, *Angkor*, 158.
55 Audric, *Angkor*, 185. See also Parmentier, *Angkor*, 70: 'The central sanctuary [of the Bayon] is a huge mass, the dark centre of which is surrounded by a narrow corridor. Numerous holes in the walls testify to the former existence of sumptuous panelling, which, under a rich roof, must have changed this rough grotto into an abode worthy of a god, glittering with gilding and lights. It was pillaged by treasure hunters and the idol was knocked into the well, which they had dug to reach the foundations.'
56 Budge, *The Gods of the Egyptians*, Vol. II, 312; Quirke, *Hieroglyphs and the Afterlife*, 122.
57 Schwaller in Tompkins, *Secrets of the Great Pyramid*, 173.
58 Ibid.
59 Ibid, 172–3.
60 E.g., Budge, *The Gods of the Egyptians*, Vol. II, 359; Lewis Spence, *Ancient Egyptian Myths and Legends*, Dover, New York, 1990, 294.
61 Budge, *The Gods of the Egyptians*, Vol. II, 359: 'On the whole the hippopotamus goddess was a beneficent creature, and she appears in the last vignette of the Theban Recension of the *Book of the Dead* as a deity of the Underworld [Duat] and a kindly guardian of the dead. She holds in her right forepaw an object which has not yet been satisfactorily explained, and her left rests upon the emblem of "protective, magical power"; on the other hand the monster Ammit, which appears in the Judgement Scene, has the hindquarters of a hippopotamus ...'
62 *Hieroglyphs and the Afterlife*, 122.
63 Budge, *The Gods of the Egyptians*, Vol. II, 359.
64 Alan W. Shorter, *The Egyptian Gods*, Routledge and Kegan Paul, London, Boston, 1981, 34.
65 Ions, *Egyptian Mythology*, 111.
66 Budge, *The Gods of the Egyptians*, Vol. II, 358.
67 Cited in Tavakar, *The Nagas*, Tavakar Prashnan, Bombay, 69.
68 Ibid.
69 Faulkner, *The Ancient Egyptian Book of the Dead*, 33.
70 Coomarswamy, *Myths of the Hindus and Buddhists*, 384.
71 Budge, *The Book of the Dead*, 598.
72 Skyglobe 3.6.
73 Edward H. Schafer, *Pacing the Void*, University of California Press, Berkeley, London, 1977, 47: Chinese mythology also depicts a ceremonial 'gateway' in the heavens – called Ch'ang-ho – which 'provides admission to exalted spirits'.

This 'great gate' is unmistakably hinged to the constellation of Draco. Indeed, its 'right pivot' is specifically described as 'pale yellow Thuban', i.e. *alpha Draconis*, the principal star of Draco.

CHAPTER TWELVE

1 *Nan Madol*, Pohnpei State Historic Preservation Office pamphlet.
2 William N. Morgan, *Prehistoric Architecture in Micronesia*, Kegan Paul International, London, 1988, 68.
3 Ibid, 68 ff.
4 David Hatcher Childress, *Lost Cities of Ancient Lemuria and the Pacific*, Adventures Unlimited Press, Stelle, Illinois, 1988, 217.
5 *Nan Madol*.
6 F. W. Christian, *The Caroline Islands*, Frank Cass and Co., London, 1967, 81.
7 *Nan Madol*.
8 See Part III.
9 Coedes, *Angkor*, 82.
10 *Nan Madol*.
11 Ibid.
12 Field research notes provided to the author by Alex McIntyre.
13 Dr Arthur Saxe, *The Nan Madol Area of Ponape: Researches into Bounding and Stabilizing an Ancient Administrative Center*, Office of the High Commissioner, Trust Territory of the Pacific, Saipan, 1980.
14 Ibid.
15 Ibid.
16 *Lost Cities of Ancient Lemuria and the Pacific*, 216–17.
17 Ibid.
18 Reymond, *The Mythical Origin of the Egyptian Temple*, 113, 109, 127.
19 *Fingerprints of the Gods*.
20 *Nature*, Vol. 234, 27 December 1971, 173–4; *New Scientist*, 6 January 1972, 7.
21 *Sunday Times*, London, 21 April 1996.
22 *Nature*, 12 February 1976.
23 Cited in John White, *Pole Shift*, ARE Press, Virginia Beach, 1994, 61.
24 *Science News*, Vol. 150, 20 July 1996, 36.
25 Charles H. Hapgood, *Maps of the Ancient Sea Kings*, Chilton Books, Philadelphia and New York, 1966, 187.
26 Cited in *Fingerprints of the Gods,* 492.
27 Albert Einstein's Foreword to Charles H. Hapgood, *Earth's Shifting Crust*, Pantheon Books, New York, 1958, 1.
28 Research notes provided to the author by John Grigsby.
29 Ibid and *Sunday Telegraph*, London, 19 May 1996.
30 Ibid.
31 Emilio Spedicato, *Apollo Objects, Atlantis and the Deluge: A Catastrophical Scenario for the End of the Last Glaciation*, Quaderni Del Dipartimento di Mathematica, Statistica, Informatica e Applicazioni, Bergamo, 1990, 10.
32 Discussed in detail in *Fingerprints of the Gods*, Part IV.

33 E.g., see *Sunday Times*, London, 10 November 1996.

34 *Sunday Times*, London 6 October 1996; *The Times*, London, 4 February 1996.

35 Spedicato, *Apollo Objects*, 14 ff.

36 *Sunday Times*, London, 21 April, 1996.

37 *Collins English Dictionary*, Collins, London, 1982, 1015.

38 *Encyclopaedia Britannica*, Micropaedia, Vol. 2, 796.

39 *Astronomy and Astrophysics*, 51, 1976, 127–35; *Science*, 10 December 1976, Vol. 1194, No. 4270, 1121–31.

40 *Science*, 1125.

41 Aveni, *Skywatchers of Ancient Mexico*, 103.

42 Childress, *Lost Cities*, 192 ff, 261 ff; Thor Heyerdahl, *The Kon-Tiki Expedition*, Unwin Paperbacks, London, 1982, 139 and 19; Thor Heyerdahl, *Early Man and the Ocean*, George Allen and Unwin, London, 1978, 91; John Macmillan Brown, *The Riddle of the Pacific*, T. Fisher Unwin, London, 1924 (reprinted 1996 by Adventures Unlimited Press), 268 ff.

43 Thor Heyerdahl, *Easter Island: The Mystery Solved*, Souvenir Press, London, 77; Heyerdahl, *Kon-Tiki*, 140 and 142; Father Sebastian Englert, *Island at the Centre of the World*, Robert Hale and Company, London, 1972, 30; Francis Maziere, *Mysteries of Easter Island*, Collins, London,, 1969, 120–22.

CHAPTER THIRTEEN

1 Maziere, *Mysteries of Easter Island*, 42.

2 Ibid, 42.

3 Ibid, 41.

4 Englert, *Island at the Centre of the World*, 45.

5 Ibid, 46–7.

6 On the navigational skills of the Polynesians, see D. Lewis, 'Voyaging Stars: Aspects of Polynesian and Micronesian Astronomy', in *Phil. Trans. R. Soc.*, London, 1974, 276, 133–148.

7 *Mysteries of Easter Island*, 41.

8 *Chile and Easter Island*, Lonely Planet Publications, 1990, 204.

9 Cited in *Island at the Centre of the World*, 46–7.

10 Cited in *Mysteries of Easter Island*, 47.

11 Cited in ibid, 47.

12 Ibid.

13 Ibid, 48.

14 Ibid, 51.

15 Heyerdahl, *Easter Island: The Mystery Solved*, 40.

16 David D. Zink, *The Ancient Stones Speak*, Paddington Press, New York, London, 1979, 165–6.

17 *Mysteries of Easter Island*, 126–8.

18 *Encyclopaedia Britannica*, Micropaedia, Vol. 4, 333; Paul Bahn and John Flenley, *Easter Island, Earth Island*, Thames and Hudson, London, 1992, 56 ff.

19 Ibid.

20 Ibid.

21 Ibid, 56 and 148–9.

22 Ibid, 149.

23 Ibid, 149.

24 Ibid, 149.

25 Ibid.

26 Ibid.

27 *Island at the Centre of the World*, 74–5.

28 Ibid, 74.

29 Ibid.

30 Ibid, 74–6.

31 Guillaume de Hevesy, *The Easter Island and Indus Valley Scripts*, Anthropos XXXIII, 1938; Alfred Metraux, *The Proto-Indian Script and the Easter Island Tablets*, Anthropos XXXIII, 1938.

32 *Island at the Centre of the World*, 74–6.

33 Heyerdahl, *Easter Island: The Mystery Solved*, 123–4.

34 Ibid, 109.

35 *Island at the Centre of the World*, 73.

36 *Easter Island: The Mystery Solved*, 157.

37 Jo Anne Van Tilburg, *Easter Island: Archaeology, Ecology and Culture*, British Museum Press, 1994, 75.

38 Ibid, 74–6.

39 Ibid.

40 *Island at the Centre of the World*, 100.

41 Scoresby Routledge, cited in ibid, 97.

42 Reymond, *The Mythical Origin of the Egyptian Temple*, 113.

43 Ibid, 113–14.

44 Ibid, 127.

45 *Island at the Centre of the World*, 98–9.

46 *Easter Island, Earth Island*, 148.

47 *Easter Island: The Mystery Solved*, 232–3.

48 *Island at the Centre of the World*, 57–8.

49 Ibid, 104.

50 Cited in Jacek Machowski, *Island of Secrets*, Robert Hale, London, 1975, 112.

51 See *Keeper of Genesis / The Message of the Sphinx*, 209.

52 *Island at the Centre of the World*, 65.

53 See *Keeper of Genesis / The Message of the Sphinx*, 209.

54 Budge, *The Egyptian Heaven and Hell*, Vol. II, 4–5.

55 Ibid, Vol. III, 38.

56 Zink, *The Ancient Stones Speak*, 165–6.

57 *Island at the Centre of the World*, 126.

58 Ibid, 125.

59 *The Ancient Stones Speak*, 165–6.

60 *Island at the Centre of the World*, 125.

61 Population figure from ibid, 108.

62 Alfred Metraux, cited in Childress, *Lost Cities of Ancient Lemuria and the Pacific*, 320.

63 *Lost Cities*, 313.

64 Maziere, *Mysteries of Easter Island*, 134.

65 Budge, *Osiris*, Vol. II, 180.

66 Harold Osborne, *Indians of the Andes*, Routledge and Kegan Paul, London, 1952, 64.

67 *Mexico: Rough Guide*, 354; *The Mythology of Mexico and Central America*, 8; J. E. Thompson, *Maya History and Religion*, University of Oklahoma Press, 1970, 340.

68 F. W. Christian, *The Caroline Islands*, 81.

CHAPTER FOURTEEN

1 Maziere, *Mysteries of Easter Island*, 134–5.

2 Ibid, 191.

3 Englert, *Island at the Centre of the World*, 108; Bahn and Flenly, *Easter Island, Earth Island*, 119.

4 *Easter Island, Earth Island*, 118.

5 Ibid, 119.

6 Ibid.

7 *Mysteries of Easter Island*, 191.

8 William Liller, *The Megalithic Astronomy of Easter Island*, History of Science, 1989, S27.

9 Ibid, S26, S37, S38.

10 Ibid, S45.

11 See Part II. And see Piankoff, *The Shrines of Tutankhamon*.

12 *The Shrines*, 128.

13 *The Megalithic Astronomy of Easter Island*, S29.

14 R. A. Jairazbhoy, *Ancient Egyptian Survivals in the Pacific*, Karnak House, London, 1990, 18.

15 Ibid, 28.

16 *The Megalithic Astronomy*, S29.

17 Ibid.

18 Ibid, S25–S26.

19 *Easter Island, Earth Island*, 192.

20 The number may be slightly less or more, as estimates vary. E.g., *Easter Island, Earth Island*, 187; *Ancient Egyptian Survivals in the Pacific*, 29; *Island at the Centre of the World*, 146.

21 *Island at the Centre of the World*, 147.

22 Ibid, 147.

23 Heyerdahl, *Easter Island: The Mystery Solved*, 145.

24 Ibid 144–5; *Island at the Centre of the World* 147 ff; *Easter Island, Earth Island*, 187 ff.

25 *Easter Island: The Mystery*, 145.

26 *Encyclopaedia Britannica*, Vol. 9, 393.

27 Ibid.

28 *Ancient Egyptian Survivals*, 31.

29 *The Ancient Egyptian Pyramid Texts*, 72.

30 Hassan, *Excavations at Giza*, 1.

31 Ibid.

32 *The Ancient Egyptian Pyramid Texts*, 72.

33 For translation of *Tangata* as 'learned' see Liller, *The Megalithic Astronomy*, S24.

34 Faulkner, *The Ancient Egyptian Book of the Dead*, 181.

35 *The Ancient Egyptian Pyramid Texts*, 284.

36 *Kon-Tiki*, 142.

37 See Part II and Budge, *Dictionary*, Vol. I, 270.

38 See Part II.

39 Budge, *Dictionary*, Vol. I, 266.

40 Budge, *The Book of the Dead*, 38.

41 *Kon-Tiki*, 142; *Easter Island: The Mystery*, 77.

42 This match was first observed by R. A. Jairazbhoy in his study *Ancient Egyptians in the Pacific*. As well as being the name of the sun-god, the word *ra* means 'the sun' in ancient Egyptian. Budge, *Dictionary*, Vol. I, 417.

43 Traditional recitation, cited in *Island of Secrets*, 112.
44 Ibid.
45 Bunsen, *The Encyclopaedia of Ancient Egypt*, 43.
46 *Hamlet's Mill*, 62.
47 Cited in *Mysteries of Easter Island*, 57.
48 Ibid, 165.
49 Ibid.
50 Zink, *The Ancient Stones Speak*, 174, citing Heyerdahl.
51 *The Megalithic Astronomy*, S25.
52 Ibid, S25.
53 *Lost Cities*, 314.
54 Ibid, 319–20.
55 Heyerdahl, *Easter Island: The Mystery*, 151.
56 Ibid, 111.
57 Ibid.
58 Robert Temple, *Genius of China*, Prion, 1991, 30.
59 *Easter Island: The Mystery*, 111.
60 Sullivan, *The Secret of the Incas*, 118–19.
61 John Michel, *At the Centre of the World*, Thames and Hudson, London, 1994, 21.
62 Mircea Eliade, *The Myth of the Eternal Return*, Princeton University Press, 1991, 16.
63 Maziere, *Mysteries*, 56–7.
64 Macmillan Brown, *Riddle of the Pacific*, opposite 40.
65 *Lost Cities*, 313.
66 See Chapter 13.
67 Heyerdahl, *Kon-Tiki*, 140.
68 *Easter Island*, Lonely Planet Publications, 226.
69 Budge, *Dictionary*, Vol. II, 217.
70 See Part II.
71 Lewis Ginzberg, *The Legends of the Jews*, The Jewish Publication Society of America, Philadelphia, 1988, Vol. I, 12.
72 Ibid, Vol. I, 350.
73 Ibid, Vol. I, 352.
74 Ibid, Vol. V, 15.
75 2 Samuel 24:16 ff and 1 Chronicles 21:26 ff. Discussed in Ginzberg, *Legends*, Vol. V, 39.
76 *Collins English Dictionary*, 1026.
77 Ibid and *Encyclopaedia Britannica*, Vol. 3, 979.
78 See discussion in *The Orion Mystery*, 201 ff.
79 *Larousse Encyclopaedia of Mythology*, 91.
80 Kenneth McCleish, *Myth*, Bloomsbury, London, 1996, 684.
81 Now in the museum on the site.
82 Ioanna K. Konstantinou, *Delphi*, Hannibal Publishing House, Athens, illustration 34.
83 *At the Centre of the World*, 21.
84 *Encyclopaedia Britannica*, Vol. 3, 979.
85 Ibid.
86 Stecchini in *Secrets of the Great Pyramid*, 298.
87 Ibid, 349.
88 See Part II.
89 Ibid.
90 *Secrets of the Great Pyramid*, 182.
91 Stecchini in ibid, 349.

92 Ibid, 182.
93 Heyerdahl, *Easter Island: The Mystery*, 77.

CHAPTER FIFTEEN

1 *Encyclopaedia Britannica*, Vol. 8, 570.
2 Skyglobe 3.6.
3 *Collins Guide to Stars and Planets*, Collins, London, 1984, 128.
4 Sullivan, *The Secret of the Incas*, 382.
5 *Collins Guide to the Stars and Planets*, 128.
6 *The Secret of the Incas*, 14–16.
7 See Tony Morrison, *Pathways to the Mountain Gods*, Book Club Associates, London, 1979, 78.
8 Correspondence from Maria Reiche to Clorinda Caller, cited in *Nazca: Lines, Clay and Mystery*, Lima, 7.
9 *Pathways to the Mountain Gods*, 55.
10 Ibid.
11 Ibid, 78.
12 David Parker is the photographer of *Broken Images*, Cornerhouse Publications, Manchester, 1992.
13 *The Mystery of the Lines*, WTW/PBS.
14 Abstract from 15th Annual SSE Meeting.
15 Ibid.
16 Ibid.
17 See Johan Reinhard, *The Nazca Lines*, Editorial Los Pinos, Lima, 1996, 9 ff.
18 See Chapter 14.
19 Maria Reiche, *Mystery on the Desert*, Stuttgart, 1989, 41.
20 All dimensions are approximate, based on ibid, 52–3.
21 *Collins Guide to the Stars and Planets*.
22 *The Secret of the Incas*, 34, 91, 183 ff.
23 Michael Moseley, *The Incas and their Ancestors*, Thames and Hudson, London, 1992, 187.
24 Reinhard, *The Nazca Lines*, 59–60.
25 Ibid.
26 Ibid.
27 Ibid, 59.
28 Emerald Tablet, cited in Maltwood, *Glastonbury's Temple of the Stars*, xix.
29 Fr Diego di Molina, cited in *The Secret of the Incas*, 118.
30 Heyerdahl, *Easter Island: The Mystery Solved*, 77.
31 Ibid.

CHAPTER SIXTEEN

1 Sir Clements Markham, *The Incas of Peru*, Smith, Elder and Co., London, 1911, 21–3.
2 Ibid, 29.
3 *The Secret of the Incas*, 119.
4 *Encyclopaedia Britannica*, Vol. 11, 752.
5 Ibid.
6 *The Incas of Peru*, 23.
7 Pedro Cieza de Leon, *Chronicle of Peru*, Hakluyt Society, London, 1864 and 1883, Part I, Chapter 87.
8 *The Incas of Peru*, 32–3.
9 Ibid, 29.

10 Ibid, 29–30.
11 *The Secret of the Incas*, 1.
12 Garcilaso de la Vega, *The Royal Commentaries of the Incas*, Orion Press, New York, 1961, 4.
13 *The Secret of the Incas*, 125.
14 *The Royal Commentaries*, 4–6.
15 *The Secret of the Incas*, 25.
16 The identification of Viracocha with sun is well attested in Arthur A. Demarest, *Viracocha: The Nature and Antiquity of the Andean High God*, Peabody Museum, Harvard, 1981.
17 *The Secret of the Incas*, 118.
18 Ibid, 119.
19 *Encyclopaedia Britannica*, Vol. 11, 803.
20 *The Secret of the Incas*, 182; *Encyclopaedia Britannica*, Vol. 11, 803.
21 Ibid.
22 *The Secret of the Incas*, 182.
23 Ibid.
24 Harold Osborne, *South American Mythology*, Paul Hamlyn, London, 1968, 74.
25 See discussion in *The Secret of the Incas*, 26 ff.
26 Ibid, 26.
27 Cited in ibid, 27.
28 Ibid.
29 *The Incas of Peru*, 43.
30 *The Secret of the Incas*, 29.
31 Harold Osborne, *Indians of the Andes*, 44.
32 Ibid.
33 Ibid.
34 Ibid.
35 *South American Mythology*, 61.
36 *The Ancient Egyptian Pyramid Texts*, 246, Utterance 600.
37 *Keeper of Genesis/The Message of the Sphinx*, 185.
38 *Peru*, Lonely Planet Publications, 182; *The Secret of the Incas*, 125–7.
39 Peter Frost, *Exploring Cuzco*, Nuevas Imagenes, Lima, 1989, 51.
40 *The Royal Commentaries*, 75–6.
41 Primary sources, cited in *The Secret of the Incas*, 121.
42 Ibid, 120.
43 Ibid.
44 *The Royal Commentaries*, 76.
45 Ibid, 76–7.
46 *Exploring Cuzco*, 35.
47 Ibid, 36.
48 See discussion in *The Secret of the Incas*, 172–7.
49 Ibid, 174–5.
50 Ibid, 176.
51 Coedes, *Angkor*, 46 ff.
52 *Royal Commentaries*, 77–8.
53 *Exploring Cuzco*, 50.
54 Ibid, 50.
55 Ibid.
56 *The Secret of the Incas*, 110.
57 Ibid.
58 Ibid, 106–7.
59 *The Secret of the Incas*.
60 Ibid, 34.

61 Ibid, 61.
62 Quechua saying, cited in ibid, 47.
63 *The Ancient Egyptian Pyramid Texts*, 68, 294.
64 See *Hermetica* and James M. Robinson, ed., *The Nag Hammadi Library*, E. J. Brill, Leiden, New York, 1988.
65 Scott, *Hermetica*, 477.
66 Scott, *Hermetica*, 477.
67 *The Secret of the Incas*, 59; Father Pablo Joseph de Arriaga, *The Extirpation of Idolatry in Peru*, University of Kentucky Press, 1968, 64.
68 Ibid.
69 *The Nag Hammadi Library*, 356.
70 Ibid, 381.
71 Ibid, 150.
72 Ibid, 150.
73 Ibid, 169.
74 *The Secret of the Incas*, 351.
75 *The Secret of the Incas*, 351–2.
76 See Tom Zuidema, 'At the King's Table: Inca Concepts of Sacred Kingship in Cuzco', *History and Anthropology*, 1989, Vol. 4, 249–74.

CHAPTER SEVENTEEN

1 Or 'satisfied falcon' – *Peru*, Lonely Planet Publications, 196.
2 *The Royal Commentaries of the Incas*, 233–5.
3 Frost, *Cuzco*, 29–30, 58.
4 Ibid, map, 59.
5 See discussion in *Keeper of Genesis/The Message of the Sphinx*.
6 Frost, *Cuzco*, 63; map, 59.
7 John Hemming, *The Conquest of the Incas*, Macmillan, London, 1993, 191.
8 Ibid.
9 As described by Vincent R. Lee, 'The Building of Sacsayhuaman', a paper presented to the Annual Meeting of the Institute of Andean Studies in Berkeley, California, 9 January 1987, 1.
10 Ibid.
11 Ibid.
12 *Royal Commentaries*, 237.
13 Frost, *Cuzco*, 63.
14 *Fingerprints of the Gods*, 52.
15 Cited in *The Secret of the Incas*, 219.
16 See, for example, Geoffrey of Monmouth, *The History of the Kings of Britain*, Penguin, London, 1987, 196 ff.
17 Zinc, *The Ancient Stones Speak*, 123–4.
18 Johan Reinhard, *Machu Picchu: The Sacred Center*, Nuevas Imagenes, 1991, 49.
19 Cited in *The Secret of the Incas*, 382.
20 Ibid.
21 *The Secret of the Incas*.
22 Ibid, 12–13.
23 Ibid, 14.
24 Ibid, 247–8.
25 See discussion in ibid, 312–13.
26 Ibid.
27 Frost, *Cuzco*, 50–51.
28 *The Secret of the Incas*, 312–13.

29 Scott, *Hermetica*, 155, 447.
30 See *Fingerprints of the Gods*.
31 *Bolivia*, Lonely Planet Publications, 157.
32 Interview with Oswaldo Rivera, May 1997.
33 *Bolivia*, Lonely Planet Publications, 157.
34 *The Secret of the Incas*, 365.
35 *Bolivia*, Lonely Planet Publications, 158.
36 Ibid.
37 Interview with Oswaldo Rivera, May 1997.
38 Emerald City Metallurgical, report to Neil Steede, 7 June 1995.
39 Daichi–Rivera interview, *BOSS Magazine*, Tokyo.
40 Emerald City Metallurgical, report to Neil Steede, 7 June 1995.
41 See *Fingerprints of the Gods*, 92.
42 Ibid, 87 ff.
43 Ibid.
44 Ibid, 81ff.
45 Interviewed in *Mysterious Origins of Man*, NBC, 1996.
46 Shun Daichi is also the translator of *Heaven's Mirror* and of *Keeper of Genesis*.
47 Daichi–Rivera interview, *BOSS Magazine*.
48 Interview with Oswaldo Rivera, May 1997.
49 See, for example, *Serpent in the Sky, Orion Mystery, Fingerprints, Keeper of Genesis/The Message of the Sphinx*.
50 *The Secret of the Incas*, 163.
51 Ibid, 313–14.
52 Interview with Oswaldo Rivera.
53 *Florentine Codex*, cited in *The Secret of the Incas*, 112.

CONCLUSION

1 Scott, *Hermetica*, 57.
2 As at Teotihuacan.
3 As in the case of Quetzalcoatl's transformation into Venus, and as repeatedly portrayed in the Pyramid Texts.
4 *The Ancient Egyptian Pyramid Texts*, 159.
5 Francis Bacon, *New Atlantis*, Kessinger Publishing Company reprint, Kila, MT, 329.
6 Ibid, 328.
7 Ibid, 324.
8 Ibid, 327.
9 Ibid, 326.
10 Ibid, 297.
11 Ibid, 304.
12 Ibid, 321.
13 Ibid, 309.
14 Ibid, 329.
15 *Encyclopaedia Britannica*, Vol. 5, 315.
16 *The Nag Hammadi Library*, 73–89.
17 Ibid, 85.
18 Ibid, 121–2.
19 Ibid, 87.
20 Ibid, 119.
21 Ibid, 387.
22 Ibid, 159.
23 Kurt Rudolph, *Gnosis: The Nature and*

History of Gnosticism, Harper, San Francisco, 1987, 116.
24 *The Nag Hammadi Library*, 194. Compare Paul, Ephesians 6:12.
25 See discussion in Francis Legge, *Forerunners and Rivals of Christianity from 330 BC to 330 AD*, University Books, New York, 1965, Vol. II, 21.
26 Normandi Ellis, *Awakening Osiris: The Ancient Egyptian Book of the Dead*, 84.
27 *The Nag Hammadi Library*, 184, 165.
28 Ibid, 185.
29 Ibid, 166.
30 Ibid, 352.
31 Ibid, 165.
32 Ibid, 340.
33 Delia Goetz and Sylvanus G. Morley, trans., *Popol Vuh: The Sacred Book of the Ancient Quiche Maya*, University of Oklahoma Press, 1991, 168–9.
34 Ibid, 169.
35 Tedlock, *Popol Vuh*, 16.
36 *Bhagavata Purana*, cited in *Atlantis: The Antediluvian World*, 88.
37 Berosus Fragments, cited in Robert K. G. Temple, *The Sirius Mystery*, Destiny Books, Vermont, 1987, 249.
38 Ibid.
39 John Greaves, *Pyramidographia*.
40 *The Nag Hammadi Library*, 325.
41 Ibid, 375.
42 Scott, *Hermetica*, 151.
43 Fax from Jacques Mayol, 16 September 1997.
44 Hinkley Allen, *Star Names*, 45.
45 Mary Caine, *The Glastonbury Zodiac*, 129–30.
46 *Star Names*, 46.
47 Hugh Harleston Jr., *El Zodiaco Maya*, Editorial Diana, Mexico City, 1991, 37 ff.
48 Ions, *Indian Mythology*, 102.
49 Wilkins, *Hindu Mythology*, 450.
50 Mackenzie, *India: Myths and Legends*, 145.
51 Caine, *The Glastonbury Zodiac*, 129 ff; Maltwood, *Glastonbury's Temple of the Stars*, 42 ff.
52 *The Ancient Egyptian Pyramid Texts*, 225–7.
53 Skyglobe 3.6.
54 *Hamlet's Mill*, 11.

INDEX